Contact Problems

SOLID MECHANICS AND ITS APPLICATIONS
Volume 155

Series Editor: G.M.L. GLADWELL
Department of Civil Engineering
University of Waterloo
Waterloo, Ontario, Canada N2L 3GI

Aims and Scope of the Series

The fundamental questions arising in mechanics are: *Why?*, *How?*, and *How much?* The aim of this series is to provide lucid accounts written by authoritative researchers giving vision and insight in answering these questions on the subject of mechanics as it relates to solids.

The scope of the series covers the entire spectrum of solid mechanics. Thus it includes the foundation of mechanics; variational formulations; computational mechanics; statics, kinematics and dynamics of rigid and elastic bodies: vibrations of solids and structures; dynamical systems and chaos; the theories of elasticity, plasticity and viscoelasticity; composite materials; rods, beams, shells and membranes; structural control and stability; soils, rocks and geomechanics; fracture; tribology; experimental mechanics; biomechanics and machine design.

The median level of presentation is the first year graduate student. Some texts are monographs defining the current state of the field; others are accessible to final year undergraduates; but essentially the emphasis is on readability and clarity.

www.springer.com/series/6557

L.A. Galin†

Author

G.M.L. Gladwell

Editor

Contact Problems

The legacy of L.A. Galin

Springer

L.A. Galin†

and

Prof. G.M.L. Gladwell
University Waterloo
Faculty Civil Engineering
Waterloo, On N2L 3GI
Canada

ISBN-13: 978-1-4020-9042-4 e-ISBN-13: 978-1-4020-9043-1

Library of Congress Control Number:

Printed on acid-free paper

9 8 7 6 5 4 3 2 1

springer.com

Table of Contents

Editor's Preface

L.A. Galin's book on contact problems is a remarkable work. Actually there are two books: the first, published in 1953 deals with contact problems in the classical theory of elasticity; this is the one that was translated into English in 1961 by Mrs. H. Moss, edited by Professor I.N. Sneddon, and published in spiral binder form by The School of Physical Sciences and Applied Mathematics at North Carolina State College. The second book, published in 1980, included the first, and then had new sections on contact problems for viscoelastic materials, and rough contact problems; this section has not previously been translated into English.

When Academician I.G. Goryacheva asked me to prepare a translation of the 1980 book, I thought that it would be a simple task: I merely had to edit the 1961 Moss translation of the 1953 book, and then get an English translation of the new part. However, when I started on the project I realised that I had to edit not only the text, but also the analysis, of both parts. The first was written in the early 1950s, and there have been many changes in style since then. Much of the analysis would be considered to be clumsy today; there were typographical errors in the 1961 translation, and there were typographical and (usually small) mathematical errors in the analysis that had not corrected from the 1953 original version. Moreover, Galin had used some results originally due to his teacher N.I. Mushkelishvili, without explaining how they were derived. Other parts of the analysis were difficult to follow, and needed supplementary material to be fully understood. The resulting book, given in the following pages, may be called a 'Ciceronian' translation: when Marcus Tullius Cicero (106–65 BC) translated a work of one of the ancient Greek scholars into Latin, he did not slavishly follow the Greek text, but inserted himself and his thoughts into the translation. This is what I have done with Galin's work: I have rearranged the order of the sections, included new explanatory parts, simplified and smoothed the analysis, omitted sections that, over the passage of time, have been found to be incorrect or unnecessary, etc. In essence I have tried to produce a version of the book that Galin could have written if he were alive today, rather than simply a translation of what he actually wrote 45 or 55 years ago.

I am grateful to Professor Goryacheva for providing draft translations of Chapter 6, and for her profile of Professor Galin.

G.M.L. Gladwell
Waterloo, Ontario
October, 2007

Biographical Sketch of L.A. Galin

Lev Alexandrovich Galin (1912–1981) was an outstanding scientist in the field of contact mechanics. This year we celebrate his 95-year birthday.

L.A. Galin was born in the town of Bogorodsk of the Nizhegorodskii region in the family of an engineer. After school, he worked first as a librarian, then entered the Moscow Technological Institute of Light Industry, graduating in 1939. At the Institute, teachers paid attention at once to his extraordinary talent, and let him study in accordance with an individual program. The first scientific work of L.A. Galin Solution of boundary value problems of elasticity by the method of point interpolation was published in the journal *Applied Mathematics and Mechanics* in the year of his graduation. He continued his research at the Institute of Mechanics of the USSR Academy of Sciences, entering as a post-graduate student at this institute in 1939. In 1942 L.A. Galin was awarded the PhD degree. His PhD thesis was devoted to methods for solving mixed problems of elasticity, and problems of elastoplastic torsion of rods with polygonal cross-section. It took him only three years to prepare the doctor thesis, during which period his scientific consultants were the well-known scientists N.E. Kochin and N.I. Muskhelishvili. In 1946 L.A. Galin obtained the doctor degree in physics and mathematics; in 1951 he became a professor; and in 1953 he was elected as a corresponding member of the USSR Academy of Sciences.

The area of scientific interest of L.A. Galin was extremely wide. His studies were devoted to various fields in the mechanics of continua – elasticity, plasticity, viscoelasticity, gas dynamics, subsoil hydromechanics, cavitation theory, and fracture mechanics.

L.A. Galin was one of those who developed an important new branch of the mechanics of solids – contact mechanics. He analyzed a large variety of two- and three-dimensional problems of contact between elastic bodies, taking into account complicated boundary conditions, anisotropy, inertia forces, and other conditions. For example, he obtained a (partial) solution for the problem of a punch indenting an elastic half-space when there are both adhesion and slip regions in the contact area (Galin's problem). His solution is based on reducing the Hilbert problem for two functions to a problem of conformal mapping., which he then solved approximately.

L.A. Galin suggested methods for solving three-dimensional contact problems; obtained the general relation for the pressure under a smooth punch of circular cross-section; analyzed the influence of an additional load applied outside the contact area on the distribution of contact pressure; considered punches of circular, elliptic, wedge, and rectangular cross-section indenting a half-space; and solved the problem for a punch pressing on an elastic plate. On the basis of the solution of the problem for a narrow punch pressing on an elastic half-space, he determined when the Zimmerman-Winkler hypothesis was applicable.

L.A. Galin developed a new direction in the theory of mixed problems – contact problems taking into account the change in shape of the surfaces during the wear process. The solutions of these problems are widely used in tribology, and also in wear and durability design of various joints.

A large number of his works concern elastoplastic problems. He found remarkable solutions for plane elastoplastic problems for a plate with a circular hole subjected to tension, and for a beam with a circular hole subjected to bending. Also, he proposed an analogy for the plane elastoplastic problem (similar to the Prandtl–Nadai analogy), which made it possible to solve many elastoplastic problems experimentally.

L.A. Galin considered a series of important contact problems taking into account the rheological properties of contacting bodies, analyzed the effect of vibration loading on structural elements (such as rods and beams) made of polymeric materials, solved the inverse problem of choosing the contour of a hole in a plate made of a glass-reinforced material. He suggested a theory of self-sustaining fracture, which describes the fracture dynamics of overstressed high-strength glasses, rock burst, and other phenomena.

L.A. Galin published about 100 works in different branches of mechanics. His monograph *Contact Problems of Elasticity* (1953) became widely known, and was translated into several languages. He was the editor of the review, *Developments in the Theory of Contact Problems* in the USSR published in 1976. The results of his investigations into contact mechanics, including problems for viscoelastic bodies and others, were presented in the monograph *Contact Problems of Elasticity and Viscoelasticity* (1980). In the last year of his life, L.A. Galin completed the monograph *Elastoplastic Problems* published in 1984 after his death. The last two monographs were awarded a USSR State Prize in 1986.

L.A. Galin was a Professor in Lomonosov Moscow State University in the Division of the Theory of Plasticitiy. I was fortunate to be a student of Professor L.A. Galin in this University, and to prepare my Master's thesis and Ph.D. thesis under his supervision. He interested me in scientific research, and showed me the beauty of analytical solutions. After my graduation, I was happy to continue my research with Professor Galin at the Institute for Problems in Mechanics of the Russian Academy of Sciences, where he was head of the laboratory.

I believe this book will be useful for researchers and students in analytical methods for contact problems. In this way my teacher L.A. Galin will continue to help future generations of scientists. I am very grateful to Professor Gladwell for his hard work in preparing this book.

Those who were fortunate to know Lev Alexandrovich remember him not only as a gifted scientist but also as a man of great kindness and principle. Apart from science, he was fond of poetry, history and literature; he himself wrote poems. He continues to live in the hearts of his students and colleagues, who remain grateful to him.

I.G. Goryacheva
Moscow, 2007

Chapter 1
A Review of Research Before 1953

1.1 Introduction

The first results on the topic of contact problems in the theory of elasticity were obtained at the end of the 19th century. After that, little work was done on these problems until the last 20 years (1933–1953). Most of the work was carried out in the Soviet Union. A number of articles published elsewhere often repeated results obtained by Soviet scientists. We shall begin by reviewing articles devoted to plane contact problems, and then to three-dimensional ones.

The solution of contact problems is significantly simplified if the friction between contacting bodies is neglected. This neglect is usually justified for machine contact problems. There is a layer of lubricant between machine parts in contact. If the speed of one part relative to the other is small, then the hydrodynamic phenomena taking place in this layer can be neglected. The presence of lubricant means that the forces of friction between the bodies are quite small, and it is possible to neglect them. Usually the size of the contact region is small compared to the radii of curvature of the bodies in contact, and we can replace one of the bodies by a semi-infinite space.

1.2 Frictionless Plane Problems

In frictionless contact between a punch (a rigid body) and an elastic body, the following boundary conditions apply. First, the normal displacement under the punch is known. If the elastic body occupies a semi-infinite plane, as is assumed in most problems, then the displacement in the direction of one of the axes (that normal to the plane) is known. Secondly, the tangential stress under the punch is taken to be zero, since there is no friction.

It is usually supposed that, outside the punch, the normal and tangential components of external force are zero, so that the surface of the body is stress-free. Some researchers consider the influence of a so-called additional load (one applied to the elastic body outside the punch) on the distribution of pressure under the punch. The known values of the normal and tangential components of external force supply two boundary conditions on the surface of the elastic body outside the punch.

Frictionless contact problems lead to the determination of one harmonic function from the solution of a Dirichlet problem. We can pose the problem differently, and determine a function of a complex variable, regular in the half plane, that is the solution of a boundary problem of mixed type.

Sadovski (1928) solved several particular frictionless contact problems for a rigid body on an elastic half-plane. He considered the pressure of a punch with a plane base on an elastic half-plane, and also the case when there is a periodic array of identical punches. The second edition of the collected works of Chaplygin (1950) contains a solution for the pressure of a punch with a plane base. This work was not published by the author; the manuscript was dated 1900, i.e. much earlier than that by Sadovski. The general solution of plane contact problems of this type was given by Muskhelishvili (1935) in the second edition of his book *Some Basic Problems in the Mathematical Theory of Elasticity*. (The book appeared in English in 1953.) Here it is supposed that the required function of a complex variable is regular everywhere, including at the ends of the segment which corresponds to the contact region. This happens when the contour of the punch is bounded by a smooth curve. Such is the case, for instance, for a punch in the shape of a circular cylinder.

The conclusions obtained in the course of solving contact problems for one punch were generalised by Begiashvili(1940) to include the case of an arbitrary number of punches. He cites, in particular, quite simple examples for several punches, each having a plane base. This enabled Lomidze (1947) to solve the problem of the pressure of a system of connected foundations on the ground. Before that, such a problem had been considered by Afonkin (1941), who obtained an approximate solution for the case of two foundations.

We should also note the work of Klubin (1938), who determined the stresses inside the elastic half plane. He assumed that the punch had a plane base. Gastev (1937) considered a problem for a half plane possessing weight. Shtaerman (1949) considered a number of frictionless contact problems for a half plane in his book *Contact Problems of the Theory of Elasticity*.

In the recently published work by Sen (1946), he shows solutions for several contact problems. However, this article does not contain any new results, since all the problems are soluble in principle by the methods developed by Muskhelishvili.

All these problems relate to an isotropic half plane. Contact problems for an anisotropic half plane, under the same assumptions on the frictional forces, were considered by Savin (1939). He gives the solution for the pressure of a punch with a plane base, and also for a base bounded by an arc of a circle, on an anisotropic half plane. These results are generalised in Savin (1940a) to include the case when the surface outside of the punch is acted on by a load; in Savin (1940b) he considers several punches. Here in particular, the following result is obtained: if the anisotropic body is orthotropic and if one of the axes of orthotropy is parallel to the boundary of the half plane, then the distribution of pressure under the punch is the same as for an isotropic body.

If the punch moves with constant speed along the boundary of an isotropic half plane, then the mixed problem which has to be solved is found to be similar to that which appears for an anisotropic half plane. This problem was considered by the author in Galin (1943a), and its solution is given in this book. If there is no friction under the punch the pressure turns out to be the same as for a stationary punch.

1.3 Plane Adhesive Contact Problems

If the coefficient of friction between the contacting bodies is large, the bodies may be assumed to be rigidly joined to one another, in so called adhesive contact. In some cases this type of boundary condition can approximate those that occur between foundations and the ground. However, in most problems on the pressure of foundations, the contact is assumed to be frictionless.

The boundary conditions for adhesive contact are as follows: if the problem concerns the contact between a punch (i.e., a rigid body) and an elastic body, then the displacement of the punch determines the displacement at its boundary with the elastic body. Thus the following are known under the punch: u, the displacement in the direction of the x-axis, and v, the displacement in the direction of the y axis (normal to the boundary of the half plane). On the remaining surface of the elastic body we are given the normal and tangential components of stress, σ_{nn} and σ_{ns}. If the surface is stress free, they are zero.

Muskhelishvili (1935) was the first to consider problems of this type, in which the displacements were given on one part of the boundary, and the stresses on the remainder. He reduced the problem to the solution of an infinite system of simultaneous linear equations. Somewhat later such a problem was considered by Florin (1936a) , who reduced it to the solution of dual integral equations, and found an approximate solution. He supposed that the pressure, and also the tangential strains under the punch, could be represented in polynomial form. Soon after that, Abramov (1937) gave an effective solution of this problem using Mellin transforms. It should be noted that Abramov's method is suitable only for the case of one punch; in the course of solving the problem, he transforms the half plane into itself by means of a rational transform, and reduces the problem to that for a plane wedge. Abramov noticed a peculiar phenomenon: both the normal and the tangential stress change sign an infinity of times as they approach the ends of the contact region. However, the points where the stresses change sign for the first time are extremely near the ends of the contact region. In addition, as a consequence of the appearance of regions of plastic deformation, the character of the distribution of stresses at the ends of the contact region is different from that obtained on the basis of the usual linear theory. The complete solution for the pressure of one or several punches on an elastic half plane was given somewhat later by Muskhelishvili (1941, 1946). He reduced it to a Riemann–Hilbert problem for an analytic function. At the same time, a similar problem stated somewhat differently was discussed by Glagolev (1942, 1943).

Sherman (1938) discussed the case when the displacements are given on one part of the body and the stresses on the remainder, and when the region occupied by the elastic body is arbitrary. Here for an arbitrary finite region, the problem was reduced to a Fredholm integral equation of the first kind. The author showed that when the region occupied by the elastic body is mapped onto a circular disc by means of a rational function, the solution is reduced to the evaluation of an integral. More effective solutions have been obtained for the case when the elastic region is bounded by a circle. Kartsivadze (1943, 1946) gave the solution of the problem for the circular disc; Mintsberg (1948) gave the solution for the outside of a circle. The mixed problem solved by Abramov (1939) was later considered by Okubo (1940), who obtained a series solution by applying special functions. We noted that Florin (1936b) gave an approximate solution for the problem of a punch rigidly connected to the base by representing the unknown pressure in polynomial form. He used the same method to solve a number of other problems encountered in calculating the strength of foundations. In particular Florin (1948) considered difficult problems concerning flexible and elastic beams of finite length lying on an elastic base. He assumed that the contact between the beam and the base was either frictionless or adhesive. Similar problems were considered by Gorbunov-Posadov (1937, 1939a) and Shekhter (1940a). The latter considered an infinitely long beam lying on an elastic layer of finite thickness.

Solutions for an anisotropic half plane when the displacements are given on one part of the boundary and the stresses on the other are given in this book.

1.4 Plane Frictional Contact Problems

When elastic bodies touch, there is often friction between them, and the coefficient of friction is some finite quantity. Two cases arise. First, one body moves with respect to the other, but the movement is so slow that dynamic effects can be neglected. In this case it can be supposed that under the action of the external forces, the punch is situated on the surface of the elastic body in a state of limiting equilibrium. In the second case, there is no displacement of the punch as a whole relative to the elastic body. At those points where the stress is less than a critical value the two bodies stick together; and where the stress exceeds the critical value, one body slips relative to the other. Thus the contact region can be partitioned into two separate parts: stick zones and slip zones. These problems present some severe mathematical difficulties.

The boundary problems for frictional contact problems are as follows: on parts of the elastic body in contact with the punch (or punches) we are given the displacement normal to the surface. In addition, since it is supposed that the friction forces obey Coulomb's law, the following relationship holds between the normal and tangential components of stress: $\sigma_{ns} = \rho\sigma_{nn}$, where ρ is the coefficient of friction. The surface of the body outside the contact region is stress free, so that both the normal and the tangential components of stress are zero. If there are stick zones in the contact region, then both the components of the displacement are known there, and the ratio of the magnitude of the tangential stress to the magnitude of the normal stress is less than the value ρ.

Muskhelishvili (1942) solved the contact problem when the friction forces obey Coulomb's law over the whole of the contact region. This problem, as well as those discussed in the preceding sections, can be reduced to finding a function of a complex variable satisfying boundary conditions of mixed type. A contact problem with the same kind of boundary conditions, but with somewhat more particular assumptions, was solved at about the same time by Glagolev (1942).

When the punch moves along the boundary of the elastic half plane, the force of friction over the whole of the contact region acts in one direction. In this connection it is interesting to consider the problem of a punch moving with constant speed along the boundary. The solution of this problem allows us to establish to what extent it is important to take into account the dynamic phenomena taking place there. This kind of problem was solved by Galin (1943a), and the results are given in this book. In addition, the frictional contact problem for an anisotropic half plane was solved in Galin (1943a). Methods used for this problem are similar to those employed for a moving punch.

When the frictional force obeys Coulomb's law, the contact region is divided, as we noted earlier, into stick and slip zones; in the former $|\sigma_{ns}| < \rho|\sigma_{nn}|$; in the latter $\sigma_{ns} = \rho\sigma_{nn}$. In general the direction of sliding will be different at different points. One problem of this type for a punch with a plane base was partly solved by Galin (1945). The partial solution is given in this book. Falkovich (1946) made different assumptions: that in some places there is no friction, while in others there is sticking.

In contact problems arising in machine construction, the contacting bodies are frequently separated by a layer of lubricant. If it is supposed that the contacting bodies are absolutely rigid, then the pressures acting on them can be determined by using the hydrodynamic theory of lubricants. The ability of bodies to become elastically deformable turns out to be important. Wherever in the lubricant layer there are large pressures, there are noticeably large deformations, and the lubricant layer becomes thicker than it would have been under the assumption that the bodies were rigid. Thus the elasticity of the contacting bodies leads to a certain evening out of the pressure; when there is a lubricant layer between elastic bodies, one must solve an elasticity problem at the same time as a lubricant hydrodynamic problem. Petrusevich (1951) considered certain problems of this type, and found some approximate solutions.

1.5 Plane Contact between Two Elastic Bodies

In the problems listed in the preceding sections it was supposed that one of the bodies in the situation is rigid. Such an assumption clearly imposes limitations, since actually both bodies will be elastic. There are, however, cases in which it is possible to neglect the elasticity of one of the bodies. This happens when the modulus of elasticity of one body is considerably larger than that of the other. For example, in foundation problems, the elastic constant of the foundation is much higher than that of the ground; the foundation may be considered to be a punch.

However, punch problems are important because some solutions for problems relating to the contact of two elastic bodies can be obtained directly from the solution of a related contact problem for a punch and an elastic body. The form of the solution remains the same; the only difference is in the values of the elastic constants.

The problem of the contact of two elastic bodies was first considered by Hertz (1895). He neglected all but the leading terms in the equations defining the contacting bodies, and in this way reduced the problem to that of two contacting elliptical paraboloids. In particular he found the solution of the plane contact problem for two parabolic cylinders whose axes are parallel. These results served as the basis of research into contact problems in the theory of elasticity. The solution of these problems was achieved by means of a semi-inverse method, using the expression for the potential for an elliptical disc.

The problem of the contact of two elastic bodies was also considered in Dinnik (1906). Some investigations devoted to the contact of elastic bodies appear in Belyaev (1917, 1924, 1929a, 1929b). He determined the position of the point inside the elastic body where the greatest stresses occur. In these works the contact was assumed to be frictionless. All these papers deal, strictly speaking, with three-dimensional contact problems. Plane problems are obtainable from these general solutions as particular cases.

Shtaerman (1941a) considered the contact of an elastic cylinder and a cylindrical hole in space whose radii are nearly equal (pin in a hole with a small clearance). This problem was considered twice by Shtaerman. His second method, slightly different from the earlier one, is cited in his book *Contact Problems in the Theory of Elasticity* (1949). Here the question has been reduced to an integral equation, analogous to that appearing in the problem of a wing of finite span.

The contact problem for two elastic bodies becomes simpler when the elastic constants of the contacting bodies are the same. Then it is possible to solve problems which would otherwise be difficult, for instance the contact problem when the contact region is subdivided into slip and stick zones, and also the contact problem for two cylinders with nearly equal radii.

Fromm (1927) considered the rolling of an elastic cylinder over an elastic half plane. He assumed that the elastic constants of both bodies are the same, and that the contact region consists of two zones, slip and stick. By making certain assumptions we can use the solution of this problem to obtain the value of the coefficient of rolling friction by reducing it to sliding friction. The rolling of elastic bodies over one another was considered by Glagolev (1945); he employed more refined methods than Fromm. This problem was considered by Glagolev in two papers with different assumptions as to the character of the distribution of zones into which the contact region is divided. It should be noticed that the boundaries of these zones are unknown; this presents one of the main difficulties of the solution.

Narodetskii (1943) considered the contact of two cylinders, with the same elastic constants, whose radii are nearly equal. He supposed that the force producing the pressure was applied at the centre of one of the cylinders. The dimensions of the contact region, and the distribution of pressure over it were determined. The solution of this problem turned out to be quite simple, much simpler than the analogous problem in which the inner cylinder is rigid.

1.6 Three-Dimensional Contact Problems

Three-dimensional contact problems present considerably greater difficulties than plane problems; this is characteristic of all spatial problems in the theory of elasticity. The premisses are as follows: The contact is frictionless. One of the elastic bodies occupies an elastic half space. That body is in contact with another, rigid or elastic, body. In the latter case it too can be replaced by an elastic half space. Only for certain problems, which will be discussed later, is it possible to find a solution of a three-dimensional axisymmetric frictional or adhesive contact problem.

Frictionless contact problems can be reduced to potential problems. Here it is necessary to find a harmonic function, zero at infinity, whose value is given on both sides of a plane contact region. Unfortunately many of the solutions of potential problems that had been found many years ago, and which could have been used in investigating contact problems, have been largely forgotten; they had to be rediscovered, and were so, but often in incomplete form. In many cases it is possible to obtain solutions of space contact problems by making use of a properly chosen system of coordinates. This system should be selected in such a way that a function satisfying Laplace's equation can be expressed as a product of three functions, each of which depends on just one variable. In addition, one of the coordinate surfaces should degenerate into a two-sided plane region coinciding with the contact region. Clearly these conditions severely restrict the class of problems for which it is possible to find an effective solution.

The first works devoted to space contact problems are due to Hertz (1895) (the date refers to the date of publication of his collected works; the research was carried out in the 1880s) and Boussinesq (1885). Hertz considered the contact of two elastic bodies which are bounded by surfaces of the second order (elliptical cylinders and elliptical paraboloids). To solve the problems he made use of an electrostatic analogy in which the harmonic function which serves to determine the pressure arising in the contact region is the potential for a certain ellipsoid. He established that the contact region was an ellipse.

Boussinesq (1885) considered the problem of the pressure of a rigid body, a punch with a circular base, pressed into an elastic half space. See also Sneddon (1944, 1946, 1947, 1948, 1951). Just as Western researchers knew little of the work of their Soviet colleagues in the immediate aftermath (1945–1953) of the Second World War, so were the Soviet researchers unfamiliar with European and U.S. work.

Most of the results on three-dimensional contact problems were obtained by Soviet scientists. Some problems were investigated by Dinnik (1906) and Belyaev (1917, 1924, 1929a). Belyaev carried out a detailed analysis of stresses arising in an elastic body. He determined the position where the tangential stress reaches its maximum value, and he found the maximum. Belyaev (1929a) applied his methods for solving elastic contact problems to the calculation of stresses in rails.

For a long time no substantial progress was made in three-dimensional contact problems. One of the first new problems that was considered was that of the pressure under an inclined circular cylinder with a plane base pressed into an elastic half space. The formula giving the pressure distribution is quite simple. However it was obtained by complicated means, by solving an integral equation. It was later shown how this result, and also some analogous ones, could be obtained by a simpler and more natural method. The problem, whose solution was first given by Abramov (1939), was later considered by Borowicka (1943).

Shtaerman (1939) gave the solution of the contact problem in which one of the contacting bodies is a paraboloid with degree higher than two. In the theory developed by Hertz it was supposed that the radii of curvature of the surfaces bounding the contacting bodies are large in comparison with the dimensions of the region. Therefore only the leading terms were preserved in the equations of these surfaces, and the problem was reduced to the contact of two bodies bounded by surfaces of the second order, namely elliptical cylinders or elliptical paraboloids. This assumption is appropriate for comparatively smooth bodies. For less smooth bodies, the unevenness of the surfaces might be of the same order as the dimensions of the contact region. In addition, we should remember that there are problems in which the contact region is large, and therefore we should take more exact expressions for the shape of the contacting bodies. Thus, problems like this take on new interest, as they concern closer contact than occurs in the usual theory of elastic contact.

Lur'e (1941) considered the solution of contact problems in which the contact region is a circle or an ellipse. He supposed that one of the contacting bodies is a paraboloid of any, including non-integer degree. In particular he obtains the solution for a rigid cone indenting an elastic half space. This had been obtained earlier by Love (1939) by other means. As to problems in which the contact region is elliptical, Lur'e (1939) also investigated the pressure of an elliptical cylinder on an elastic half space, and considered the cases when base of the elliptical cylinder is either plane, or a surface whose equation in rectangular coordinates is a polynomial of the second degree.

Shtaerman (1941b, 1941c) also considered cases in which the contact region is elliptical. In the first paper he used Lamé functions to determine the stresses between the bodies; in the other he investigates the problem in which there is closer contact than in the usual theory. Leonov (1939) gives a more general solution of axisymmetric problems than was found by Shtaerman or Lur'e. One of the contacting bodies is an arbitrary solid of revolution. The problem is reduced to the solution of a pair of integral equations, and the pressure arising under the punch is obtained by integration. Recently Harding and Sneddon (1945) obtained substantially the same results.

Leonov (1940) considered the axisymmetric bending of a circular plate on an elastic half space. He reduced the problem to a Fredholm integral equation of the first kind. Reissner and Sagoci (1944) gave the solution to an axisymmetric problem in which a punch, rigidly connected to an elastic body, rotates about the axis of symmetry.

A number of investigations into three-dimensional problems, due to the author, are contained in this book. We give a further generalisation of the problem of the pressure of a punch of circular cross-section indenting an elastic half space. Here we suppose that the base of the punch is bounded by a surface which can be represented by a function of two coordinates, as in Galin (1946). We investigate the case in which the punch is a solid of revolution. We determine the relation between the displacement of the punch and the force acting on it. In addition, we consider axisymmetric frictional punch problems. In this case, the punch, which is pressed against the elastic body, rotates about its own axis, and the frictional forces have axial symmetry. It appears that the stresses produced by these friction forces can be superimposed on those arising in the frictionless case. We also consider the influence of a load acting outside the circular punch on the distribution of pressure under the punch.

In Galin (1947a) we investigated punches of elliptical cross-section. Here we showed that if the equation of the base of the punch is a polynomial, then the pressure under the punch is also a polynomial, of the same degree, divided by a simple algebraic function. In Galin (1948a) we obtained an estimate for the force which should act on a punch of an arbitrary cross-section to produce a given displacement. In Galin (1947b) we gave an approximate solution for the problem of the pressure of a wedge-shaped punch on an elastic half space. This solution establishes a law for the distribution of pressure near a vertex of a punch of polygonal, say rectangular, cross-section.

In Galin (1943b) we investigated a punch with slender cross-section. Such a problem arises when a slender beam exerts pressure on an elastic half space. Here, in the limit as the width of the beam tends to zero, the mean pressure across the width is proportional to the corresponding displacement.

In all these problems it was supposed that the elastic body on which the punch exerts pressure is sufficiently large to be represented as an elastic half space. There is an interesting limiting case in which the punch presses into a lamina (Galin, 1948b). If, as usual, we preserve only the leading terms in the equation of the surface bounding the punch (this is equivalent to assuming that the punch is an elliptical paraboloid) then the contact region is elliptical, as it is when the elastic body is a half space.

Problems relating to the bending of beams and plates on an elastic half space also belong among contact problems. As we do not discuss such problems in this book, we restrict ourselves to a short review of the principal results. Gorbunov-Posadov (1949) discusses such questions in his book Beams and Plates on Elastic Foundation, and gives an extensive bibliography. Proktor (1922) seems to have been the first to consider a beam on an elastic half space, rather than a simple so-called Winkler foundation. Kuznetsov (1938, 1939, 1940) considered the bending of beams on an elastic half space. The pressure under the beam was obtained from integral and integro-differential equations, which were solved numerically. The pressure was found as a polynomial whose coefficients were determined from a system of simultaneous equations.

Gorbunov-Posadov (1939b, 1941, 1946a, 1946b, 1946c, 1948) wrote many papers devoted to plates on elastic half spaces; they are listed below. He considered a number of problems related to the bending of infinite or semi-infinite beams and plates, and he studied punches of rectangular cross-section pressed into an elastic half space. He supposed that the pressure under the beam or plate could be represented as a polynomial, and determined the coefficients by solving simultaneous linear equations. Shekhter (1940b) considered some problems concerning a plate situated on a layer of finite thickness. Filippov (1942) also investigated the problem of a beam of infinite length resting on an elastic half space.

Zhemochkin (1937, 1938) and Zhemochkin and Sinitsin (1948) considered the bending of beams on an elastic half space. They subdivided the beam into segments, and supposed that the pressure was constant in each segment. The force causing the deformation of the beam was applied at a point. Their method combines those used in the theory of elasticity and in strength of materials. Ishkova (1947) considered the bending of a circular plate on an elastic half space.

Problems concerning beams and plates on an elastic half space can be reduced to integral and integro-differential equations. In all the papers listed above it is really the approximate solution of these equations that is discussed. Sometimes these approximate methods give satisfactory results, however the investigation of closed form solutions of these equations, or at least bounds for the errors in the approximate solutions, requires further attention.

Chapter 2
Plane Elasticity Theory

2.1 The Fundamental Equations

The state of stress in a plane elastic body is determined by three components of stress: $\sigma_{xx}, \sigma_{xy}, \sigma_{yy}$.

Recall that σ_{xx}, σ_{xy} *are the components, in the* x*- and* y*-directions, of the force per unit area exerted, at* (x, y)*, on the plane normal to the* x*-axis, applied from the* $x+$ *side to the* $x-$ *side, as shown in Figure 2.1.1.*

Fig. 2.1.1 (a) Stress $\sigma_{xx}\hat{\imath} + \sigma_{xy}\hat{\jmath}$ acts on a plane with normal $\hat{\imath}$. (b) Stress $\sigma_{xy}\hat{\imath} + \sigma_{yy}\hat{\jmath}$acts on a plane with normal $\hat{\jmath}$.

Similarly, $\sigma_{yx} \equiv \sigma_{xy}$*, and* σ_{yy} *are the components of the force per unit area on the plane normal to the* y*-axis, again applied from the side* $y+$ *to the side* $y-$*.*

They satisfy the *equilibrium equations*

$$\frac{\partial \sigma_{xx}}{\partial x} + \frac{\partial \sigma_{xy}}{\partial y} = 0 = \frac{\partial \sigma_{xy}}{\partial x} + \frac{\partial \sigma_{yy}}{\partial y}. \tag{2.1.1}$$

The deformation of the elastic body can be expressed by the relative extensions $\varepsilon_{xx}, \varepsilon_{yy}$ in the directions of the x- and y-axes, and by the angular rotation ε_{xy}.

The *elastic strains* $\varepsilon_{xx}, \varepsilon_{yy}, \varepsilon_{xy}$ are related to the components u, v of the *elastic displacement vector* (u, v) by the equations

$$\varepsilon_{xx} = \frac{\partial u}{\partial x}, \quad \varepsilon_{xy} = \frac{1}{2}\left(\frac{\partial u}{\partial y} + \frac{\partial v}{\partial x}\right), \quad \varepsilon_{yy} = \frac{\partial v}{\partial y}. \tag{2.1.2}$$

Since the *three* strain components are expressed in terms of *two* displacement components, there should be some relationship between them. This is called the *strain compatibility condition*:

$$\frac{\partial^2 \varepsilon_{xx}}{\partial y^2} + \frac{\partial^2 \varepsilon_{yy}}{\partial x^2} = \frac{2\partial^2 \varepsilon_{xy}}{\partial x \partial y}. \tag{2.1.3}$$

There are two variants of plane elastic problems: plane strain *and* plane stress. *Plane strain is an idealisation of the elastic state in an infinitely long cylinder with its axis being the z-axis, and acted upon by forces in the* x, y*-plane that are independent of z. In plane strain, u and v are functions of* x, y *only, while w, the elastic displacement in the z-direction, is zero:*

$$u = u(x, y), \quad v = v(x, y), \quad w = 0 \tag{2.1.4}$$

so that the strains

$$\varepsilon_{xz} = \frac{1}{2}\left(\frac{\partial u}{\partial z} + \frac{\partial w}{\partial x}\right), \quad \varepsilon_{yz} = \frac{1}{2}\left(\frac{\partial v}{\partial z} + \frac{\partial w}{\partial y}\right), \quad \varepsilon_{zz} = \frac{\partial w}{\partial z}, \tag{2.1.5}$$

are all zero.

Plane stress *is an idealisation of the state of stress in a thin plate acted on by forces in its plane; in plane stress* $\sigma_{xz} = 0 = \sigma_{yz} = \sigma_{zz}$.

In *plane strain*, the components of strain are related to the components of stress by the *stress-strain equations*

$$\sigma_{xx} = \lambda\theta + 2\mu\varepsilon_{xx}, \quad \sigma_{yy} = \lambda\theta + 2\mu\varepsilon_{yy}, \quad \sigma_{xy} = 2\mu\varepsilon_{xy}. \tag{2.1.6}$$

Here $\theta = \varepsilon_{xx} + \varepsilon_{yy}$ is the relative increase of volume, or *dilatation*, and λ, μ are *Lamé's constants*. The strains, in their turn, may be related to the stresses by the equations

$$2\mu\varepsilon_{xx} = \sigma_{xx} - \nu(\sigma_{xx} + \sigma_{yy}), \quad 2\mu\varepsilon_{yy} = \sigma_{yy} - \nu(\sigma_{xx} + \sigma_{yy}), \tag{2.1.7}$$
$$2\mu\varepsilon_{xy} = \sigma_{xy}$$

The *Young's modulus*, E, and *Poisson's ratio*, ν, are related to λ and μ by

$$E = \frac{\mu(3\lambda + 2\mu)}{\lambda + \mu}, \quad \nu = \frac{\lambda}{2(\lambda + \mu)}, \tag{2.1.8}$$

and the inverse equations

$$\lambda = \frac{E\nu}{(1+\nu)(1-2\nu)}, \qquad \mu = \frac{E}{2(1+\nu)}. \tag{2.1.9}$$

In plane stress we start with the full stress-strain equations *for an isotropic elastic body*:

$$\sigma_{xx} = \lambda(\varepsilon_{xx}+\varepsilon_{yy}+\varepsilon_{zz}) + 2\mu\varepsilon_{xx}, \quad \sigma_{yy} = \lambda(\varepsilon_{xx}+\varepsilon_{yy}+\varepsilon_{zz}) + 2\mu\varepsilon_{yy},$$
$$\sigma_{zz} = \lambda(\varepsilon_{xx}+\varepsilon_{yy}+\varepsilon_{zz}) + 2\mu\varepsilon_{zz},$$
$$\sigma_{xz} = 2\mu\varepsilon_{xz}, \quad \sigma_{yz} = 2\mu\varepsilon_{yz}, \quad \sigma_{xy} = 2\mu\varepsilon_{xy}. \tag{2.1.10}$$

If $\sigma_{zz} = 0$*, then* $\lambda(\varepsilon_{xx}+\varepsilon_{yy}+\varepsilon_{zz})+2\mu\varepsilon_{zz} = 0$*. Thus* $\varepsilon_{zz} = -\lambda(\varepsilon_{xx}+\varepsilon_{yy})/(\lambda+2\mu)$ *and* $\varepsilon_{xx}+\varepsilon_{yy}+\varepsilon_{zz} = [1-\lambda/(\lambda+2\mu)](\varepsilon_{xx}+\varepsilon_{yy})$*. Thus*

$$\sigma_{xx} = \lambda^*\theta + 2\mu\varepsilon_{xx}, \quad \sigma_{yy} = \lambda^*\theta + 2\mu\varepsilon_{yy}, \quad \sigma_{xy} = 2\mu\varepsilon_{xy}, \tag{2.1.11}$$

where

$$\lambda^* = \lambda[1-\lambda/(\lambda+2\mu)] = 2\mu\lambda/(\lambda+2\mu). \tag{2.1.12}$$

The effective Poisson's ratio for plane stress is therefore

$$\nu^* = \frac{\lambda^*}{2(\lambda^*+\mu)} = \frac{\nu}{1+\nu}. \tag{2.1.13}$$

We conclude that the equations for plane stress may be derived from those of plane strain by replacing μ, λ *by* μ, λ^**; or equivalently,* μ, ν *by* μ, ν^*.

If the expressions (2.1.7) are substituted in the compatibility equation (2.1.3), we find

$$\frac{\partial^2\sigma_{xx}}{\partial y^2} - \frac{\nu\partial^2}{\partial y^2}(\sigma_{xx}+\sigma_{yy}) + \frac{\partial^2\sigma_{yy}}{\partial x^2} - \frac{\nu\partial^2}{\partial x^2}(\sigma_{xx}+\sigma_{yy}) = \frac{2\partial^2\sigma_{xy}}{\partial x\partial y}.$$

The equilibrium equations (2.1.1) yield

$$\frac{2\partial^2\sigma_{xy}}{\partial x\partial y} = -\frac{\partial^2\sigma_{xx}}{\partial x^2} - \frac{\partial^2\sigma_{yy}}{\partial y^2}$$

so that

$$\left(\frac{\partial^2}{\partial x^2} + \frac{\partial^2}{\partial y^2}\right)(\sigma_{xx}+\sigma_{yy}) = 0. \tag{2.1.14}$$

We write this as

$$\Delta(\sigma_{xx}+\sigma_{yy}) = 0. \tag{2.1.15}$$

The equilibrium equations (2.1.1) are satisfied if

$$\sigma_{xx} = \frac{\partial^2 U}{\partial y^2}, \quad \sigma_{xy} = -\frac{\partial^2 U}{\partial x\partial y}, \quad \sigma_{yy} = \frac{\partial^2 U}{\partial x^2} \tag{2.1.16}$$

where $U(x, y)$ is called the *Airy stress function*. Substituting (2.1.16) into (2.1.15) we find that

$$\frac{\partial^4 U}{\partial x^4} + \frac{2\partial^4 U}{\partial x^2 \partial y^2} + \frac{\partial^4 U}{\partial y^4} = 0. \tag{2.1.17}$$

(Galin uses Φ for the Airy stress function, and then later uses Φ for one of the complex potentials. We use U here to avoid confusion.) This equation, called the *biharmonic equation*, is written

$$\Delta^2 U = 0; \tag{2.1.18}$$

We say that U is *biharmonic.*

In solving plane contact problems we shall make use of *complex variable methods*.

We now give a brief account of a method of formulating the basic equations for plane problems in the theory of elasticity. The detailed exposition of these matters may be found in Muskhelishvili (1953) or Gladwell (1980).

If we substitute the expressions (2.1.6) for $\sigma_{xx}, \sigma_{xy}, \sigma_{yy}$ into the equations (2.1.16), and express $\varepsilon_{xx}, \varepsilon_{xy}, \varepsilon_{yy}$ in terms of u, v we find

$$\lambda \left(\frac{\partial u}{\partial x} + \frac{\partial v}{\partial y}\right) + 2\mu \frac{\partial u}{\partial x} = \frac{\partial^2 U}{\partial y^2}, \tag{2.1.19}$$

$$\lambda \left(\frac{\partial u}{\partial x} + \frac{\partial v}{\partial y}\right) + 2\mu \frac{\partial v}{\partial y} = \frac{\partial^2 U}{\partial x^2}, \tag{2.1.20}$$

$$\mu \left(\frac{\partial u}{\partial y} + \frac{\partial v}{\partial x}\right) = -\frac{\partial^2 U}{\partial x \partial y}. \tag{2.1.21}$$

We may determine $\frac{\partial u}{\partial x}$ and $\frac{\partial v}{\partial y}$ from the first two equations. Introducing the notation

$$\frac{\partial^2 U}{\partial x^2} + \frac{\partial^2 U}{\partial y^2} = \Delta U = P \tag{2.1.22}$$

we find

$$2\mu \frac{\partial u}{\partial x} = -\frac{\partial^2 U}{\partial x^2} + \frac{\lambda + 2\mu}{2(\lambda + \mu)} P, \tag{2.1.23}$$

$$2\mu \frac{\partial v}{\partial y} = -\frac{\partial^2 U}{\partial y^2} + \frac{\lambda + 2\mu}{2(\lambda + \mu)} P, \tag{2.1.24}$$

since U is biharmonic, $P = \Delta U$ is harmonic.

$P(x, y)$ and $Q(x, y)$ are said to be conjugate if they satisfy the Cauchy–Riemann equations

$$\frac{\partial P}{\partial x} = \frac{\partial Q}{\partial y}, \quad \frac{\partial P}{\partial y} = -\frac{\partial Q}{\partial x}. \tag{2.1.25}$$

If P is harmonic, so is Q, because

$$\frac{\partial^2 Q}{\partial x^2} + \frac{\partial^2 Q}{\partial y^2} = -\frac{\partial^2 P}{\partial x \partial y} + \frac{\partial^2 P}{\partial y \partial x} = 0.$$

Moreover, $P + iQ$ is a function of the complex variable $z = x + iy$. Put $\bar{z} = x - iy$ and $f(x, y) = P(x, y) + iQ(x, y)$, then $x = (z + \bar{z})/2$, $y = (z - \bar{z})/2i$ and we can write f as a function of z and $\bar{z}$; $\partial x/\partial \bar{z} = 1/2$, $\partial y/\partial \bar{z} = -1/2i = i/2$. Thus

$$\frac{\partial f}{\partial \bar{z}} = \frac{\partial f}{\partial x}\frac{\partial x}{\partial \bar{z}} + \frac{\partial f}{\partial y}\frac{\partial y}{\partial \bar{z}} = \frac{1}{2}\left(\frac{\partial f}{\partial x} + \frac{i\partial f}{\partial y}\right)$$

$$= \frac{1}{2}\left(\frac{\partial P}{\partial x} + \frac{i\partial Q}{\partial x} + \frac{i\partial P}{\partial y} - \frac{\partial Q}{\partial y}\right) = 0.$$

We conclude that f is a function of z only.

Thus, if Q is conjugate to P then

$$P(x, y) + iQ(x, y) = f(z). \tag{2.1.26}$$

Let us now find the expressions for the displacements u, v. Introduce two more conjugate harmonic functions p, q, where

$$\phi(z) = p + iq = \frac{1}{4}\int f(z)dz, \tag{2.1.27}$$

then since

$$\frac{\partial x}{\partial z} = \frac{1}{2}, \quad \frac{\partial y}{\partial z} = -\frac{i}{2},$$

we have

$$\phi'(z) = \frac{\partial}{\partial z}(p + iq) = \frac{1}{2}\frac{\partial}{\partial x}(p + iq) - \frac{i}{2}\frac{\partial}{\partial y}(p + iq),$$

$$= \frac{1}{2}\frac{\partial p}{\partial x} + \frac{i}{2}\frac{\partial q}{\partial x} - \frac{i\partial p}{2\partial y} + \frac{1}{2}\frac{\partial q}{\partial y}.$$

The Cauchy–Riemann, equations are

$$\frac{\partial p}{\partial x} = \frac{\partial q}{\partial y}, \quad \frac{\partial p}{\partial y} = -\frac{\partial q}{\partial x}$$

so that

$$\phi'(z) = \frac{\partial p}{\partial x} + \frac{i\partial q}{\partial x} = \frac{1}{4}(P + iQ),$$

and

$$\frac{\partial p}{\partial x} = \frac{\partial q}{\partial y} = \frac{1}{4}P, \quad \frac{\partial q}{\partial x} = -\frac{\partial p}{\partial y} = \frac{1}{4}Q. \tag{2.1.28}$$

This means that equations (2.1.23), (2.1.24) may be written

$$2\mu\frac{\partial u}{\partial x} = -\frac{\partial^2 U}{\partial x^2} + \frac{2(\lambda+2\mu)}{\lambda+\mu}\frac{\partial p}{\partial x},$$

$$2\mu\frac{\partial v}{\partial y} = -\frac{\partial^2 U}{\partial y^2} + \frac{2(\lambda+2\mu)}{\lambda+\mu}\frac{\partial q}{\partial y}.$$

After integration, we get

$$2\mu u = -\frac{\partial U}{\partial x} + \frac{2(\lambda+2\mu)}{\lambda+\mu}p + f_1(y), \tag{2.1.29}$$

$$2\mu v = -\frac{\partial U}{\partial y} + \frac{2(\lambda+2\mu)}{\lambda+\mu}q + f_2(x), \tag{2.1.30}$$

and after substitution into (2.1.21) we find

$$f_1'(y) + f_2'(x) = 0.$$

Since each term in this equation must be constant, we find

$$f_1'(y) = 2\mu\gamma, \quad f_2'(x) = -2\mu\gamma$$

so that

$$f_1(y) = 2\mu(\gamma y + \alpha), \quad f_2(x) = 2\mu(-\gamma x + \beta).$$

They correspond to a rigid body displacement and rotation:

$$u = \gamma y + \alpha, \quad v = -\gamma x + \beta$$

and we can omit them.

Let us now form the function $U - px - qy$. It is harmonic because equation (2.1.28) gives

$$\Delta(U - px - qy) = P - \frac{2\partial p}{\partial x} - \frac{2\partial q}{\partial y} = 0.$$

Since $U - px - qy$ is harmonic, it is the real part of a function $\chi(z)$, i.e.,

$$2(U - px - qy) = \chi(z) + \overline{\chi(z)}.$$

But

$$\begin{aligned}2(px + qy) &= (x - iy)(p + iq) + (x + iy)(p - iq)\\ &= \bar{z}\phi(z) + z\overline{\phi(z)}\end{aligned}$$

so that

$$2U = \bar{z}\phi(z) + z\overline{\phi(z)} + \chi(z) + \overline{\chi(z)} \tag{2.1.31}$$

and

$$\frac{\partial U}{\partial x} + i\frac{\partial U}{\partial y} = \frac{2\partial U}{\partial \bar{z}} = \phi(z) + z\overline{\phi'(z)} + \overline{\psi(z)} \tag{2.1.32}$$

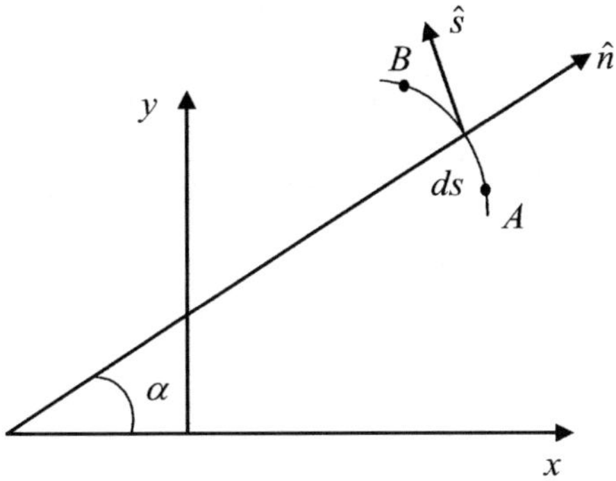

Fig. 2.1.2 The directions $\hat{n}$ and $\hat{s}$.

where

$$\psi(z) = \chi'(z). \tag{2.1.33}$$

Now, using (2.1.29), (2.1.30) we find

$$2\mu(u + iv) = -\left(\frac{\partial U}{\partial x} + i\frac{\partial U}{\partial y}\right) + 2\frac{(\lambda + 2\mu)}{\lambda + \mu}(p + iq);$$

we have neglected $f_1(y)$ and $f_2(x)$. Substituting from (2.1.27), (2.1.32) we find

$$2\mu(u + iv) = \kappa\phi(z) - z\overline{\phi'(z)} - \overline{\psi(z)}, \tag{2.1.34}$$

where

$$\kappa = \frac{\lambda + 3\mu}{\lambda + \mu} = 3 - 4\nu. \tag{2.1.35}$$

Note that for plane stress,

$$\kappa^* = 3 - 4\nu^* = (3 - \nu)/(1 + \nu). \tag{2.1.36}$$

If $\nu = 0{\cdot}3$, then $\kappa = 1{\cdot}8$, $\kappa^* = 2{\cdot}08$.

We now express $\sigma_{xx}, \sigma_{xy}, \sigma_{yy}$, and also certain complex combinations of these quantities, in terms of $\phi(z)$ and $\psi(z)$.

Consider the arc AB situated in the region occupied by the elastic body (Figure 2.1.2) and denote the length of the arc measured in the positive direction from A to B by ds.

We denote the normal to the arc AB by $\hat{n}$; we take as positive the direction along the normal lying to the right of an observer moving along the arc from A to B. We denote the components of force acting on ds from the direction of the outside normal, i.e., in Figure 2.1.2, from the upper right to the lower left, by $X_n ds$ and $Y_n ds$. In terms of the stress components we have

$$\sigma_{xx} = X_x, \quad \sigma_{xy} = X_y = Y_x, \quad \sigma_{yy} = Y_y. \tag{2.1.37}$$

The components X_n, Y_n are

$$X_n = \sigma_{xx}\cos\alpha + \sigma_{xy}\sin\alpha, \qquad Y_n = \sigma_{xy}\cos\alpha + \sigma_{yy}\sin\alpha,$$

so that on introducing the Airy stress function U and noting that

$$\frac{d}{ds} = (\hat{s}\cdot\nabla) = -\sin\alpha\frac{\partial}{\partial x} + \cos\alpha\frac{\partial}{\partial y},$$

we find

$$X_n = \cos\alpha\frac{\partial^2 U}{\partial y^2} - \sin\alpha\frac{\partial^2 U}{\partial x\partial y} = \frac{d}{ds}\left(\frac{\partial U}{\partial y}\right),$$

$$Y_n = -\cos\alpha\frac{\partial^2 U}{\partial x\partial y} + \sin\alpha\frac{\partial^2 U}{\partial x^2} = -\frac{d}{ds}\left(\frac{\partial U}{\partial x}\right),$$

so that

$$(X_n + iY_n)ds = \frac{d}{ds}\left(\frac{\partial U}{\partial y} - i\frac{\partial U}{\partial x}\right)ds = -i\frac{d}{ds}\left(\frac{\partial U}{\partial x} + i\frac{\partial U}{\partial y}\right)ds.$$

Substituting from (2.1.32) we have

$$(X_n + iY_n)ds = -i\frac{d}{ds}\left(\phi(z) + z\overline{\phi'}(\bar{z}) + \overline{\psi(z)}\right)ds. \tag{2.1.38}$$

Take $\hat{n}$ in the direction $\hat{j}$, then $\alpha = \frac{\pi}{2}$, and $X_n = \sigma_{xy}, \quad Y_n = \sigma_{yy}$,

$$\frac{d}{ds} = -\frac{\partial}{\partial x} = -\frac{\partial}{\partial z} - \frac{\partial}{\partial \bar{z}},$$

and

$$\sigma_{xy} + i\sigma_{yy} = i(\phi'(z) + \bar{\phi}'(\bar{z}) + z\bar{\phi}''(\bar{z}) + \bar{\psi}'(\bar{z})),$$

or

$$\sigma_{yy} - i\sigma_{xy} = \phi'(z) + \bar{\phi}'(\bar{z}) + z\bar{\phi}''(\bar{z}) + \bar{\psi}'(\bar{z}).$$

Similarly, taking $\hat{n}$ on the $\hat{\imath}$ direction, so that $\alpha = 0$,

$$\sigma_{xx} + i\sigma_{xy} = \phi'(z) + \bar{\phi}'(\bar{z}) - z\bar{\phi}''(\bar{z}) - \bar{\psi}'(\bar{z}).$$

Introduce the notation

$$\phi'(z) = \Phi(z), \qquad \psi'(z) = \Psi(z) \tag{2.1.39}$$

then these equations may be written

$$\sigma_{xx} + i\sigma_{xy} = \Phi(z) + \bar{\Phi}(\bar{z}) - z\bar{\Phi}'(\bar{z}) - \bar{\Psi}(\bar{z}), \tag{2.1.40}$$

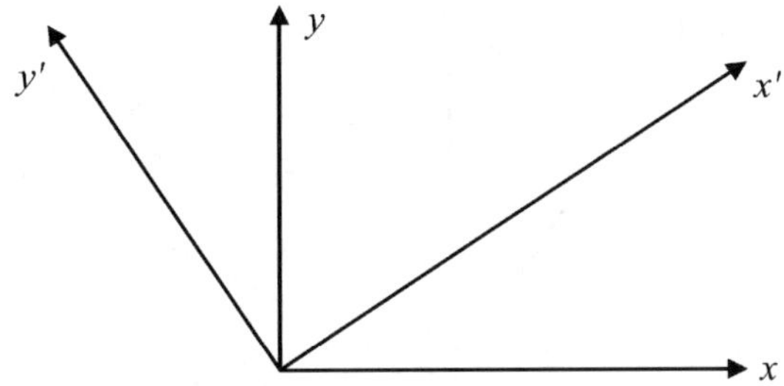

Fig. 2.1.3 The axes x', y'.

$$\sigma_{yy} - i\sigma_{xy} = \Phi(z) + \bar{\Phi}(\bar{z}) + z\bar{\Phi}'(\bar{z}) + \bar{\Psi}(\bar{z}). \tag{2.1.41}$$

These may be combined to give

$$\sigma_{xx} + \sigma_{yy} = 2[\Phi(z) + \bar{\Phi}(\bar{z})], \tag{2.1.42}$$

$$\sigma_{yy} - \sigma_{xx} + 2i\sigma_{xy} = 2[\bar{z}\Phi'(z) + \Psi(z)]. \tag{2.1.43}$$

The three stress components $\sigma_{xx}, \sigma_{yy}, \sigma_{xy}$ are (the only three non-zero) components of the stress tensor; they are the components $\sigma_{11}, \sigma_{22}, \sigma_{12}$ of the rank-two symmetric tensor with components σ_{ij}, $i, j = 1, 2, 3$ Under a change of axes, the components change according to the usual tensor law. In particular, if x', y' are axes as shown in Figure 2.1.3, then

$$\sigma_{x'x'} = \cos^2\alpha\sigma_{xx} + \sin^2\alpha\sigma_{yy} + 2\cos\alpha\sin\alpha\sigma_{xy} \tag{2.1.44}$$

$$\sigma_{y'y'} = \sin^2\alpha\sigma_{xx} + \cos^2\alpha\sigma_{yy} - 2\cos\alpha\sin\alpha\sigma_{xy} \tag{2.1.45}$$

$$\sigma_{x'y'} = -\cos\alpha\sin\alpha(\sigma_{xx} - \sigma_{yy}) + (\cos^2\alpha - \sin^2\alpha)\sigma_{xy}. \tag{2.1.46}$$

These may be rewritten as

$$\sigma_{x'x'} = \frac{\sigma_{xx} + \sigma_{yy}}{2} + \cos 2\alpha\left(\frac{\sigma_{xx} - \sigma_{yy}}{2}\right) + \sin 2\alpha\sigma_{xy} \tag{2.1.47}$$

$$\sigma_{y'y'} = \frac{\sigma_{xx} + \sigma_{yy}}{2} - \cos 2\alpha\left(\frac{\sigma_{xx} - \sigma_{yy}}{2}\right) - \sin 2\alpha\sigma_{xy} \tag{2.1.48}$$

$$\sigma_{x'y'} = -\sin 2\alpha\left(\frac{\sigma_{xx} - \sigma_{yy}}{2}\right) + \cos 2\alpha\sigma_{xy}. \tag{2.1.49}$$

The combinations $\sigma_{xx} + \sigma_{yy}$ and $\sigma_{yy} - \sigma_{xx} + 2i\sigma_{xy}$ are convenient for the investigation of the state of stress in an elastic body. The sum of the stresses, $\sigma_{xx} + \sigma_{yy}$, is an invariant: equations (2.1.47), (2.1.48) show that

$$\sigma_{xx} + \sigma_{yy} = \sigma_{x'x'} + \sigma_{y'y'}. \tag{2.1.50}$$

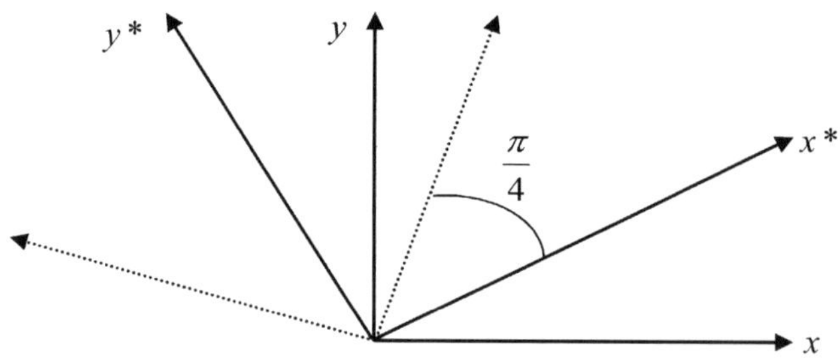

Fig. 2.1.4 The principal axes x^*, y^* and the directions corresponding to maximum shearing stress.

The principal directions of stress *at a given point, are those for which* $\sigma_{x'y'} = 0$. *Equation (2.1.49) shows that these are given by*

$$\sin 2\alpha \left(\frac{\sigma_{xx} - \sigma_{yy}}{2}\right) = \cos 2\alpha \sigma_{xy}. \tag{2.1.51}$$

We may choose two roots of this equation: α *and* $\alpha + \frac{\pi}{2}$. *Denote these two directions by* x^*, y^* *as in Figure 2.1.4, then*

$$\cos 2\alpha = \frac{\sigma_{xx} - \sigma_{yy}}{2\tau}, \quad \sin 2\alpha = \frac{\sigma_{xy}}{\tau} \tag{2.1.52}$$

where

$$\tau = \left\{\left(\frac{\sigma_{xx} - \sigma_{yy}}{2}\right)^2 + \sigma_{xy}^2\right\}^{\frac{1}{2}}. \tag{2.1.53}$$

When α *satisfies (2.1.51), then*

$$\cos 2\alpha \left(\frac{\sigma_{xx} - \sigma_{yy}}{2}\right) + \sin 2\alpha \sigma_{xy} = \tau$$

so that, from equations (2.1.47) and (2.1.48)

$$\sigma_{x^*x^*} = \frac{\sigma_{xx} + \sigma_{yy}}{2} + \tau$$

$$\sigma_{y^*y^*} = \frac{\sigma_{xx} + \sigma_{yy}}{2} - \tau$$

The maximum shearing stress *at the point occurs for the angles* β *given by*

$$\cos 2\beta = \frac{\sigma_{xy}}{\tau}, \quad \sin 2\beta = -\left(\frac{\sigma_{xx} - \sigma_{yy}}{2\tau}\right). \tag{2.1.54}$$

Clearly, combining (2.1.52), (2.1.54) we find

$$\cos 2\alpha \cos 2\beta + \sin 2\alpha \sin 2\beta = 0$$

so that $\cos(2\alpha - 2\beta) = 0$, *or* $\alpha - \beta = \pm\frac{\pi}{4}$. *This means that the directions of maximum shearing stress bisect the angles between the principal directions, as shown in Figure 2.1.4. The maximum shearing stress is*

$$\sigma_{x'y'} = -\sin 2\beta \left(\frac{\sigma_{xx} - \sigma_{yy}}{2}\right) + \cos 2\beta \sigma_{xy} = \tau :$$

the maximum shearing stress is τ.

The magnitude of $\sigma_{yy} - \sigma_{xx} + 2i\sigma_{xy}$ is twice the maximum shearing stress, τ, at the given point. The principal stresses at the point are (the values of $\sigma_{x^*x^*}, \sigma_{y^*y^*}$ respectively)

$$\sigma_1 = \frac{\sigma_{xx} + \sigma_{yy}}{2} + \tau, \tag{2.1.55}$$

$$\sigma_2 = \frac{\sigma_{xx} + \sigma_{yy}}{2} - \tau. \tag{2.1.56}$$

Now equations (2.1.42), (2.1.43) give

$$\tau = |\bar{z}\Phi'(z) + \Psi(z)| \tag{2.1.57}$$

$$\sigma_1 = \Phi(z) + \bar{\Phi}(\bar{z}) + |\bar{z}\Phi'(z) + \Psi(z)|, \tag{2.1.58}$$

$$\sigma_2 = \Phi(z) + \bar{\Phi}(\bar{z}) - |\bar{z}\Phi'(z) + \Psi(z)|. \tag{2.1.59}$$

2.2 Stresses and Displacements in a Semi-Infinite Elastic Plane

Usually, the linear dimensions of the area of contact are small compared with the radii of curvature of the touching bodies. Therefore, we assume when considering plane contact problems, that the elastic body which is subjected to the pressure of the punch is semi-infinite. For plane problems, we assume that the elastic body occupies a semi-infinite plane. This assumption somewhat distorts the picture of the state of stress. However, this distortion is appreciable only fairly far away from the contact region.

In this chapter we give the solutions for a number of plane contact problems. Some results appear for the first time, others were given earlier, in particular, in the third edition of Muskhelishvili (1953).

However, we employ a slightly different method for solving these problems. In Muskhelishvili (1953), the problem is reduced to the determination of the functions $\Phi(z)$ and $\Psi(z)$ in (2.1.39). In this section, we introduce the functions $w_1(z)$ and $w_2(z)$ which are integrals of Cauchy type, whose densities are the normal pressure and tangential load acting on the boundary. $\Phi(z)$ and $\Psi(z)$, from which the state of stress in an elastic half-plane can be found, are easily determined from $w_1(z)$ and $w_2(z)$.

The functions $w_1(z)$ and $w_2(z)$ have many advantages: anisotropic contact problems, problems for a moving punch, and also more complicated problems (those with zones of various types on the contact region) can be reduced to mixed boundary value problems for these functions.

Thus, we shall proceed to determine the stresses and displacements in a half-plane on whose boundary normal pressure is applied and tangential stress is distributed.

We shall make use of equation (2.1.41):

$$\sigma_{yy} - i\sigma_{xy} = \Phi(z) + \bar{\Phi}(\bar{z}) + z\bar{\Phi}'(\bar{z}) + \bar{\Psi}(\bar{z}).$$

We consider this complex combination of stresses for the half-plane under the assumption that the stresses tend to zero at infinity. This implies the following behaviour at infinity, i.e., for large values of $|z|$:

$$\Phi(z) = \frac{\gamma_1}{z} + o\left(\frac{1}{z}\right), \quad \Psi(z) = \frac{\gamma_2}{z} + o\left(\frac{1}{z}\right)$$

$$\Phi'(z) = -\frac{\gamma_1}{z^2} + o\left(\frac{1}{z^2}\right)$$

We recall the elements of the theory of Cauchy integrals, see Gladwell (1980) for a fuller version.

We start with the definition of a holomorphic *function of a complex variable z. The function $f(z)$ is said to be holomorphic (sometimes the term regular is used) in a finite region Sof the complex plane if it is single-valued in S, and its complex derivative $f'(z)$ exists at every point in S. The condition that $f(z)$ have a complex derivative is so strong that it may be proved that if $f(z)$ is holomorphic in S, then it will possess complex derivatives $f^{(n)}(z)$ of any finite order at every point in S, so that each such derivative will also be holomorphic in S. (Note the contrast with functions of a real variable, where the existence of $f''(x)$ by no means follows from the existence of $f'(x)$.) Further, it may be expanded in a series*

$$f(x) = \sum_{n=0}^{\infty} a_n(z - z_0)^n$$

about any point $z_0 \in S$. If the region S is infinite, then $f(z)$ is said to be holomorphic at infinity *if $f(1/z)$ is holomorphic at the origin. This means that, for large $|z|$, $f(z)$ may be expanded in the form*

$$f(z) = \sum_{n=0}^{\infty} b_n z^{-n}.$$

If $f(z)$ is holomorphic in the entire complex plane, except the point at infinity, then it must be a polynomial in z. If, in addition it is holomorphic at infinity, then it must be a constant.

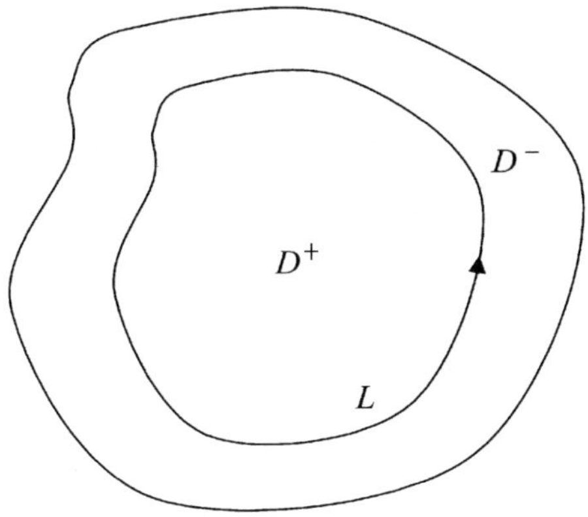

Fig. 2.2.1 The contour L divides S into D^+ and D^-.

Now we introduce

Theorem 1 (Cauchy's Theorem). *If L is a simple closed contour lying wholly in a region S in which the function $f(z)$ is holomorphic, then*

$$\int_L f(t)dt = 0, \tag{2.2.1}$$

where we shall use t to denote the generic point of the contour L.

The contour L divides S into two parts, D^+ lying to the left, the inside of L, and D^- on the right, the outside, as shown in Figure 2.2.1.

Apply this theorem to the function $f(z) = 1/(z - z_0)$, which is holomorphic in any region excluding z_0. If z lies outside L, i.e., in D^-, then Cauchy's Theorem gives

$$\int_L \frac{dt}{t - z_0} = 0 \text{ for } z_0 \in D^-.$$

If z_0 lies inside L, i.e., $z_0 \in D^+$, then we construct the contour $L + C_1 + C_\varepsilon + C_2$, as shown in Figure 2.2.2, so that again z_0 lies outside the contour, and

$$\left\{\int_L + \int_{C_1} + \int_{C_\varepsilon} + \int_{C_2}\right\} \frac{dt}{t - z_0} = 0.$$

But the integrals along C_1, C_2 are equal and opposite, and the integral around C_ε may be evaluated by writing $t = z_0 + \varepsilon \exp(i\theta)$, $dt = i\varepsilon \exp(i\theta)$ so that, since C_ε is traversed clockwise,

$$-\int_{C_\varepsilon} \frac{dt}{t - z_0} = \int_0^{2\pi} \frac{i\varepsilon \exp(i\theta)}{\varepsilon \exp(i\theta)} d\theta = 2\pi i,$$

and therefore

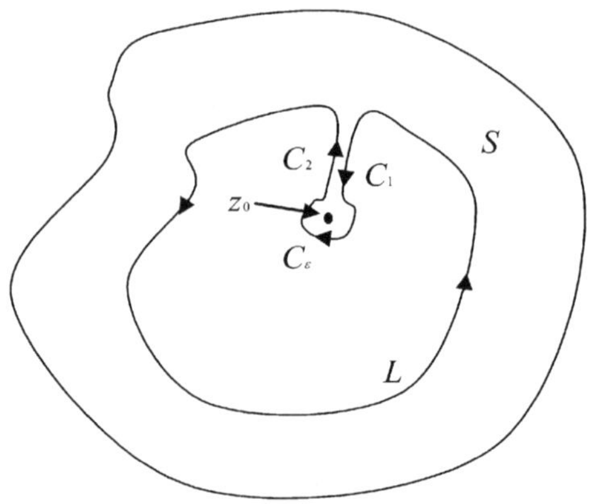

Fig. 2.2.2 The point z_0 lies outside the contour $L + C_1 + C_2 + C_\varepsilon$.

$$\frac{1}{2\pi i}\int_L \frac{dt}{t - z_0} = 1 \text{ for } z_0 \in D^+.$$

Now write

$$\int_L \frac{f(t)dt}{t - z_0} = \int_L \frac{f(t) - f(z_0)}{t - z_0}dt + f(z_0)\int_L \frac{dt}{t - z_0}.$$

If L is a closed contour lying in a region in which $f(z)$ is holomorphic, then $(f(z) - f(z_0))/(z - z_0)$ will also be holomorphic, so that the first integral will be zero, giving

$$\frac{1}{2\pi i}\int_L \frac{f(t)dt}{t - z_0} = \begin{cases} f(z_0), & \text{if } z_0 \in D^+ \\ 0, & \text{if } z_0 \in D^- \end{cases}. \tag{2.2.2}$$

We emphasize that this equation holds provided that $f(z)$ is holomorphic in S.

Now let L again be a simple closed contour, and let $f(t)$ be a function given and continuous on L; it need be defined only on L, not as a function in S. The equation

$$F(z) = \frac{1}{2\pi i}\int_L \frac{f(t)dt}{t - z} \tag{2.2.3}$$

defines a function which may easily be shown to be holomorphic everywhere except on L. Such a function is called a Cauchy integral. *If $f(t)$ happens to be the boundary value of a function $f(z)$ holomorphic in S then, according to (2.2.2),*

$$F(z) = \begin{cases} f(z), & \text{if } z \in D^+ \\ 0, & \text{if } z \in D^- \end{cases}. \tag{2.2.4}$$

Note, however, that $F(z)$ may be defined by (2.2.3) provided only that $f(t)$ is continuous on L. (Even this condition may be relaxed.)

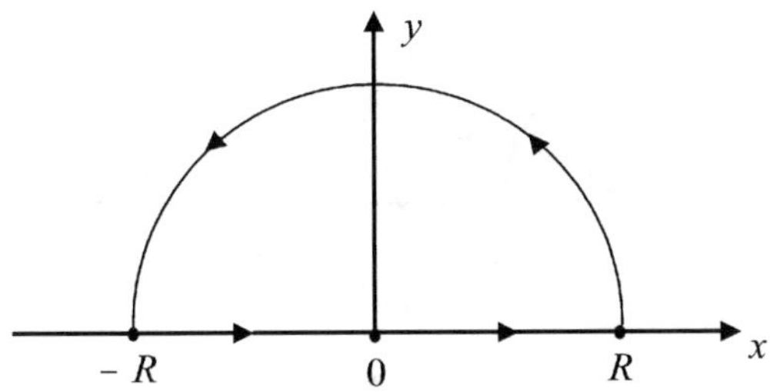

Fig. 2.2.3 A semi-circle in the upper half-plane.

We need to extend these results to the case in which L is the whole x-axis. Consider the contour shown in Figure 2.2.3 consisting of a semi-circle of radius R and the segment $(-R, R)$. If $f(z)$ is holomorphic in the upper half plane and

$$f(z) = \frac{\gamma}{z} + o\left(\frac{1}{z}\right)$$

at infinity, then equation (2.2.2) gives

$$\frac{1}{2\pi i}\int_{C_R} \frac{f(t)dt}{t-z} + \frac{1}{2\pi i}\int_{-R}^{R} \frac{f(t)dt}{t-z} = \begin{cases} f(z), & \text{if } z \in D^+ \\ 0, & \text{if } z \in D^- \end{cases}.$$

For large R, we write

$$\frac{1}{t-z} = \frac{1}{t} + \frac{z}{t^2} + \cdots$$

and evaluate the integral around C_R. Here $t = R\exp(i\theta)$, $dt = iR\exp(i\theta)d\theta$, so that the leading term in the expansion has the form

$$\gamma\int_{C_R} \frac{dt}{t^2} = \gamma\int_0^{\pi} \frac{iR\exp(i\theta)}{R^2\exp(2i\theta)}d\theta = O\left(\frac{1}{R}\right).$$

Thus, letting $R \to \infty$, we find that the integral around C_R vanishes, and so

$$\frac{1}{2\pi i}\int_{-\infty}^{\infty} \frac{f(t)dt}{t-z} = \begin{cases} f(z), & \text{if } z \in S^+ \\ 0, & \text{if } z \in S^- \end{cases} \tag{2.2.5}$$

where, in the limit, S^+ and S^- are the upper and lower half-planes respectively.

If the point $z = x + iy$ is in the upper half-plane, i.e., $y > 0$, then $\zeta = \bar{z} = x - iy$ is in the lower half-plane and $\bar{f}(\zeta)$ is holomorphic in the lower half-plane. Thus, applying the second line of (2.2.5) to the lower half-plane, we deduce that

$$\frac{1}{2\pi i}\int_{-\infty}^{\infty} \frac{\bar{f}(t)dt}{t-z} = 0, \quad z \in S^+ \tag{2.2.6}$$

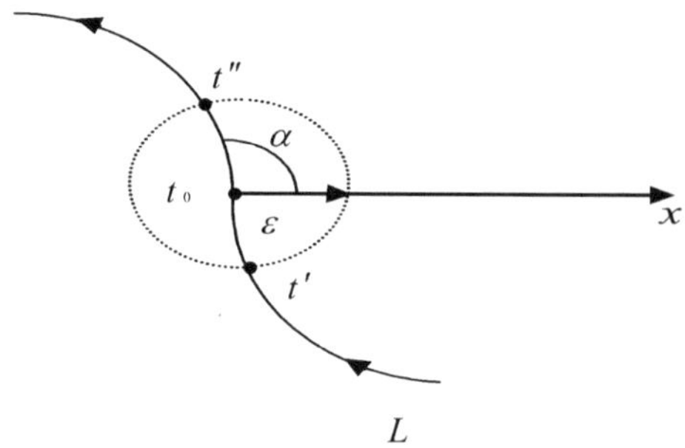

Fig. 2.2.4 The points t', $t''$are on L, equidistant from t_0.

where again S^+ denotes the upper half-plane.

Now return to (2.2.3), and assume that $f(t)$ is defined and continuous on L. $F(z)$ is holomorphic everywhere except on L. We compute the limiting values $F^+(t)$ and $F^-(t)$ as z approaches a point t of L from D^+ or D^- respectively.

To do this, we assume that, in addition to being continuous on L, $f(t)$ satisfies a so-called Hőlder condition. *The function $f(t)$ is said to satisfy a* Hőlder condition *on L if there exist parameters A, λ, where $0 < \lambda < 1$ such that, for every two points t_1, t_2 of L we have*

$$|f(t_2) - f(t_1)| < A|t_2 - t_1|^\lambda. \qquad (2.2.7)$$

The function $f(t)$ is said to satisfy a Hőlder condition in the neighbourhood of a point $t_0 \in L$ if (2.2.6) holds for all t_1, t_2 sufficiently near t_0. Under this condition, we shall show that $F(t)$ in (2.2.3) may be given a meaning when $z \in L$, and $F(z)$ tends to definite limits $F^+(t)$, $F^-(t)$ as $z \to t \in L$ from D^+ or D^-.

Let $t_0 \in L$, and suppose $f(t)$ satisfies a Hőlder condition in the neighbourhood of t_0. Let t', t'' be two points on L on either side of t_0, such that

$$|t_0 - t'| = t_0 - t''| = \varepsilon$$

as shown in Figure 2.2.4. The Cauchy Principal Value *of the integral (2.2.3) at t_0 is defined to be*

$$\frac{1}{2\pi i}\oint_L \frac{f(t)dt}{t - t_0} = \lim_{\varepsilon \to 0} \frac{1}{2\pi i}\int_{L-\ell} \frac{f(t)dt}{t - t_0}$$

where ℓ is the arc $t't''$. The integral may be written

$$\frac{1}{2\pi i}\int_{L-\ell} \frac{f(t) - f(t_0)}{t - t_0}dt + \frac{f(t_0)}{2\pi i}\int_{L-\ell} \frac{dt}{t - t_0}.$$

Since $\frac{|f(t)-f(t_0)|}{|t-t_0|} < A|t - t_0|^{\lambda-1}$, the limit of the first integral exists in the ordinary sense, i.e., provided only that t', t'' tend to t_0; it is not necessary for $|t_0 - t'|, |t_0 - t''|$ to be equal. The second integral is

$$\int_{L-\ell} \frac{dt}{t-t_0} = [\ell n(t-t_0)]_{t''}^{t'}$$

where we have taken a branch of ℓnz that is continuous on $L-\ell$. Now

$$t' = t_0 + \varepsilon \exp[i(\alpha+\pi)], \quad t'' = t_0 + \varepsilon \exp(i\alpha)$$

so that

$$\ell n(t'-t_0) - \ell n(t''-t_0) = i\pi$$

and the Cauchy Principal Value of the integral is

$$\frac{1}{2\pi i}\oint_L \frac{f(t)dt}{t-t_0} = \frac{1}{2\pi i}\int_L \frac{f(t)-f(t_0)}{t-t_0}dt + \frac{1}{2}\cdot f(t_0).$$

This is the meaning that will be attached to the integral (2.2.3) when $z \in L$; thus

$$F(t_0) = \frac{1}{2\pi i}\oint_L \frac{f(t)}{t-t_0} = \frac{1}{2}f(t_0) + \frac{1}{2\pi i}\int_L \frac{f(t)-f(t_0)}{t-t_0}dt. \tag{2.2.8}$$

Now return to equation (2.2.3) and write

$$F(z) = \frac{1}{2\pi i}\int_L \frac{f(t)-f(t_0)}{t-z}dt + \frac{f(t_0)}{2\pi i}\int_L \frac{dt}{t-z}$$

where $t_0 \in L$. It may be proved that the first integral tends to

$$\frac{1}{2\pi i}\int_L \frac{f(t)-f(t_0)}{t-t_0}dt$$

as $z \to t_0$, from whichever side of L. The second integral has, by the argument used before, the values

$$\frac{f(t_0)}{2\pi i}\int_L \frac{dt}{t-z} = \begin{cases} f(t_0), & \text{if } z \in D^+ \\ 0, & \text{if } z \in D^- \end{cases}.$$

Thus, the limits of $F(z)$ as $z \to t_0$, from D^+ and D^- are respectively

$$F^+(t_0) = \frac{1}{2\pi i}\int_L \frac{f(t)-f(t_0)}{t-t_0}dt + f(t_0)$$

$$F^-(t_0) = \frac{1}{2\pi i}\int_L \frac{f(t)-f(t_0)}{t-t_0}dt.$$

Now, returning to the definition of the Cauchy Principal Value of the integral in (2.2.3) we have

$$F^+(t_0) = \frac{1}{2}f(t_0) + \frac{1}{2\pi i}\oint_L \frac{f(t)}{t-t_0}dt \tag{2.2.9}$$

$$F^-(t_0) = -\frac{1}{2}f(t_0) + \frac{1}{2\pi i}\oint_L \frac{f(t)}{t - t_0}dt \tag{2.2.10}$$

These equations, called the Plemelj formulae, *are often written in the form*

$$F^+(t_0) - F^-(t_0) = f(t_0), \quad t_0 \in L \tag{2.2.11}$$

$$F^+(t_0) + F^-(t_0) = \frac{1}{\pi i}\oint_L \frac{dt}{t - t_0}, \quad t_0 \in L. \tag{2.2.12}$$

When L is the real axis, then these results still hold if $f(t)$ is finite and integrable along any finite part of the axis, and satisfies the condition

$$f(t) = f(\infty) + O(|t|^{-\lambda}) \quad \lambda > 0$$

for large $|t|$. We then define the Cauchy integral (2.2.3) as

$$F(z) = \lim_{N\to\infty} \frac{1}{2\pi i}\int_{-N}^{N} \frac{f(t)}{t - z}$$

and find

$$F(z) = \pm\frac{1}{2}f(\infty) + \frac{1}{2\pi i}\int_{-\infty}^{\infty} \frac{f(t) - f(\infty)}{t - z}dt$$

where the sign is $\pm$ according to whether $z \in S^+$ or $z \in S^-$. Further details may be found in Muskhelishvili (1953) or Gladwell (1980).

We now return to the text. Galin assumes that the elastic body occupies the lower *half-plane. While this is perhaps appealing to an engineer – a punch is pressed* down *on a medium, it complicates the mathematics. Also, this section in the original version is made complicated by the chosen notation; we have therefore changed the notation and rearranged the analysis.*

Suppose that the elastic body, occupying the upper half-plane, is subject to normal and shear stresses

$$\sigma_{yy}(x, 0) = -p(x), \quad \sigma_{xy}(x, 0) = -q(x) \tag{2.2.13}$$

as shown in Figure 2.2.5. Remember the convention regarding these stresses shown in Figure 2.1.1.

Equation (2.1.41) gives

$$-p(x) + iq(x) = \{\Phi(z) + \bar{\Phi}(\bar{z}) + \bar{z}\bar{\Phi}'(\bar{z}) + \bar{\Psi}(\bar{z})\}|_{y=0}, \tag{2.2.14}$$

where in the third term on the right, we have replaced z by $\bar{z}$ ($z = \bar{z}$ on the x-axis). Taking the complex conjugate of this equation, we find

$$-p(x) - iq(x) = \{\bar{\Phi}(\bar{z}) + \Phi(z) + z\Phi'(z) + \Psi(z)\}|_{y=0}. \tag{2.2.15}$$

Multiply each of these equations by $1/(2\pi i(x - z))$ and integrate over $(-\infty, \infty)$, using equations (2.2.5), (2.2.6) and making use of the fact that both $\Phi(z)$ and $\Phi(z)+$

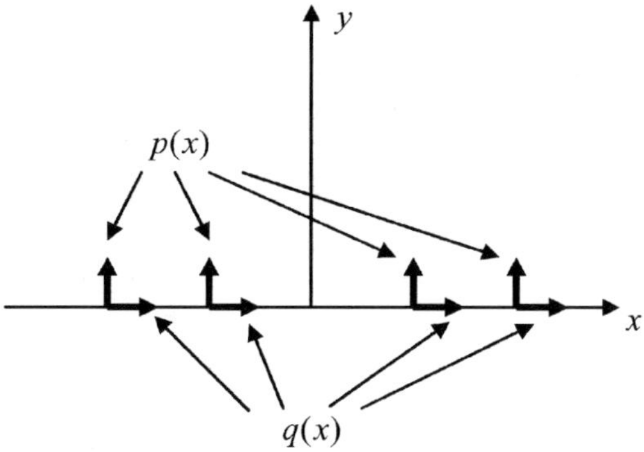

Fig. 2.2.5 The upper half-plane is subjected to distributed forces on the boundary.

$z\Phi'(z) + \Psi(z)$ are holomorphic in the upper half-plane. We find

$$\frac{1}{2\pi i}\int_{-\infty}^{\infty} \frac{-p(x) + iq(x)}{x - z} dx = \Phi(z), \tag{2.2.16}$$

$$\frac{1}{2\pi i}\int_{-\infty}^{\infty} \frac{-p(x) - iq(x)}{x - z} dx = \Phi(z) + z\Phi'(z) + \Psi(z). \tag{2.2.17}$$

Now turn to equation (2.1.34) for the displacements

$$2\mu(u(x,0) + iv(x,0)) = \kappa\phi(z) - \bar{z}\bar{\phi}'(\bar{z}) - \bar{\psi}(\bar{z})\}|_{y=0}$$

where again, in the second term, we have replaced z by $\bar{z}$. Differentiating w.r.t. x and using (2.1.39), we find

$$2\mu(u'(x,0) + iv'(x,0)) = \kappa\Phi(z)|_{y=0} - \{\bar{\Phi}(\bar{z}) + \bar{z}\bar{\Phi}'(\bar{z}) + \bar{\Psi}(\bar{z})\}|_{y=0}. \tag{2.2.18}$$

Now $\Phi(z)$ is given by (2.2.16) and $\Phi(z)+z\Phi'(z)+\Psi(z)$ by (2.2.17). Thus, according to (2.2.9), the value of $\Phi^+(x)$ is

$$\Phi^+(x) = -\frac{1}{2}(p(x) - iq(x)) - \frac{1}{2\pi i}\oint_{-\infty}^{\infty} \frac{p(t) - iq(t)}{t - x} dt$$

and similarly

$$\Phi(x) + x\Phi'(x) + \Psi(x)|^+ = -\frac{1}{2}(p(x) + iq(x)) - \frac{1}{2\pi i}\oint_{-\infty}^{\infty} \frac{p(t) + iq(t)}{t - x} dt$$

where these integrals are interpreted as Cauchy principal values. Inserting these into (2.2.18), we find

$$2\mu(u'(x,0)+iv'(x,0)) = -\frac{\kappa-1}{2}(p(x)-iq(x)) - \frac{\kappa+1}{2\pi i}\oint_{-\infty}^{\infty}\frac{p(t)-iq(t)}{t-x}dt.$$

Separating the real and imaginary parts, we find

$$2\mu u'(x,0) = -\frac{\kappa-1}{2}p(x) + \frac{\kappa+1}{2\pi}\oint_{-\infty}^{\infty}\frac{q(t)dt}{t-x}, \tag{2.2.19}$$

$$2\mu v'(x,0) = \frac{\kappa-1}{2}q(x) + \frac{\kappa+1}{2\pi}\oint_{-\infty}^{\infty}\frac{p(t)dt}{t-x}. \tag{2.2.20}$$

Introducing the parameters

$$\beta = \frac{\kappa-1}{\kappa+1}, \quad \vartheta = \frac{\kappa+1}{4\mu}, \tag{2.2.21}$$

we may write

$$\frac{u'(x)}{\vartheta} = -\beta p(x) + \frac{1}{\pi}\oint_{-\infty}^{\infty}\frac{q(t)dt}{t-x}, \tag{2.2.22}$$

$$\frac{v'(x)}{\vartheta} = \frac{1}{\pi}\oint_{-\infty}^{\infty}\frac{p(t)dt}{t-x} + \beta q(x). \tag{2.2.23}$$

Note that the integrals must be interpreted as Cauchy principal values. If the stresses are applied over a finite interval $(-a, b)$, then the integrals will have limits $-a$ and b.

Suppose the stresses act over a finite interval $(-a, b)$, then we may integrate (2.2.22), (2.2.23) w.r.t. x and find

$$\frac{u(x,0)}{\vartheta} = -\beta\int_{-a}^{x}p(t)dt - \frac{1}{\pi}\oint_{-a}^{b}q(t)\ell n|t-x|dt + C_1 \tag{2.2.24}$$

$$\frac{v(x,0)}{\vartheta} = -\frac{1}{\pi}\oint_{-a}^{b}p(t)\ell n|t-x|dt + \beta\int_{-a}^{x}q(t)dt + C_2 \tag{2.2.25}$$

where C_1, C_2 are arbitrary constants. The equations are due to Muskhelishvili (1953).

If we use Young's modulus, E, and Poisson's ratio ν, instead of μ, $\kappa+1$ and β, we have

$$\kappa = 3-4\nu, \quad 2\mu = E/(1+\nu) \tag{2.2.26}$$

so that

$$\frac{1}{\vartheta} = \frac{4\mu}{\kappa+1} = \frac{2E}{(1+\nu)(4-4\nu)} = \frac{E}{2(1-\nu^2)}, \quad \beta = \frac{1-2\nu}{2(1-\nu)}. \tag{2.2.27}$$

We now introduce two functions holomorphic in the upper half-plane:

$$w_1(z) = \frac{1}{2\pi i} \oint_{-\infty}^{\infty} \frac{p(t)dt}{t-z} = u_1 + iv_1, \tag{2.2.28}$$

$$w_2(z) = \frac{1}{2\pi i} \oint_{-\infty}^{\infty} \frac{q(t)dt}{t-z} = u_2 + iv_2. \tag{2.2.29}$$

(Note that Galin omits the factor $1/(2\pi i)$ in the definitions of w_1 and w_2. The analysis is neater if it is included.) Using equation (2.2.9), we see that the upper boundary values of these functions are

$$w_1^+(x) = \frac{1}{2}p(x) + \frac{1}{2\pi i} \oint_{-\infty}^{\infty} \frac{p(t)dt}{t-x} = u_1^+(x) + iv_1^+(x), \tag{2.2.30}$$

$$w_2^+(x) = \frac{1}{2}q(x) + \frac{1}{2\pi i} \oint_{-\infty}^{\infty} \frac{q(t)dt}{t-x} = u_2^+(x) + iv_2^+(x), \tag{2.2.31}$$

so that

$$u_1^+(x) = \frac{1}{2}p(x), \quad v_1^+(x) = -\frac{1}{2\pi} \oint_{-\infty}^{\infty} \frac{p(t)dt}{t-x}, \tag{2.2.32}$$

$$u_2^+(x) = \frac{1}{2}q(x), \quad v_2^+(x) = -\frac{1}{2\pi} \oint_{-\infty}^{\infty} \frac{q(t)dt}{t-x}, \tag{2.2.33}$$

and we may write equations (2.2.22), (2.2.23) as

$$\frac{u'(x,0)}{2\vartheta} = -\beta u_1^+(x) - v_2^+(x), \tag{2.2.34}$$

$$\frac{v'(x,0)}{2\vartheta} = -v_1^+(x) + \beta u_2^+(x). \tag{2.2.35}$$

We now establish certain properties of the functions $w_1(z)$ and $w_2(z)$. Equations (2.2.32), (2.2.33) show that the real parts of these functions are related to the normal pressure and shear stress acting on the surface $y = 0$. These quantities can become infinite at certain points. We now investigate the character of the singularities that $w_1(z)$ and $w_2(z)$ can have.

If a concentrated force is applied to the boundary of the half-plane, this can be pictured as the transmission of pressure (and shear stress) by means of an extremely narrow punch. In this case, the functions $w_1(z)$ and $w_2(z)$ possess poles of the first order.

When, on the other hand, the pressure and shear stress is transmitted by means of a punch of finite width, there can be no concentrated forces under the punch, even at the ends. It follows that the real parts of $w_1(z)$ and $w_2(z)$ can have only integrable singularities on the real axis. This condition is satisfied if the functions of $w_1(z)$ and $w_2(z)$, which are integrals of Cauchy type, have singularities of the form $(z-c)^{-\theta}$, where $0 < \theta < 1$.

To obtain the limiting forms of $w_1(z)$, $w_2(z)$ as $z \to \infty$, we return to equations (2.2.28), (2.2.29):

$$w_1(z) \to \frac{iP}{2\pi z}, \qquad w_2(z) \to \frac{iQ}{2\pi z} \tag{2.2.36}$$

where

$$P = \int_{-a}^{b} p(t)dt, \qquad Q = \int_{-a}^{b} q(t)dt \tag{2.2.37}$$

are the resultants of the forces applied by the punch. If the normal pressure and shear are distributed over a finite number of intervals of finite length, then $w_1(z)$, $w_2(z)$ will still have the form (2.2.36) at infinity. In the contact problems discussed in this book, $w_1(z)$ and $w_2(z)$ will always possess these properties.

We now express the functions $\phi'(z) \equiv \Phi(z)$ and $\psi'(z) \equiv \Psi(z)$, which serve as the basis for determining the sresses, in terms of $w_1(z)$ and $w_2(z)$. Equations (2.2.16), (2.2.17) give

$$\Phi(z) = -w_1(z) + iw_2(z), \tag{2.2.38}$$

$$\Psi(z) = -2iw_2(z) + zw_1'(z) - izw_2'(z), \tag{2.2.39}$$

from which the stresses $\sigma_{xx}, \sigma_{xy}, \sigma_{yy}$ may be found by using (2.1.42), (2.1.43).

Chapter 3
Plane Static Isotropic Contact Problems

3.1 Boundary Conditions for Plane Contact Problems

In the contact problems that are the subject of this book, we shall deal with cases in which the external forces creating the state of stress are applied by different means at different parts of the surface. However, we can establish several fairly general types of boundary conditions, to various combinations of which most contact problems can be reduced. External forces acting on an elastic body can be applied directly to the surface, for instance by hydrostatic pressure, or through the presence of another body, rigid or elastic. In this section, we shall consider the case in which the bodies transmitting the forces to the elastic body are rigid.

If the forces acting on an elastic body are applied to it directly, then the values of the normal and tangential stresses, σ_{nn} and σ_{ns}, are given on the surface. If the forces are transmitted through a rigid body, then the elastic body can either be rigidly connected to the external rigid body or it can move relative to the external rigid body. In the former case, the surface values of the elastic displacements u and v are given. In the latter case, the value of the displacement normal to the surface is given, and also the relationship between the normal and tangential stress, indicating the presence of friction obeying *Coulomb's law*: $\sigma_{ns} + \rho\sigma_{nn} = 0$. The sign of the coefficient of friction, ρ, depends on the displacement on the surface of the elastic body relative to the rigid body. If the contact is frictionless, the boundary condition takes the form $\sigma_{ns} = 0$.

Thus, the principal types of boundary conditions in contact problems of the theory of elasticity are as follows:

Type I: Part of the surface of the elastic body is free of stress:

$$\sigma_{nn} = 0 = \sigma_{ns}.$$

This is a particular case of

Type II: External forces are given on part of the surface of the elastic body:

$$\sigma_{nn} = f_1(s), \quad \sigma_{ns} = f_2(s).$$

Type III: Part of the surface of the elastic body is rigidly connected to a rigid body that is being displaced. In this case, the displacements along the coordinate axes are given on that part:

$$u = g_1(s), \quad v = g_2(s).$$

Type IV: The elastic body is pressed against a frictionless punch. The normal displacement is known and the shear stress is zero:

$$v_n = h(s) \quad \sigma_{ns} = 0.$$

This is a particular case of

Type V: On part of the surface of the elastic body, where the rigid body is pressed against it, there are frictional forces. The normal displacement is known, as is the relation between the stresses:

$$v_n = h(s), \quad \sigma_{ns} + \rho\sigma_{nn} = 0.$$

In all these boundary conditions, $f_1(s)$, $f_2(s)$, $g_1(s)$, $g_2(s)$, $h(s)$ are given functions of the distance parameter s on the boundary. (Note that Galin introduced just four types, treating our I, IV as special cases.)

We conclude that boundary conditions for contact problems relating to the upper half-plane are such that, on each part of the boundary, we are given either σ_{yy} and σ_{xy}, or a combination of stresses $\sigma_{xy} + \rho\sigma_{yy} = 0$, or displacement derivatives $u'(x)$ and $v'(x)$. As shown in the previous section, each of these quantities may be represented as a linear combination of the functions $w_1(z) = u_1 + iv_1$ and $w_2(z) = u_2 + iv_2$.

Thus, taking equations (2.2.32)–(2.2.35) into account, we can write the principal types of boundary conditions as follows:

$$\begin{array}{ll}
\text{I} & u_1^+(x) = 0 = u_2^+(x) \\
\text{II} & u_1^+(x) = \frac{1}{2}p(x), \quad u_2^+(x) = \frac{1}{2}q(x) \\
\text{III} & -\beta u_1^+(x) - v_2^+(x) = g_1'(x)/(2\vartheta) \\
 & -v_1^+(x) + \beta u_2^+(x) = g_2'(x)/(2\vartheta) \\
\text{IV} & -v_1^+(x) + \beta u_2^+(x) = h'(x)/(2\vartheta) \\
 & u_2^+(x) = 0 \\
\text{V} & -v_1^+(x) + \beta u_2^+(x) = h'(x)/(2\vartheta) \\
 & u_2^+(x) + \rho u_1^+(x) = 0
\end{array}$$

In these equations, β is given by equation (2.2.21).

This means that, in general, the basic contact problems for the half-plane lead to the problem of finding two holomorphic functions $w_1(z)$ and $w_2(z)$ satisfying the following boundary conditions: on various segments of the x-axis the values of two different linear combinations of real and imaginary parts of the functions $w_1(z)$

and $w_2(z)$ are known. Recall that $w_1(z)$,$w_2(z)$ are defined as Cauchy integrals in equations (2.2.28), (2.2.29). In a number of cases, the problem can be reduced to that of finding a single holomorphic function satisfying a boundary condition of mixed type; this is called the *Riemann–Hilbert problem.*

3.2 The Riemann–Hilbert Problem for the Half-Plane

One version of the Riemann–Hilbert problem is this: find a function $w(z) = u + iv$, holomorphic in a region D, and satisfying the condition

$$c(s)u^+(s) + d(s)v^+(s) = f(s)$$

on the boundary of the region. For the upper half-plane, the boundary is the x-axis and the condition is

$$c(x)u^+(x) + d(x)v^+(x) = f(x). \tag{3.2.1}$$

We take $c(x), d(x)$ and $f(x)$ to be real functions. It is unfortunate that Galin uses u, v for the components of the elastic displacement as well as for the real and imaginary parts of the function $w(z)$; we shall follow his practice and trust that it will not cause confusion. We shall, however, use $c(x), d(x)$, instead of Galin's $a(x), b(x)$, to avoid confusion with the limits $-a, b$, on the contact region. In this condition, $c(x), d(x)$ are taken to be piecewise-continuous functions. We assume that $c(x), d(x)$ are not simultaneously zero. An examination of the Types I-V shows that $c(x), d(x)$ are usually piecewise *constant.*

From our earlier examination of the peculiarities of contact problems, we see that $w(z)$ should satisfy the following conditions:

(a) It is holomorphic in the upper half-plane, excluding the boundary $y = 0$; on the boundary, at points of discontinuity of $c(x)$ and $d(x)$, it can have singularities of the form $(x - c)^{-\theta}$, where $0 < \theta < 1$.
There are problems in which $w(z)$ is required to be bounded at points of discontinuity of $c(x)$ and $b(x)$. This imposes certain conditions on the function $f(x)$.

(b) At infinity, $w(z)$ behaves like z^{-1}.
A detailed investigation of the Riemann–Hilbert problem is given in Muskhelishvili (1953b). Here we will touch upon the solution of this problem.

With the stated conditions, the function $w(z)$ may not be unique; it may contain several arbitrary constants.

We will represent $w(z)$ in the form

$$w(z) = w_0(z) + w_*(z) \tag{3.2.2}$$

where $w_0(z)$ is the general solution of the *homogeneous problem* with boundary condition

$$c(x)u_0^+(x) + d(x)v_0^+(x) = 0, \tag{3.2.3}$$

and $w_*(z)$ is a particular solution of the nonhomogeneous problem.

To find the function $w_0(z)$, we consider the function

$$\omega(x) = \arctan[d(x)/c(x)]. \tag{3.2.4}$$

We define $\omega(x)$ to be the angle α satisfying $-\frac{\pi}{2} < \alpha \leq \frac{\pi}{2}$. If $c(x)$ and/or $d(x)$ are discontinuous at a point x_0, $\omega(x)$ may also be discontinuous there.

Equation (2.2.9) shown that the function $\Omega(z)$ holomorphic in the upper half-plane, and whose real part is equal to $\omega(x)$ on the x-axis, meaning $\Omega^+(x) = \omega(x)$, is

$$\Omega(z) = \frac{1}{2\pi i}\int_{-\infty}^{\infty}\frac{2\omega(t)dt}{t-z} + ic,$$

where c is an arbitrary constant. Note, however, that if we require $\Omega(z) \to 0$ as $z \to \infty$, then since the Cauchy integral tends to zero as $z \to \infty$, c must be taken to be zero. Thus, we repeat, equation (2.2.30) states that

$$\Omega^+(x) = \omega(x) + \frac{1}{2\pi i}\oint_{-\infty}^{\infty}\frac{2\omega(t)dt}{t-x} = \Omega_1^+(x) + i\Omega_2^+(x). \tag{3.2.5}$$

Now return to equation (3.2.4); if

$$\frac{d(x)}{c(x)} = \tan\alpha$$

then

$$c(x) = \sqrt{c^2(x)+d^2(x)}\cos\alpha, \quad d(x) = \sqrt{c^2(x)+d^2(x)}\sin\alpha$$

so that we can write equation (3.2.3) as

$$\cos\alpha u_0^+(x) + \sin\alpha v_0^+(x) = 0.$$

But $\cos\alpha - i\sin\alpha = \exp(-i\alpha)$ and $u_0^+(x) + iv_0^+(x) = w_0^+(x)$ so that

$$\exp(-i\alpha)w_0^+(x) = (\cos\alpha - i\sin\alpha)(u_0^+(x) + iv_0^+(x)),$$

and taking the real parts, we find

$$\mathrm{Re}(\exp(-i\alpha)w_0^+(x)) = \cos\alpha u_0^+(x) + \sin\alpha v_0^+(x) = 0. \tag{3.2.6}$$

But $\alpha = \omega(x)$, and, as equation (3.2.5) shows, $\omega(x)$ is the real part of $\Omega^+(x)$. Multiplying (3.2.6) by $\exp(\Omega_2^+(x))$ we see that

$$\begin{aligned}
&\exp(\Omega_2^+(x))\,\mathrm{Re}(\exp(-i\Omega_1^+(x))w_0^+(x)\\
&\quad= \mathrm{Re}(\exp(-i\Omega^+(x))w_0^+(x)\\
&\quad= \mathrm{Re}(\exp\left(-i\cdot\frac{1}{2\pi i}\int_{-\infty}^{\infty}\frac{2\omega(t)dt}{t-z}\right)w_0(z))|_+ = 0.
\end{aligned}$$

This equation states that the function of z inside the bracket takes imaginary values on the real axis. Since $w_0(z)$ is holomorphic in the upper half-plane, the function must be a rational function with imaginary coefficients such that its poles are situated on the real axis. Thus

$$\exp\left[-\frac{1}{\pi}\int_{-\infty}^{\infty}\frac{\omega(t)dt}{t-z}\right]w_0(z) = \frac{iN(z)}{D(z)}$$

where $N(z)$ is a polynomial with real coefficients, and $D(z) = \prod_n (z-\beta_n)$, where the β_n are real. Thus,

$$w_0(z) = \exp\left[\frac{1}{\pi}\int_{-\infty}^{\infty}\frac{\omega(t)dt}{t-z}\right]\frac{iN(z)}{D(z)}. \tag{3.2.7}$$

Since $w_0(z)$ can have poles only at the points of discontinuity of $c(x)$ and $d(x)$, the β_n must coincide with these points. Note, however, that the number of poles can be less than the number of discontinuities.

Thus, the solution of the homogeneous Riemann–Hilbert problem is not unique: there are several arbitrary constants in the polynomial $N(z)$. The number of these constants can be reduced if we prescribe that the function $w_0(z)$ has a definite form at discontinuities of $c(x)$ and $d(x)$, and also a definite form, z^{-1}, at infinity. Note that if $c(x), d(x)$ have discontinuities at just two points, corresponding to the contact problem of a single punch, the satisfaction of these requirements completely determines the function being sought.

In the particular case when the rational function appearing in (3.2.7) is constant, we have

$$w_0(z) = iA\exp\left[\frac{1}{\pi}\int_{-\infty}^{\infty}\frac{\omega(t)dt}{t-z}\right].$$

Now return to the non-homogeneous Riemann–Hilbert problem (3.2.1), namely

$$c(x)u^+(x) + d(x)v^+(x) = f(x). \tag{3.2.8}$$

Take a particular homogeneous solution $w_p(z)$ satisfying

$$c(x)u_p^+(x) + d(x)v_p^+(x) = 0. \tag{3.2.9}$$

Now consider

$$F(z) = \frac{w(z)}{iw_p(z)}.$$

We have

$$F(z) = \frac{(c-id)w}{i(c-id)w_p} = \frac{(c-id)(u+iv)}{i(c-id)(u_p+iv_p)}$$
$$= \frac{cu+dv+i(cv-du)}{i(cu_p+dv_p)+(du_p-cv_p)}.$$

Hence, using (3.2.8) and (3.2.9), we find

$$F^+(x) = \frac{f(x)+i(cv^+(x)-du^+(x))}{du_p^+(x)-cv_p^+(x)}$$

and

$$\mathrm{Re}(F^+(x)) = \frac{f(x)}{du_p^+(x)-cv_p^+(x)} = \frac{f(x)}{i(c-id)w_p^+(x)}. \tag{3.2.10}$$

Equation (2.2.9) shows that the function $F(z)$ holomorphic in the upper half-plane, and with real part of $F^+(x)$ given by (3.2.10) is

$$F(z) = \frac{1}{\pi i}\int_{-\infty}^{\infty} \frac{f(x)dx}{i(c-id)w_p^+(x)(x-z)},$$

i.e.,

$$w_*(z) = \frac{w_p(z)}{\pi i}\int_{-\infty}^{\infty} \frac{f(x)dx}{(c(x)-id(x)w_p^+(x)(x-z)}. \tag{3.2.11}$$

Note that while Galin takes the Riemann–Hilbert problem in the form (3.2.1), an equation linking the real and imaginary parts of $w(z)$, Muskhelishvili takes it to be an equation linking the limiting values $\Omega^+(t)$ and $\Omega^-(t)$ on the two sides of a contour L, or of the real axis:

$$c(t)\Omega^+(t) + d(t)\Omega^-(t) = f(t), \quad t \in L. \tag{3.2.12}$$

He introduces a solution $X(z)$ of the homogeneous problem

$$c(t)X^+(t) + d(t)X^-(t) = 0, \quad t \in L$$

and then writes (3.2.12) as

$$\frac{\Omega^+(t)}{X^+(t)} - \frac{\Omega^-(t)}{X^-(t)} = \frac{f(t)}{c(t)X^+(t)},$$

so that on writing

$$F(z) = \frac{\Omega(z)}{X(z)},$$

we have

$$F^+(t) - F^-(t) = \frac{f(t)}{c(t)X^+(t)}, \quad t \in L$$

and the Plemelj formula (2.2.11) gives

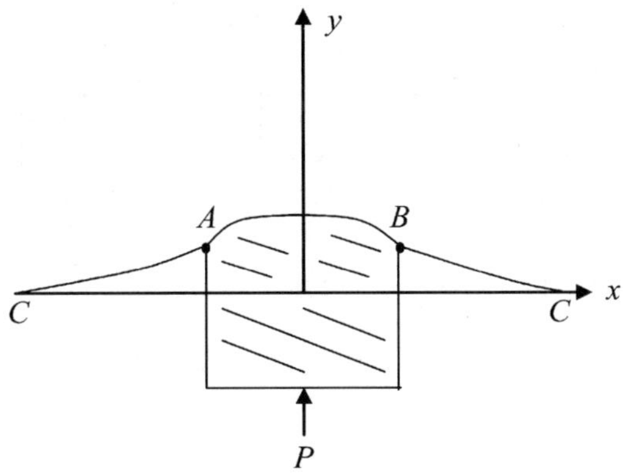

Fig. 3.3.1 A punch indents the half-plane.

$$F(z) = \frac{1}{2\pi i} \int_L \frac{f(t)dt}{c(t)X^+(t)(t-z)},$$

and

$$\Omega(z) = \frac{X(z)}{2\pi i} \int_L \frac{f(t)dt}{c(t)X^+(t)(t-z)}. \tag{3.2.13}$$

3.3 Frictionless Punch Problems – Introduction

As we stated in Section 3.1, in these problems, the boundary condition is of Type I on the free surface, and Type IV under the punch or punches.

Thus, for the punch shown in Figure 3.3.1, we have the boundary conditions

$$\sigma_{yy} = 0 = \sigma_{xy} \text{ on } CA \text{ and } BC$$

$$v = h(x) + c_0, \quad \sigma_{xy} = 0 \text{ on } AB.$$

Here $h(x)$ is the function specifying the shape of the punch, and c_0 is an arbitrary constant. Here we suppose that the punch exerting the pressure moves parallel to the y-axis, and does not rotate. The punch is acted upon by a force P which is the resultant of the pressure produced over the contact region. The force P can be arbitrary, but its point of application is determined by the condition that the punch does not rotate.

If there are several punches pressing against the half-plane, as shown in Figure 3.3.2 for three punches, we have the following boundary conditions:

I $\quad \sigma_{yy} = 0 = \sigma_{xy}$ on $CA_1, B_1A_2, B_2A_3, B_3C$

IV $\quad v = h_n(x) + c_n, \quad \sigma_{xy} = 0$ on $A_nB_n, \quad n = 1, 2, 3.$

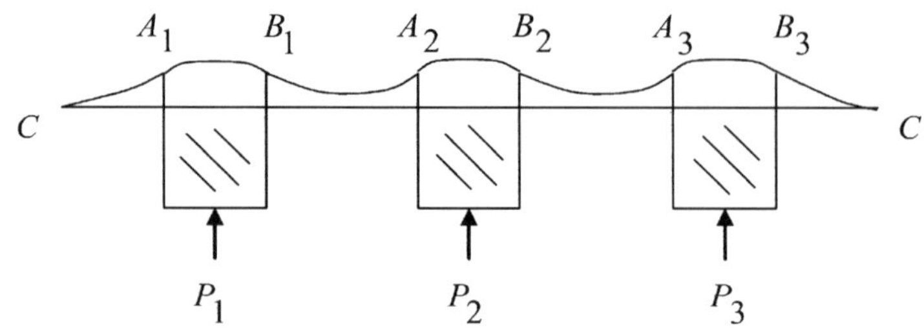

Fig. 3.3.2 Three punches indent the half-plane.

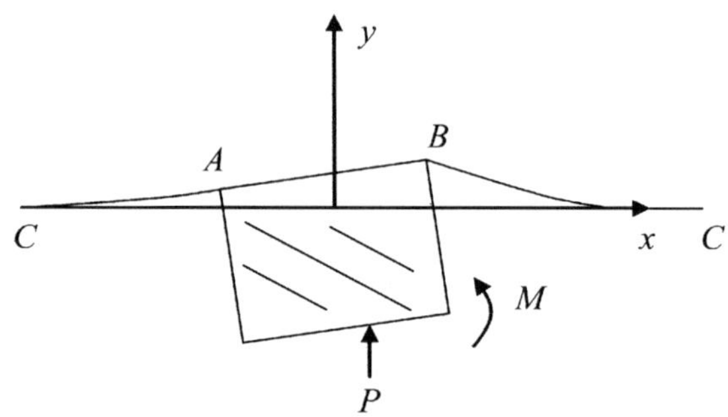

Fig. 3.3.3 A force and a moment are applied to the punch.

Here $h_n(x)$ is the function defining the shape of the nth punch. As before, we suppose that each punch moves parallel to the y-axis and does not rotate. Each punch is acted upon by a force, whose magnitude may be arbitrary, but whose point of application is determined. In this problem, each punch acts independently of the others. We can pose another problem in which the punches are linked to each other by rigid connectors. In this case, the ratios of the forces acting on each punch, and their points of application, are determined from the solution of the problem, but the magnitude of the resultant can be arbitrary. Both problems lead to the same Riemann–Hilbert problem, but since the solution is not unique, and involves various constants, these constants may be chosen to lead to solutions of each of the two problems.

In all the problems discussed to this point, the forces on the punches were applied in such a way that the punches moved parallel to the y-axis without rotation. In the general case, there will be translation and rotation, as shown, exaggerated, in Figure 3.3.3.

Now the boundary conditions will be:

I $\quad \sigma_{yy} = 0 = \sigma_{xy}$ on CA and BC
IV $\quad v = h(x) + \alpha x + c_0, \quad \sigma_{xy} = 0$ on AB.

The resultant force and moment applied to the punch will be

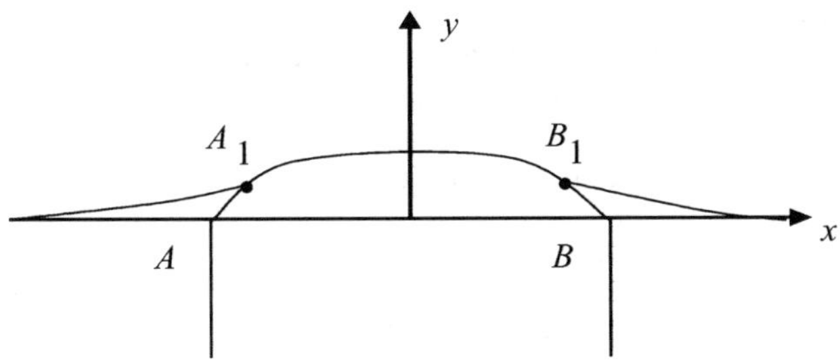

Fig. 3.3.4 A smooth frictionless punch indents the half-space.

$$P = \int_{AB} p(x)dx, \quad M = \int_{AB} xp(x)dx.$$

These conditions allow us to determine the (small) angle α of rotation of the punch. One problem of this type will be discussed in Section 3.4.

Until now, we have assumed that the dimensions of the contact region are given. In this case, if the punch is acted upon by an arbitrary force, there will in general be infinitely large stresses at the ends (corners) of the punch. However, there are *frictionless punch* problems where this is not the case. Consider the pressure of a punch bounded by a smooth surface, as shown in Figure 3.3.4.

Consider what happens as the force P increases from zero. Initially, the punch touches the elastic half-plane at a point. As the force increases, the size of the contact region increases. Until the edges of the contact region reach the points A and B, the pressure remains bounded everywhere, including the points A_1 and B_1 where it is zero. The condition of boundedness leads to the unique determination of the pressure acting on the punch, and hence to the determination of the force corresponding to that contact region.

3.4 Frictionless Punch Problems – Theory

In frictionless punch problems, as we have seen in Section 3.3, the boundary conditions are of Type I and IV. Thus, for the system in Figure 3.3.1, the boundary conditions expressed in terms of the functions $w_1(z)$ and $w_2(z)$ are

$$u_1^+(x) = 0 = u_2^+(x) \text{ on } CA \text{ and } BC$$

$$v_1^+(x) - \beta u_2^+(x) = f(x), \quad u_2^+(x) = 0 \text{ on } AB$$

where

$$f(x) = -h'(x)/(2\vartheta). \tag{3.4.1}$$

First, consider the function $w_2(z)$. It is holomorphic in the upper half-plane, and its real part is zero on (the upper side of) the whole real axis; it is zero. This means that we have to find only $w_1(z)$; this satisfies

$$u_1^+(x) = 0 \text{ on } CA \text{ and } BC \tag{3.4.2}$$

$$v_1^+(x) = f(x) \text{ on } AB. \tag{3.4.3}$$

Recalling that the Riemann–Hilbert problem is

$$c(x)u_1^+ + d(x)v_1^+(x) = f(x)$$

we see that

$$\begin{aligned} c(x) &= 1, \quad d(x) = 0, \quad f(x) = 0 \text{ on } CA \text{ and } BC \\ c(x) &= 0, \quad d(x) = 1, \qquad\qquad \text{on } AB. \end{aligned}$$

Recalling equations (3.2.7), (3.2.11), we note that $w_0(z)$ is

$$w_0(z) = \exp\left[\frac{1}{\pi}\int_{-\infty}^{\infty}\frac{\omega(t)dt}{t-z}\right]\frac{iN(z)}{D(z)} \tag{3.4.4}$$

and

$$w_*(z) = \frac{w_p(z)}{\pi i}\int_{-\infty}^{\infty}\frac{f(x)dx}{(c(x)-id(x))w_p^+(x)(x-z)} \tag{3.4.5}$$

where $w_p(z)$ is a particular solution given by (3.4.4).

First find $\omega(t)$:

$$\omega(t) = \arctan[d(t)/c(t)]$$

so that equations (3.4.2), (3.4.3) give

$$\omega(t) = \begin{cases} 0 \text{ on } CA \text{ and } BC \\ \frac{\pi}{2} \text{ on } AB \end{cases} \tag{3.4.6}$$

and

$$\frac{1}{\pi}\int_{-\infty}^{\infty}\frac{\omega(t)dt}{t-z} = \frac{1}{2}\int_{-a}^{b}\frac{dt}{t-z} = \frac{1}{2}\ell n\left(\frac{z-b}{z+a}\right),$$

$$w_0(z) = \left(\frac{z-b}{z+a}\right)^{\frac{1}{2}}\frac{iN(z)}{D(z)}. \tag{3.4.7}$$

In order to define $w_0(z)$ we must consider the definition of the complex function $((z+a)(z-b))^{1/2}$, and the more general $(z+a)^{\theta}(z-b)^{1-\theta}$, $0 < \theta < 1$, that we will encounter in frictional punch problems. We must define it in a *cut plane*; we choose to cut along the real axis from $-a$ to b, as in Figure 3.4.1.
In this cut plane

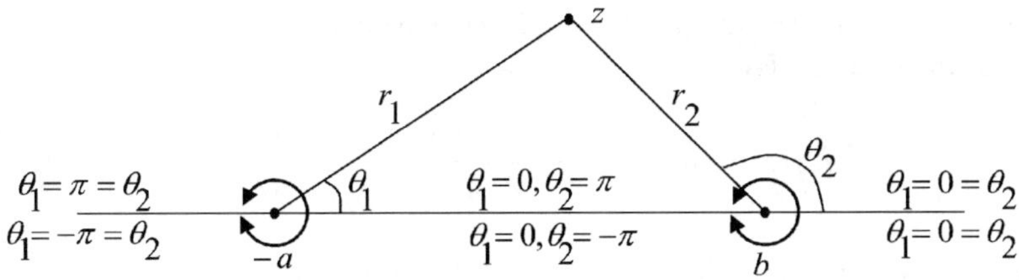

Fig. 3.4.1 The plane cut along $(-a, b)$.

$$z + a = r_1 \exp(i\theta_1), \quad z - b = r_2 \exp(i\theta_2)$$

where $-\pi \leq \theta_1 \leq \pi, \quad -\pi \leq \theta_2 \leq \pi$.

On the real axis to the right of b, $\theta_1 = 0 = \theta_2$ so that

$$(z + a)^\theta (z - b)^{1-\theta} = r_1^\theta r_2^{1-\theta} = (x + a)^\theta (x - b)^{1-\theta}. \tag{3.4.8}$$

On the upper side of $(-a, b)$, $\theta_1 = 0, \theta_2 = \pi$, so that

$$\begin{aligned}(z + a)^\theta (z - b)^{1-\theta} &= r_1^\theta \cdot r_2^{1-\theta} \exp[i(1 - \theta)\pi] \\ &= \exp[i(1 - \theta)\pi](x + a)^\theta (b - x)^{1-\theta}.\end{aligned} \tag{3.4.9}$$

On the lower side of $(-a, b)$, $\theta_1 = 0, \theta_2 = -\pi$, so that

$$(z + a)^\theta (z - b)^{1-\theta} = \exp[-i(1 - \theta)\pi](x + a)^\theta (b - x)^{1-\theta}. \tag{3.4.10}$$

On the real axis to the left of $-a$, $\theta_1 = \pi = \theta_2$ on the upper side, $\theta_1 = -\pi = \theta_2$ on the lower; on either side therefore

$$\begin{aligned}(z + a)^\theta (z - b)^{1-\theta} &= \exp(\pm i\pi)(-a - x)^\theta (b - x)^{1-\theta} \\ &= -(-a - x)^\theta (b - x)^{1-\theta}\end{aligned} \tag{3.4.11}$$

Thus, the function is continuous across the real axis except for the segment $(-a, b)$. On the upper side of $(-a, b)$, taking $\theta = 1/2$, we find

$$w_0^+(x) = \frac{A}{(x + a)^{\frac{1}{2}} (b - x)^{\frac{1}{2}}} \cdot \frac{N(x)}{D(x)}.$$

We note that in finding one solution, $w_*(z)$ of the nonhomogeneous problem, we can use a particular solution $w_p(z)$ of the homogeneous problem, namely

$$w_p(z) = \frac{1}{(z + a)^{\frac{1}{2}} (z - b)^{\frac{1}{2}}} \cdot \frac{N_p(z)}{D_p(z)}, \tag{3.4.12}$$

and choose $N_p(z)$, $D_p(z)$ suitably to simplify the evaluation of the integral in (3.4.5) by contour integration. Then

$$w_1(z) = w_*(z) + \frac{iA}{(z+a)^{\frac{1}{2}}(z-b)^{\frac{1}{2}}} \frac{N(z)}{D(z)}. \tag{3.4.13}$$

We must choose $N(z)$ and $D(z)$ to make $w_1(z) = O(z^{-1})$ for large z.

Having found $w_1(z)$, we find the pressure distribution from $p(x) = 2u_1^+(x)$, the total pressure P, and the potentials $\Phi(z)$, $\Psi(z)$ from equations (2.2.38), (2.2.39), viz.

$$\Phi(z) = -w_1(z), \quad \Psi(z) = zw_1'(z). \tag{3.4.14}$$

The equations (2.1.40), (2.1.41) now give the stresses $\sigma_{xx}, \sigma_{yy}, \sigma_{xy}$; the maximum shearing stress and principal stresses are given by (2.1.53), (2.1.55), (2.1.56).

From $\Phi(z)$, $\Psi(z)$ we may find $\phi(z)$, $\psi(z)$, and hence the elastic displacements u and v from equation (2.1.34). Since $w_1(z)$, $w_2(z)$ have the form (2.2.36) at infinity, so do $\Phi(z)$, $\Psi(z)$. Thus $\phi(z)$, $\psi(z)$, and hence the elastic displacements, have logarithmic singularities at infinity: the displacements at infinity are infinite! This phenomenon occurs in other plane problems in which the half-space is acted on by a force distribution with non-zero resultant: $P \neq 0$ or $Q \neq 0$. The plane problem of the theory of elasticity is an idealisation, since in reality the pressure is exerted by one finite body on another. We can determine stresses from the solution of a plane problem, but we cannot determine displacements.

3.5 Examples of Frictionless Problems

For a punch with a plane face, $h'(x)$ in equation (3.4.1) is zero; $f(x) = 0$, and equation (3.4.13) gives

$$w_1(z) = \frac{iA}{\sqrt{(z+a)(z-b)}}. \tag{3.5.1}$$

For large z,

$$w_1(z) = \frac{iA}{z}$$

so that equation (2.2.36) gives $A = P/(2\pi)$ and thus $p(x) = 2u_1^+(x)$ is

$$p(x) = \frac{P}{\pi\sqrt{(x+a)(b-x)}}. \tag{3.5.2}$$

After giving the equation (3.5.2) for the pressure under a plane punch, Galin briefly considers an inclined punch, and one with a circular base, and then embarks on a long and rather complicated analysis of the inclined punch problem. We have rearranged and simplified his analysis.

The major simplication arises by noting that we can obtain a particular solution of equations (3.4.2), (3.4.3) by inspection, without using the equation (3.4.5), when the given function $f(x)$ in (3.4.3) is a polynomial – and this is not a severe constraint:

$$w_*(z) = if(z). \tag{3.5.3}$$

This solution will not have the appropriate form $O(z^{-1})$ for large z, so we must add a solution of the homogeneous problem:

$$w_1(z) = if(z) + i\left(\frac{z-b}{z+a}\right)^{\frac{1}{2}}\frac{N(z)}{D(z)}. \tag{3.5.4}$$

It is now unnecessary to evaluate any integrals, only to find $N(z)$, $D(z)$ so that $w_1(z) = O(z^{-1})$ for large z, and has, or has not, the required singularities at the ends, $-a$ and b, of the contact region. We now follow this procedure for the various cases.

Consider a punch with inclination α, as in Figure 3.3.3;

$$v = \alpha x + d_0, \quad h'(x) = \alpha, \quad f(x) = f(x) = -\frac{\alpha}{2\vartheta} = -\delta. \tag{3.5.5}$$

Equation (3.5.4) gives

$$w_1(z) = -i\delta + i\left(\frac{z-b}{z+a}\right)^{\frac{1}{2}}\frac{N(z)}{D(z)}, \tag{3.5.6}$$

and we must take $N(z) = \delta(z+c), \quad D(z) = z - b$. For large z

$$w_1(z) = \frac{i\delta}{z}\left(c - \frac{a}{2} + \frac{b}{2}\right) + O(z^{-2}),$$

so that, on using (2.2.36),

$$\frac{P}{2\pi} = \delta\left(c - \frac{a}{2} + \frac{b}{2}\right), \tag{3.5.7}$$

and

$$p(x) = 2u_1^+(x) = 2\delta\left(\frac{b-x}{x+a}\right)^{\frac{1}{2}}\frac{(x+c)}{b-x}. \tag{3.5.8}$$

We introduce the notation

$$\ell = \frac{a+b}{2}, \quad g = \frac{b-a}{2}, \tag{3.5.9}$$

so that the contact width is 2ℓ, its mid point is g, and

$$\frac{P}{2\pi} = \delta(c+g). \tag{3.5.10}$$

First consider what happens when $\delta \to 0$; equation (3.5.10) shows that $\delta c \to P/(2\pi)$ so that (3.5.8) gives

$$p(x) = \frac{P}{\pi(x+a)^{\frac{1}{2}}(b-x)^{\frac{1}{2}}}. \tag{3.5.11}$$

agreeing with (3.5.2).

Now examine the pressure as α, the inclination, changes from zero. At $\alpha = 0$, the punch is flat, and $p(x)$ is given by equation (3.5.2); $p(x)$ is positive throughout $(-a, b)$ and is singular at the ends, $-a$ and b. As α increases, ($\alpha > 0$) this situation still holds: the pressure is positive throughout $(-a, b)$ and singular at the ends $-a$ and b; c is large and $c+x > 0$ for $-a < x < b$. The situation continues until $c = a$, at which point

$$\frac{P}{2\pi} = \delta(a+g) = \delta\ell. \tag{3.5.12}$$

The critical value of α is

$$\alpha = \alpha_0 = 2\vartheta\delta = \vartheta P/(\pi\ell). \tag{3.5.13}$$

At $\alpha = \alpha_0$ the contact pressure is

$$p(x) = \frac{P}{\pi\ell}\left(\frac{x+a}{b-x}\right)^{\frac{1}{2}}. \tag{3.5.14}$$

When α exceeds the value α_0, the punch loses contact with the half-plane at $x = -c$; $p(x) = 0$ when $x = -c$; the contact region is $(-c, b)$, and

$$p(x) = \frac{P}{\pi\ell^*}\left(\frac{x+c}{b-x}\right)^{\frac{1}{2}}, \quad \ell^* = \frac{c+b}{2}. \tag{3.5.15}$$

We may examine the situation in which α decreases from zero in a similar way. Put $\alpha = -\alpha^*$, $\delta = -\delta^*$,

$$w_1(z) = i\delta^* - i\delta^* \frac{(z+c)}{(z+a)^{\frac{1}{2}}(z-b)^{\frac{1}{2}}} \tag{3.5.16}$$

$$\frac{P}{2\pi} = -\delta^*(c+g). \tag{3.5.17}$$

When δ^* is small, c is large negative, and $x + c < 0$ for $x \in (-a, b)$. The critical situation is $c = -b$, when

$$\frac{P}{2\pi} = -\delta^*(-b+g) = \delta^*\ell. \tag{3.5.18}$$

At this critical value, $\alpha = -\alpha_0$, where α_0 is given by

$$p(x) = \frac{P}{\pi \ell} \left(\frac{b-x}{x+a} \right)^{\frac{1}{2}}. \tag{3.5.19}$$

When $\alpha < -\alpha_0$, the contact region is $(-c, b)$, and

$$p(x) = \frac{P}{\pi \ell^*} \left(\frac{b-x}{x+c} \right)^{\frac{1}{2}}, \quad \ell^* = \frac{c+b}{2}. \tag{3.5.20}$$

We may link the inclination α to the moment applied to the punch. At $\alpha = 0$ the pressure is given by (3.5.2) and the moment about the centre, $x = g$, is zero. When $0 < \alpha < \alpha_0$, $p(x)$ is given by (3.5.8) and the moment M about the centre is

$$M = \int_{-a}^{b} (x-g) p(x) dx = \pi \ell^2 \delta = \frac{P\ell^2}{2(c+g)}. \tag{3.5.21}$$

When α reaches the critical value α_0, $c = a$, and

$$M_0 = \frac{P\ell^2}{2(a+g)} = \frac{P\ell}{2}. \tag{3.5.22}$$

When $\alpha > \alpha_0$, $p(x)$ is given by (3.5.15), so that

$$\begin{aligned} M &= \int_{-c}^{b} (x-g) p(x) dx = \int_{-c}^{b} (x - g^* + g^* - g) p(x) dx \\ &= \frac{P\ell^*}{2} + P(g^* - g) = \frac{P(c+b)}{4} + \frac{P(a-c)}{2} \\ &= P\ell - \frac{P(b+c)}{4}. \end{aligned} \tag{3.5.23}$$

Theoretically therefore, the punch will overturn when $c = -b$, i.e., $M = P\ell$; although, of course, the linear theory can hope to simulate only small, actually infinitesimal deflections.

Now consider a punch with a circular base; $h(x) = c_0 - x^2/(2R)$, where R is the radius of curvature of the base, so that

$$f(x) = x/(2\vartheta R) = \delta x, \quad \delta = 1/(2\vartheta R); \tag{3.5.24}$$

and we assume that the punch is in contact over $(-a, a)$, as in Figure 3.5.1.

Now

$$w_1(z) = i\delta z + iA \left(\frac{z-a}{z+a} \right)^{\frac{1}{2}} \frac{N(z)}{D(z)}. \tag{3.5.25}$$

We must choose A, $N(z)$, $D(z)$ so that $w_1(z) = 0(z^{-1})$ for large z, and $w_1(z)$ is finite at $z = \pm a$:

$$w_1(z) = i\delta(z - (z^2 - a^2)^{\frac{1}{2}}). \tag{3.5.26}$$

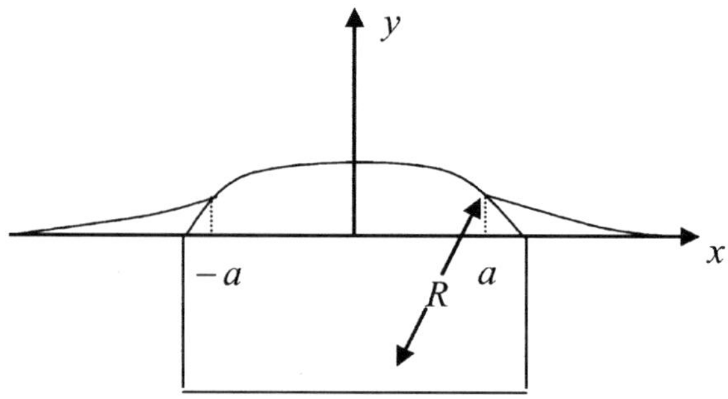

Fig. 3.5.1 The punch is in contact with the half-plane over $(-a, a)$.

Now

$$p(x) = 2\delta(a^2 - x^2)^{\frac{1}{2}}, \tag{3.5.27}$$

and P, which may be obtained from (2.2.36), is

$$P = \frac{\pi a^2}{2\vartheta R}. \tag{3.5.28}$$

Finally, consider a punch with a base that is given by $h(x) = c_0 - kx^4$, so that

$$f(x) = \frac{2kx^3}{\vartheta} = \delta x^3, \quad \delta = 2k/\vartheta. \tag{3.5.29}$$

Proceeding as before, we have

$$w_1(z) = i\delta z^3 + iA(z^2 + a_1 z + a_2)(z^2 - a^2)^{\frac{1}{2}}$$

where we have used the fact that $w_1(z)$ is finite at $z = \pm a$. For large z,

$$(z^2 + a_1 z + a_2)(z^2 - a^2)^{\frac{1}{2}} = z(z^2 + a_1 z + a_2)\left(1 - \frac{a^2}{2z^2} - \frac{a^4}{8z^4}\cdots\right)$$

so that

$$\begin{aligned} w_1(z) &= i\delta z^3 \\ &\quad + iA\left\{z^3 + a_1 z^2 + \left(a_2 - \frac{a^2}{2}\right)z - a_1\frac{a^2}{2} - \left(\frac{a^2}{2}a_2 + \frac{a^4}{8}\right)\frac{1}{z} + o\left(\frac{1}{z^2}\right)\right\} \end{aligned}$$

and $A = -\delta$, $a_1 = 0$, $a_2 = a^2/2$. For large z,

$$w_1(z) = \frac{3i\delta a^4}{8z} = \frac{iP}{2\pi z}$$

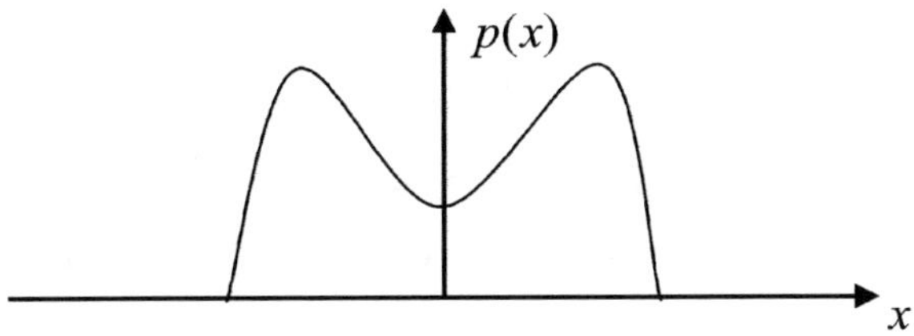

Fig. 3.5.2 The pressure has one local minimum and two local maxima.

so that

$$P = \frac{3\pi\delta a^4}{4} \tag{3.5.30}$$

and

$$w_1(z) = i\delta\left\{z^3 - \left(z^2 + \frac{a^2}{2}\right)(z^2 - a^2)^{\frac{1}{2}}\right\} \tag{3.5.31}$$

and

$$p(x) = \delta(2x^2 + a^2)(a^2 - x^2)^{\frac{1}{2}} \tag{3.5.32}$$

as in Figure 3.5.2. The pressure has a local minimum at $x = 0$, and local maxima at $x = \pm a/\sqrt{2}$.

3.6 Frictional Punch Problems – Theory

We shall solve problems relating to the state of stress arising in an elastic body when one or more punches press into it, and there is friction between the punches and the elastic body. We shall assume that the punches are in a state of limiting equilibrium, when each of them is subjected to the action of a force equal to the product of the coefficient of friction and the force pressing the punch.

It will be shown later that if the punches move along the surface of the elastic body with a speed that is small compared to the speed of sound in the medium, then in solving this problem, we can neglect the dynamic phenomena, and regard the problem as in this chapter.

An account of the results contained in this section was given in Galin (1943a).

We consider a punch in contact with an elastic half-plane over the segment $(-a, b)$.

The boundary conditions are of Type I outside the punch, and Type V under the punch:

$$u_1^+(x) = 0 = u_2^+(x) \text{ for } x \in (-\infty, -a) \cup (b, \infty)$$

$$-v_1^+(x) + \beta u_2^+(x) = -f(x), \quad u_2^+(x) + \rho u_1^+(x) = 0, \text{ for } x \in (-a, b)$$

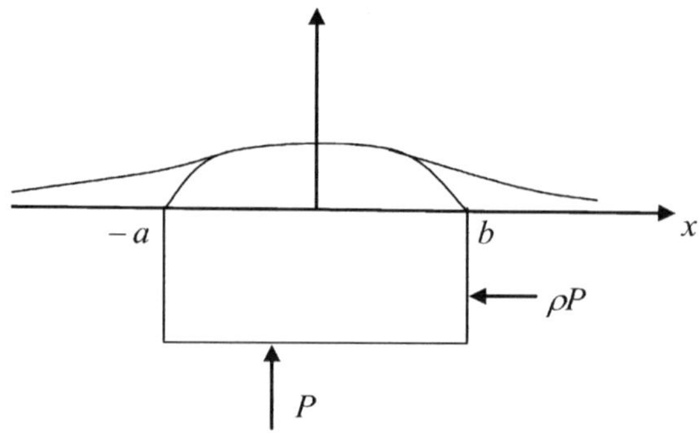

Fig. 3.6.1 A punch with limiting friction presses against the upper half-plane.

where

$$f(x) = -\frac{h'(x)}{2\vartheta}.$$

These can be combined to give a boundary value problem for $w_1(z)$:

$$u_1^+(x) = 0 \text{ for } x \in (-\infty, -a) \cup (b, \infty), \tag{3.6.1}$$

$$v_1^+(x) + \rho\beta u_1^+(x) = f(x), \text{ for } x \in (-a, b). \tag{3.6.2}$$

Once $w_1(z)$ has been found, then $w_2(z)$ is found from

$$u_2^+(x) = 0 \text{ for } x \in (-\infty, -a) \cup (b, \infty),$$

$$u_2^+(x) = -\rho u_1^+(x), \text{ for } x \in (-a, b).$$

Let us denote $\rho\beta = \gamma$. In the notation introduced for the Riemann–Hilbert problem in Section 3.2,

$$c(x) = 1, \quad d(x) = 0 \text{ for } x \in (-\infty, -a) \cup (b, \infty),$$

$$c(x) = \gamma, \quad d(x) = 1 \text{ for } x \in (-a, b).$$

Thus, in equation (3.2.4)

$$\omega(x) = \begin{cases} 0 \quad \text{for } x \in (-\infty, -a) \cup (b, \infty), \\ \arctan(1/\gamma) \quad \text{for } x \in (-a, b), \end{cases}$$

and in equation (3.2.7)

$$w_0(z) = \exp\left[\frac{1}{\pi} \arctan(1/\gamma) \int_{-a}^{b} \frac{dt}{t - z}\right] \frac{iN(z)}{D(z)}.$$

Again, as in Section 3.4

$$w_0(z) = \left(\frac{z-b}{z+a}\right)^{\theta} \frac{iN(z)}{D(z)} \tag{3.6.3}$$

where

$$\theta = \frac{1}{\pi} \arctan\left(\frac{1}{\gamma}\right) = \frac{1}{2} - \eta, \tag{3.6.4}$$

where η is given by

$$\eta = \frac{1}{\pi} \arctan \gamma. \tag{3.6.5}$$

A particular solution of the boundary value problem is provided by equation (3.4.5), but again we note that if $f(x)$ in equation (3.6.2) is a polynomial, then we can take

$$w_*(z) = if(z); \tag{3.6.6}$$

it satisfies $u_1^+(x) = 0$, $v_1^+(x) = f(x)$ everywhere on the real axis. Again, realising this allows us to simplify Galin's analysis.

3.7 Frictional Punch Problems – Examples

First, consider a punch with plane face, pressed normally into the half-plane, in limiting equilibrium. A similar problem was considered by Lvin (1950). Now $h(x) = \text{const.}$, so that $f(x) = 0$; there is just the homogeneous solution $w_0(z)$. There will be singularities in the normal pressure and shearing stress at both ends of the punch, so that

$$w_1(z) = \frac{iA}{(z+a)^{\frac{1}{2}-\eta}(z-b)^{\frac{1}{2}+\eta}}. \tag{3.7.1}$$

For large z, $w_1(z) = iAz^{-1}$, so that $A = P/(2\pi)$, and

$$p(x) = \frac{P}{\pi} \frac{\cos \pi\eta}{(a+x)^{\frac{1}{2}-\eta}(b-x)^{\frac{1}{2}+\eta}}. \tag{3.7.2}$$

Since $\eta > 0$, we note that the order of the singularity at $x = b$, namely $(1/2) + \eta$, is greater than that at $x = a$, namely $(1/2) - \eta$. The pressure has its minimum value at $x = g - 2\ell\eta$, just to the left of the mid-point, $x = g$, of the punch.

Now consider a punch with a plane face pressed at an angle α into the half-plane, as shown in Figure 3.7.1.

First, take $\alpha > 0$ as in Figure 3.7.1a

$$f(x) = -\alpha/(2\vartheta) = -\delta \tag{3.7.3}$$

and

$$w_1(z) = -i\delta + iA\left(\frac{z-b}{z+a}\right)^{\frac{1}{2}+\eta} \frac{N(z)}{D(z)}$$

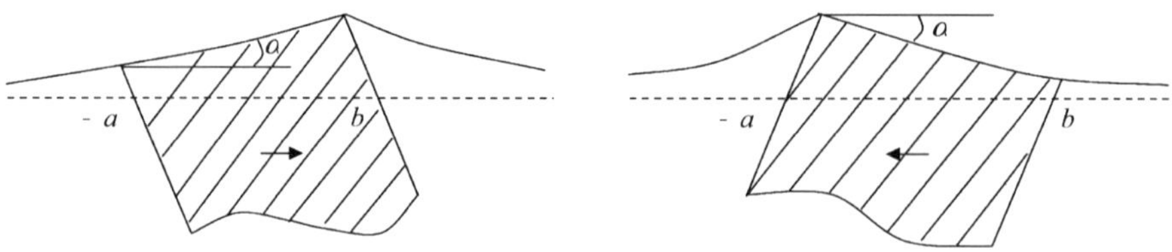

Fig. 3.7.1 A punch is pressed into the half-plane at a small, positive or negative, angle α.

and $N(z) = z + c$, $D(z) = z - b$, $A = \delta$, so that

$$w_1(z) = -i\delta + \frac{i\delta(z+c)}{(z+a)^{\frac{1}{2}-\eta}(z-b)^{\frac{1}{2}+\eta}}. \tag{3.7.4}$$

For large z,

$$w_1(z) = \frac{i\delta}{z}\{c + g + 2\eta\ell\}$$

so that

$$\frac{P}{2\pi} = \delta(c + g + 2\eta\ell). \tag{3.7.5}$$

As $\delta \to 0$, $\delta c \to P/(2\pi)$, in agreement with (3.7.1).

Equation (3.7.4) gives the contact pressure

$$p(x) = \frac{2\delta\cos(\pi\eta)(x+c)}{(x+a)^{\frac{1}{2}-\eta}(b-x)^{\frac{1}{2}+\eta}}. \tag{3.7.6}$$

This agrees with equation (3.5.8) when $\eta = 0$.

Now examine this pressure as α increases from zero. For α small, c is large positive, and $x + c > 0$ for $x \in (-a, b)$; $p(x)$ is positive throughout $(-a, b)$. As α increases, c decreases until $c = a$, at which point $p(-a) = 0$, and

$$p(x) = 2\delta_0 \cos\pi\eta \left(\frac{x+a}{b-x}\right)^{\frac{1}{2}+\eta} \tag{3.7.7}$$

where

$$\frac{P}{2\pi} = \delta_0(a + g + 2\eta\ell) = \delta_0\ell(1 + 2\eta). \tag{3.7.8}$$

As α increases further, $c < a$, $p(-c) = 0$ and the contact region is $(-c, b)$ with

$$p(x) = 2\delta\cos\pi\eta\left(\frac{x+c}{b-x}\right)^{\frac{1}{2}+\eta} \tag{3.7.9}$$

where δ is related P by (3.7.8) in which a is replaced by c, i.e.,

$$\frac{P}{2\pi} = \delta\left\{c + \frac{b-c}{2} + \eta\left(\frac{c+b}{2}\right)\right\} = \delta\ell^*(1+2\eta) \tag{3.7.10}$$

where $\ell^* = (c+b)/2$.

We may treat the situation in which α decreases from zero likewise.

Put $\alpha = -\alpha^*$, so that

$$f(x) = \alpha^*/(2\vartheta) = \delta^* \tag{3.7.11}$$

and

$$w_1(z) = i\delta^* + iA\left(\frac{z-b}{z+a}\right)^{\frac{1}{2}-\eta}\frac{N(z)}{D(z)}.$$

Now $N(z) = z + c$, $D(z) = z - b$, $A = -\delta^*$, so that

$$w_1(z) = i\delta^* - \frac{i\delta^*(z+c)}{(z+a)^{\frac{1}{2}-\eta}(z-b)^{\frac{1}{2}+\eta}}. \tag{3.7.12}$$

For large z

$$w_1(z) = -i\delta^*\{c + g + 2\eta\ell\}/z$$

so that

$$\frac{P}{2\pi} = -\delta^*(c + g + 2\eta\ell). \tag{3.7.13}$$

As $\alpha \to 0-$, $-\delta^* c \to P/(2\pi)$: and $w_1(z)$ tends to the value in (3.7.1). For small negative α, c is large negative. Put $c = -c^*$

$$p(x) = \frac{2\delta^*\cos(\pi\eta)(c^* - x)}{(x+a)^{\frac{1}{2}-\eta}(b-x)^{\frac{1}{2}+\eta}}. \tag{3.7.14}$$

The critical point is $c^* = b$, at which

$$\frac{P}{2\pi} = -\delta_0^*(-b + g + 2\eta\ell) = \delta_0^*\ell(1-2\eta) \tag{3.7.15}$$

where

$$p(x) = 2\delta_0^*\cos\pi\eta\left(\frac{b-x}{x+a}\right)^{\frac{1}{2}-\eta}. \tag{3.7.16}$$

As α^* increases further, the contact region is $(-a, c^*)$,

$$p(x) = 2\delta^*\cos\pi\eta\left(\frac{c^*-x}{x+a}\right)^{\frac{1}{2}-\eta} \tag{3.7.17}$$

where

$$\frac{P}{2\pi} = \delta^*\ell^{**}(1-2\eta), \quad \ell^{**} = (a+c^*)/2. \tag{3.7.18}$$

We now consider a punch that is smooth near both end points, $-a$ and b. In this case, $w_1(z)$ must be bounded at $-a$ and b, and as a result, $p(x)$ is bounded, actually zero, there. Now

$$w_1(z) = if(z) + iA\left(\frac{z-b}{z+a}\right)^{\frac{1}{2}-\eta}\frac{N(z)}{D(z)}, \tag{3.7.19}$$

where we must choose A, $N(z)$ and $D(z)$ so that $w_1(z) = 0(z^{-1})$ for large z, and is bounded at $-a$ and b. Consider a punch with a circular profile with its apex at $x = c$.

$$h(x) = d_0 - (x-c)^2/(2R), \quad f(x) = (x-c)/(2\vartheta R) = \delta(x-c).$$

Now $N(z) = z + a$, $D(z) = 1$, so that $A = -\delta$ and

$$w_1(z) = i\delta(z-c) - i\delta(z-b)^{\frac{1}{2}-\eta}(z+a)^{\frac{1}{2}+\eta}. \tag{3.7.20}$$

For large z,

$$w_1(z) = i\delta\left\{z - c - z\left[1 - \left(\frac{1}{2}-\eta\right)\frac{b}{z} + \left(\frac{1}{2}+\eta\right)\frac{a}{z} + O(z^{-2})\right]\right\}$$

so that $-c + (\frac{1}{2}-\eta)b - (\frac{1}{2}+\eta)a = 0$, or

$$c = g - 2\ell\eta. \tag{3.7.21}$$

When there is no friction, $\eta = 0$; the apex, $x = c$, is at the centre of the contact region. When there is friction, $\eta > 0$, and the apex shifts to the left, in the direction of the friction force. The total force P may be obtained by using (2.2.3);

$$P = \frac{2\pi\left(\frac{1}{4}-\eta^2\right)\ell^2}{\vartheta R}. \tag{3.7.22}$$

When $\eta = 0$, this reduces to (3.5.20)

We obtained the condition (3.7.19) by demanding that $w_1(z)$ be $O(z^{-1})$ for large z, and finite at both ends, $-a$ and b. Galin obtains the condition that must hold in the general case, when $f(x)$ is not a polynomial; it is

$$\int_{-a}^{b}(t+a)^{-\frac{1}{2}-\eta}(b-t)^{-\frac{1}{2}+\eta}f(t)dt = 0. \tag{3.7.23}$$

When this condition holds, $w_1(z)$ may be written

$$w_1(z) = \frac{1}{\pi i(1+\gamma^2)^{\frac{1}{2}}}(z-b)^{\frac{1}{2}-\eta}(z+a)^{\frac{1}{2}+\eta}$$
$$\times\int_{-a}^{b}(t+a)^{-(1/2)-\eta}(b-t)^{-(1/2)+\eta}\frac{f(t)dt}{t-z}. \tag{3.7.24}$$

3.8 Sliding Contact with Coulomb Friction

This section is an addendum to Galin's consideration of frictional contact in Section 3.7 derived with the help of I.G. Goryacheva.

In Sections 3.6 and 3.7, the friction law was the one-term law $q(x) = -\rho p(x)$. It was found that there was a range of α values $(-\alpha_0^*, \alpha_0)$ for which an inclined punch had contact with an elastic half-plane over $(-a, b)$. The critical angles α_0, α_0^* may be found by combining equations (3.7.3) and (3.7.8) to give

$$\alpha_0 = 2\vartheta\delta_0 = \frac{\vartheta P}{\pi\ell(1+2\eta)},$$

and equations (3.7.11), (3.7.15) to give

$$\alpha_0^* = 2\vartheta\delta_0^* = \frac{\vartheta P}{\pi\ell(1-2\eta)}.$$

If we define the dimensionless quantity

$$\kappa = \frac{\ell\alpha}{\vartheta P}, \tag{3.8.1}$$

then we can state that κ satisfies

$$-\kappa_1 < \kappa < \kappa_2, \tag{3.8.2}$$

where

$$\kappa_1 = \frac{1}{\pi(1-2\eta)}, \quad \kappa_2 = \frac{1}{\pi(1+2\eta)}. \tag{3.8.3}$$

Now consider the general, Coulomb, friction law

$$q(x) = -\tau_0 - \rho p(x) \tag{3.8.4}$$

where again we suppose that the punch slips in the x-direction. The boundary conditions (3.6.1), (3.6.2) change to

$$u_1^+(x) = 0 \text{ for } x \in (-\infty, -a) \cup (b, \infty), \tag{3.8.5}$$

$$v_1^+(x) + \rho\beta u_1^+ = f(x), \tag{3.8.6}$$

where now

$$f(x) = \frac{-h'(x)}{2\vartheta} - \frac{\beta\tau_0}{2}. \tag{3.8.7}$$

For an inclined flat punch, $h'(x) = \alpha$, so that

$$f(x) = \frac{-\alpha}{2\vartheta} - \frac{\beta\tau_0}{2} = -\delta. \tag{3.8.8}$$

This means that the analysis of Section 3.7 holds with the δ of equation (3.7.3) replaced by that of (3.8.8); κ in equation (3.8.1) is replaced by

$$\kappa = \frac{\ell}{\vartheta P}(\alpha + \beta\vartheta\tau_0) \tag{3.8.9}$$

and κ satisfies (3.8.2).

The contact pressure is given by equation (3.7.6). Near the left-hand end, $x = -a + \xi$ and

$$p(-a+\xi) = \frac{2\delta\cos(\pi\eta)(c-a)}{\xi^{\frac{1}{2}-\eta}(2\ell)^{\frac{1}{2}+\eta}} + 0\left(\xi^{\frac{1}{2}+\eta}\right) \tag{3.8.10}$$

while near the right-hand end, $x = b - \xi$, and

$$p(b-\xi) = \frac{2\delta\cos(\pi\eta)(b+c)}{(2\ell)^{\frac{1}{2}-\eta}\xi^{\frac{1}{2}+\eta}} + 0\left(\xi^{\frac{1}{2}-\eta}\right) \tag{3.8.11}$$

where c is given by equation (3.7.5). After some algebraic manipulation, we find that (3.8.10) may be written

$$p(-a+\xi) = \frac{P}{\pi}\cdot\frac{\cos(\pi\eta)\cdot(1-\kappa/\kappa_2)}{\xi^{\frac{1}{2}-\eta}(2\ell)^{\frac{1}{2}+\eta}} + 0\left(\xi^{\frac{1}{2}+\eta}\right) \tag{3.8.12}$$

while (3.8.11) may be written

$$p(b-\xi) = \frac{P}{\pi}\frac{\cos(\pi\eta)}{(2\ell)^{\frac{1}{2}-\eta}}\cdot\frac{(1+\kappa/\kappa_1)}{\xi^{\frac{1}{2}+\eta}} + 0\left(\xi^{\frac{1}{2}-\eta}\right) \tag{3.8.13}$$

We note that the singularity at $-a$ disappears when $\kappa = \kappa_2$, while that at b disappears when $\kappa = -\kappa_1$.

Now return to the contact pressure $p(x)$ in equation (3.7.6). We may write this as

$$p(x) = \frac{P\kappa}{\ell}\frac{\cos(\pi\eta)(x+c)}{(x+a)^{\frac{1}{2}-\eta}(b-x)^{\frac{1}{2}+\eta}}, \tag{3.8.14}$$

where c is given by equation (3.7.5); that equation may be written in two ways:

$$c - a = \frac{\ell}{\kappa}\left(\frac{1}{\kappa} - \frac{1}{\kappa_2}\right); \quad c + b = \frac{\ell}{\kappa}\left(\frac{1}{\kappa} + \frac{1}{\kappa_1}\right), \tag{3.8.15}$$

showing that $c = a$ when $\kappa = \kappa_2$, $c = -b$ when $\kappa = -\kappa_1$.

To display $p(x)$ dimensionlessly, we put $x = g + \ell\xi$, so that $-1 \le \xi \le 1$, and find

$$\frac{\ell p(x)}{P} = \frac{\cos\pi\eta}{2\pi}\cdot\frac{\{(1+\kappa/\kappa_1)(1+\xi) + (1-\kappa/\kappa_2)(1-\xi)\}}{(1+\xi)^{\frac{1}{2}-\eta}(1-\xi)^{\frac{1}{2}+\eta}}. \tag{3.8.16}$$

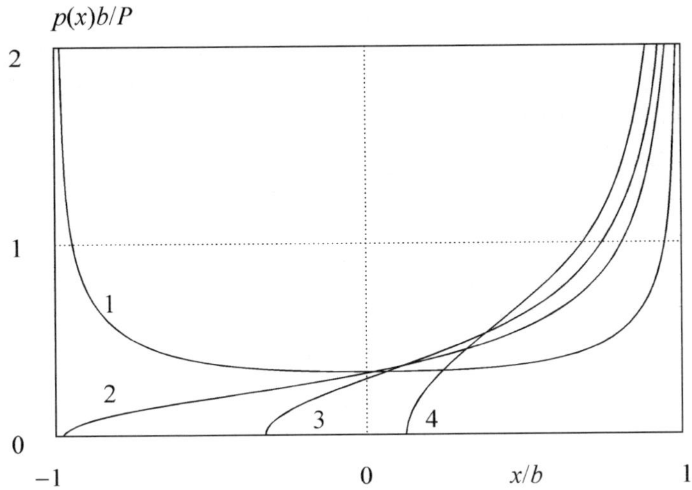

Fig. 3.8.1 Contact pressure under a flat inclined punch sliding on an elastic half-plane.

We note that this agrees with (3.7.7) when $\kappa = \kappa_2$ with (3.7.16) when $\kappa = -\kappa_1$, and with (3.7.2) when $\kappa = 0$. We note that η is defined in (3.6.5) so that

$$\cos \pi\eta = 1/\sqrt{1 + \tan^2 \pi\eta} = 1/\sqrt{1 + \gamma^2}. \tag{3.8.17}$$

The contact pressure distribution for different values of κ are shown in Figure 3.8.1.

The curves are shown for $a = b$, $\rho = 0.2$, $\nu = 0.3$, so that $\gamma = 0.0057$, $\eta = 0.0018$, $\kappa_1 = 0.139$, $\kappa_2 = 0.317$. Curve 1 corresponds to $\kappa = 0$; the pressure is infinite at both ends. Curve 2 corresponds to $\kappa = \kappa_2$, so that $p(-a) = 0$. Curve 3 and 4 correspond to $\kappa = -0.5$ and $\kappa = -0.75$; there is partial contact.

The parameter κ depends on the inclination α, according to equation (3.8.9). The inclination α may be found by using the equilibrium conditions for the punch. The force P can be assumed to act at $x = h$, where

$$\int_{-a}^{b} (x - h)p(x)dx = 0. \tag{3.8.18}$$

The force Q is assumed to act at $y = -d$. Thus

$$P = \int_{-a}^{b} p(x)dx, \quad Q = \int_{-a}^{b} q(x)dx = -2\ell\tau_0 - \rho P$$

and

$$(h - g)P - Qd = 0. \tag{3.8.19}$$

After some manipulation, we find

$$2P\ell\left\{\eta+\kappa\pi\left(\frac{1}{4}-\eta^2\right)\right\}=(2\ell\tau_0+\rho P)d. \tag{3.8.20}$$

If $\kappa > 0$, there is full contact over $(-a, b)$ when $\kappa < \kappa_2$, and equation (3.8.14) shows that this occurs when $d < d_2$, where

$$d_2=\frac{P\ell\left(\frac{1}{2}+\eta\right)}{2\ell\tau_0+\rho P}. \tag{3.8.21}$$

If $\kappa < 0$ there is full contact over $(-a, b)$ when $\kappa > -\kappa_1$, and equation (3.8.14) shows that this occurs when $d > -d_1$, where

$$d_1=\frac{P\ell\left(\frac{1}{2}-\eta\right)}{2\ell\tau_0+\rho P}. \tag{3.8.22}$$

Equation (3.8.14), when combined with equation (3.8.9), gives an explicit expression for the inclination α in terms of τ_0 and d.

3.9 A Two-Punch Problem

Consider two identical plane punches, rigidly linked together, and situated at the same height, pressing into an elastic body, as shown in Figure 3.9.1. The force P is directed along the axis of symmetry of the punches. When such a rigid body presses into an elastic body, the surface of the elastic body is displaced relative to it, and slides along it. We shall assume that there are no sectors with linkage (i.e., stick) in the contact region.

The boundary conditions for the potential $w_1(z)$ are as follows:

$$\begin{aligned}
u_1 &= 0 \quad x\in(-\infty,-b)\\
v_1+\rho\beta u_1 &= 0 \quad x\in(-b,-a)\\
u_1 &= 0 \quad x\in(-a,a)\\
v_1-\rho\beta u_1 &= 0 \quad x\in(a,b)\\
u_1 &= 0 \quad x\in(b,\infty)
\end{aligned} \tag{3.9.1}$$

Here $w_1(z) \to iP/(2\pi z)$ as $z \to \infty$. The potential $w_1(z)$ may have singularities of the form $(z-c)^\alpha$, where $c = \pm a, \pm b$, and $0 < \alpha < 1$. From symmetry, the singularities at $x = +a$ and $-a$ must be the same, as for those at $x = +b$ and $-b$.

Recall that the function $\omega(x)$ is given by equation (3.2.4), namely

$$\omega(x) = \arctan[d(x)/c(x)]$$

where $c(x), d(x)$ characterise the boundary condition

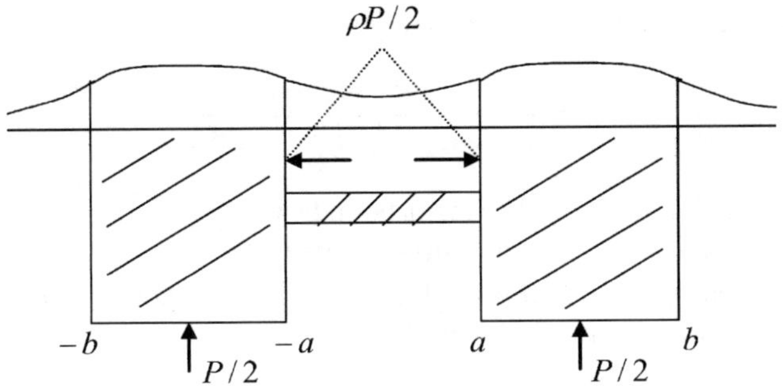

Fig. 3.9.1 Two identical frictional punches indent the half plane.

$$c(x)u^+(x) + d(x)v^+(x) = 0.$$

Equation (3.8.1) show that

$$\begin{aligned}
&c(x) = 1, \quad d(x) = 0, \ x \in (-\infty, -b) \cup (-a, a) \cup (b, \infty)\\
&c(x) = \gamma, \quad d(x) = 1, \ x \in (-b, -a)\\
&c(x) = -\gamma \ d(x) = 1 \ \ x \in (a, b)
\end{aligned}$$

Thus,

$$\omega(x) = \begin{cases} 0 \text{ for } x \in (-\infty, -b) \cup (-a, a) \cup (b, \infty) \\ \arctan(1/\gamma) \text{ for } x \in (-b, -a) \\ -\arctan(1/\gamma) \text{ for } x \in (a, b). \end{cases}$$

Thus, in equation (3.4.4),

$$\begin{aligned}
\frac{1}{\pi}\int_{-\infty}^{\infty} \frac{\omega(t)dt}{t-z} &= \left(\frac{1}{2} - \eta\right)\left\{\int_{-b}^{-a} \frac{dt}{t-z} - \int_a^b \frac{dt}{t-z}\right\},\\
&= \left(\frac{1}{2} - \eta\right)\left(\ell n\left(\frac{z^2-a^2}{z^2-b^2}\right)\right),
\end{aligned}$$

so that

$$w_1(z) = \left(\frac{z^2-a^2}{z^2-b^2}\right)^{\left(\frac{1}{2}-\eta\right)} \frac{iN(z)}{D(z)},$$

where, as before, $N(z)$ is a polynomial with real coefficients, $D(z)$ has possible zeros at $\pm a$, $\pm b$ and η is given by equation (3.6.5). Since $w_1(z) = iP/(2\pi z)$ for large $|z|$, clearly $D(z)$ can have no zero at $\pm b$, and thus $D(z) = z^2 - a^2$, consequently, $N(z) = Pz/(2\pi)$, and

$$w_1(z) = \frac{iPz}{2\pi} \cdot \frac{1}{(z^2-a^2)^{\frac{1}{2}+\eta}(z^2-b^2)^{\frac{1}{2}-\eta}}; \tag{3.9.2}$$

and the pressure exerted by the right hand punch is

$$p(x) = \frac{(P\cos\pi\eta)x}{\pi(x^2 - a^2)^{\frac{1}{2}+\eta}(b^2 - x^2)^{\frac{1}{2}-\eta}}. \tag{3.9.3}$$

Since $\eta > 0$, the singularity at the inner boundary is greater than that at the outer one. The pressure has a local minimum when $x^2 = (\frac{1}{2} - \eta)a^2 + (\frac{1}{2} + \eta)b^2$. When there is no friction, $\eta = 0$, and the minimum is at x_0, where $x_0^2 = (a^2 + b^2)/2$. When there is friction, the minimum is at

$$x^2 = x_0^2 + \eta(b^2 - a^2). \tag{3.9.4}$$

Since $\eta > 0$, the effect of friction is to shift the minimum away from the origin. Begiashvili (1940) considered the frictionless case.

Chapter 4
Moving Punches, and Anisotropic Media

4.1 Introduction

The two topics considered in this chapter, problems related to moving punches, and to anisotropic media, while unrelated, have much in common: they may each be formulated in terms of functions of two complex variables z_1, z_2.

In Chapter 3 we considered problems relating to the pressing of a rigid punch into a semi-infinite plane when there were frictional forces present. These frictional forces, having one direction over the whole contact region, can arise when a rigid body slides over the surface of an elastic one. Thus, problems with friction of this type presuppose, generally speaking, movement of one body relative to the other. We assumed in Chapter 3 that if the speed of the rigid body relative to the elastic one is 'small' (in some sense) then the dynamic character of the phenomenon could be neglected. The expressions (2.2.22), (2.2.23) deduced for the case of a static load were used for the determination of the displacements of a point on the surface of the elastic body.

However, there are problems that arise in practice in which the speed of one body relative to the other is quite large, and we therefore need to investigate whether it is necessary to take the dynamic character of the problem into account.

In Section 4.2, we develop a complex variable formulation for the state of stress in a half-plane when a load moves with constant speed over the plane bounding it. In Section 4.3, we consider the solution of the boundary value problems relating to a moving punch. In Sections 4.4 and 4.5, we consider a complex formulation for an anisotropic plane medium, and the boundary value problem relating to a punch indenting an anisotropic semi-infinite plane.

4.2 Dynamic Plane Isotropic Elasticity Theory

As in static elasticity theory, there are three sets of equations: the strain-displacement equations (2.1.2), the stress-strain equations (2.1.6), and the equilibrium equations. It is only the latter that have to be changed, from (2.1.1) to

$$\frac{\partial \sigma_{xx}}{\partial x_1} + \frac{\partial \sigma_{xy}}{\partial y_1} = \delta \frac{\partial^2 u}{\partial t^2}, \tag{4.2.1}$$

$$\frac{\partial \sigma_{xy}}{\partial x_1} + \frac{\partial \sigma_{yy}}{\partial y_1} = \delta \frac{\partial^2 v}{\partial t^2}, \tag{4.2.2}$$

where δ, assumed constant, is the density of the elastic material. We use x_1, y_1 for the coordinates because most of the analysis will be carried out in a moving frame, and we reserve x, y for the coordinates with respect to that frame.

Expressing the stresses in terms of the strains, and the strains in terms of the displacements u, v, we find

$$(\lambda + \mu) \frac{\partial \theta}{\partial x_1} + \mu \Delta u - \delta \frac{\partial^2 u}{\partial t^2} = 0 \tag{4.2.3}$$

$$(\lambda + \mu) \frac{\partial \theta}{\partial y_1} + \mu \Delta v - \delta \frac{\partial^2 v}{\partial t^2} = 0 \tag{4.2.4}$$

where, as in Section 2.1,

$$\theta = \varepsilon_{xx} + \varepsilon_{yy} = \frac{\partial u}{\partial x_1} + \frac{\partial v}{\partial y_1} \tag{4.2.5}$$

is the dilatation. Subsituting for θ from (4.2.5) we find

$$\frac{\lambda + \mu}{\mu} \left(\frac{\partial^2 u}{\partial x_1^2} + \frac{\partial^2 v}{\partial x_1 \partial y_1} \right) + \frac{\partial^2 u}{\partial x_1^2} + \frac{\partial^2 u}{\partial y_1^2} - \frac{\delta}{\mu} \frac{\partial^2 u}{\partial t^2} = 0, \tag{4.2.6}$$

$$\frac{\lambda + \mu}{\mu} \left(\frac{\partial^2 u}{\partial x_1 \partial y_1} + \frac{\partial^2 v}{\partial y_1^2} \right) + \frac{\partial^2 v}{\partial x_1^2} + \frac{\partial^2 v}{\partial y_1^2} - \frac{\delta}{\mu} \frac{\partial^2 v}{\partial t^2} = 0. \tag{4.2.7}$$

In (4.2.6), put

$$u = -\frac{\partial^2 U}{\partial x_1 \partial y_1}, \tag{4.2.8}$$

then (4.2.6) is satisfied if

$$\left(\frac{\lambda + \mu}{\mu} \right) v = \frac{\lambda + 2\mu}{\mu} \frac{\partial^2 U}{\partial x_1^2} + \frac{\partial^2 U}{\partial y_1^2} - \frac{\delta}{\mu} \frac{\partial^2 U}{\partial t^2}. \tag{4.2.9}$$

When these expressions are substituted into (4.2.7) we find

$$\left(\frac{\partial^2}{\partial x_1^2}+\frac{\partial^2}{\partial y_1^2}-\frac{1}{c_1^2}\frac{\partial^2}{\partial t^2}\right)\left(\frac{\partial^2}{\partial x_1^2}+\frac{\partial^2}{\partial y_1^2}-\frac{1}{c_2^2}\frac{\partial^2}{\partial t^2}\right)U=0, \tag{4.2.10}$$

where

$$c_1=\sqrt{\frac{\lambda+2\mu}{\delta}},\quad c_2=\sqrt{\frac{\mu}{\delta}} \tag{4.2.11}$$

are the speeds of propagation of *longitudinal* and *transverse* waves in the elastic medium.

Now consider the case in which the state of stress in the medium moves with constant speed c along the line bounding the semi-infinite plane. We introduce a system of coordinates

$$x=x_1-ct,\quad y_1=y, \tag{4.2.12}$$

in the moving frame, and suppose that the stresses, and hence U, are functions of x and y, i.e.,

$$U(x_1,y_1,t)=V(x,y)$$

so that

$$\frac{\partial^2 U}{\partial x_1^2}=\frac{\partial^2 V}{\partial x^2},\quad \frac{\partial^2 U}{\partial t^2}=c^2\frac{\partial^2 V}{\partial x^2},$$

and (4.2.10) becomes

$$\left(\frac{\partial^2}{\partial x^2}+\frac{1}{k_1^2}\frac{\partial^2}{\partial y^2}\right)\left(\frac{\partial^2}{\partial x^2}+\frac{1}{k_2^2}\frac{\partial^2}{\partial y^2}\right)V=0, \tag{4.2.13}$$

where

$$k_1^2=1-\frac{c^2}{c_1^2},\quad k_2^2=1-\frac{c^2}{c_2^2}. \tag{4.2.14}$$

This equation shows clearly that the character of the stress state will depend on whether c, the speed of the punch, is larger or smaller than c_1 and c_2. We suppose $c<c_2$, and thus $c<c_1$ also.

Introduce the complex variables

$$z_1=x+ik_1y,\quad z_2=x+ik_2y, \tag{4.2.15}$$

so that equation (4.2.13) becomes

$$\frac{\partial^2}{\partial z_1\partial\bar{z}_1}\left(\frac{\partial^2 V}{\partial z_2\partial\bar{z}_2}\right)=0, \tag{4.2.16}$$

with general (real) solution

$$V=\Phi(z_1)+\bar{\Phi}(\bar{z}_1)+\Psi(z_2)+\bar{\Psi}_2(\bar{z}_2); \tag{4.2.17}$$

this yields the displacements

$$u = -ik_1\{\Phi''(z_1) - \bar{\Phi}''(\bar{z}_1)\} - ik_2\{\Psi''(z_2) - \bar{\Psi}''(\bar{z}_2)\}, \tag{4.2.18}$$

$$v = k_1^2\{\Phi''(z_1) + \bar{\Phi}''(\bar{z}_1)\} + \Psi''(z_2) + \bar{\Psi}''(\bar{z}_2). \tag{4.2.19}$$

These yield

$$u' = -ik_1\{\phi(z_1) - \bar{\phi}(\bar{z}_1)\} - ik_2\{\psi(z_2) - \bar{\psi}(\bar{z}_2)\}, \tag{4.2.20}$$

$$v' = k_1^2\{\phi(z_1) + \bar{\phi}(\bar{z}_1)\} + \psi(z_2) + \bar{\psi}(\bar{z}_2), \tag{4.2.21}$$

$$\sigma_{xy}/\mu = 2k_1^2\{\phi(z_1) + \bar{\phi}(\bar{z}_1)\} + (1 + k_2^2)\{\psi(z_2) + \bar{\psi}(\bar{z}_2)\}, \tag{4.2.22}$$

$$\sigma_{yy}/\mu = i(1 + k_2^2)k_1\{\phi(z_1) - \bar{\phi}(\bar{z}_1)\} + 2ik_2\{\psi(z_2) - \bar{\psi}(\bar{z}_2)\}, \tag{4.2.23}$$

where

$$\phi(z_1) = \Phi'''(z_1), \quad \psi(z_2) = \Psi'''(z_2). \tag{4.2.24}$$

4.3 Displacement – Stress Relations

Equations (4.2.20)–(4.2.23) yield the stresses and $\partial u/\partial x$ and $\partial v/\partial x$ in terms of the potentials $\phi(z_1)$ and $\psi(z_2)$. We now proceed as in Section 2.2, and suppose that the elastic body, occupying the upper half-plane, is subject to normal and shear stresses

$$\sigma_{yy}(x, 0) = -p(x), \quad \sigma_{xy}(x, 0) = -q(x) \tag{4.3.1}$$

where now, remember, x is the coordinate $x_1 - ct$ in the moving frame. Now form combinations of σ_{yy} and σ_{xy} from which $\psi(z_2)$ and $\phi(z_1)$ are eliminated, in turn:

$$\begin{aligned}\{-(1 + k_2^2)p(x) + 2ik_2q(x)\}/\mu &= i\{(1 + k_2^2)^2 - 4k_1k_2\}k_1\phi(z_1)\\ -i\{(1 + k_2^2)^2 + 4k_1k_2\}k_1\bar{\phi}(\bar{z}_1) &- 4ik_2(1 + k_2^2)\bar{\psi}(\bar{z}_2)|_{y=0},\end{aligned} \tag{4.3.2}$$

$$\begin{aligned}\{-2k_1^2p(x) + i(1 + k_2^2)k_1q(x)\}/\mu &= -i\{(1 + k_2^2)^2 - 4k_1k_2\}k_1\psi(z_2)\\ -4ik_1^3(1 + k_2^2)\bar{\phi}(\bar{z}_1) &- i\{(1 + k_2^2)^2 + 4k_1k_2\}k_1\bar{\psi}(\bar{z}_2)|_{y=0}.\end{aligned} \tag{4.3.3}$$

Multiplying these equations by $1/(2\pi i(x - z))$, making use of the fact that, on $y = 0, z_1 = z_2 = x - ct$; and $\phi(z_1)$, $\psi(z_2)$ are holomorphic in the upper half-plane, so that equation (2.2.5) holds:

$$ik_1K\phi(z_1) = \frac{1}{2\pi i}\int_{-\infty}^{\infty}\frac{\{-(1 + k_2^2)p(x) + 2ik_2q(x)\}dx}{x - z_1}, \tag{4.3.4}$$

$$ik_1K\psi(z_2) = \frac{1}{2\pi i}\int_{-\infty}^{\infty}\frac{\{2k_1^2p(x) - i(1 + k_2^2)k_1q(x)\}dx}{x - z_2}. \tag{4.3.5}$$

Introducing $w_1(z)$, $w_2(z)$ as before, viz.

$$w_1(z) = \frac{1}{2\pi i} \oint_{-\infty}^{\infty} \frac{p(t)dt}{t-z} = u_1 + iv_1, \tag{4.3.6}$$

$$w_2(z) = \frac{1}{2\pi i} \oint_{-\infty}^{\infty} \frac{q(t)dt}{t-z} = u_2 + iv_2, \tag{4.3.7}$$

we find

$$k_1 K\phi(z_1) = i(1+k_2^2)w_1(z_1) + 2k_2 w_2(z_1), \tag{4.3.8}$$

$$K\psi(z_2) = -2ik_1 w_1(z_2) - (1+k_2^2)w_2(z_2), \tag{4.3.9}$$

where

$$K = \mu\{(1+k_2^2)^2 - 4k_1k_2\}. \tag{4.3.10}$$

Using (2.2.9) we see that the upper boundary values of w_1 and w_2 are

$$w_1^+(x) = \frac{1}{2}p(x) + \frac{1}{2\pi i} \oint_{-\infty}^{\infty} \frac{p(t)dt}{t-z} = u_1^+(x) + iv_1^+(x), \tag{4.3.11}$$

$$w_2^+(x) = \frac{1}{2}q(x) + \frac{1}{2\pi i} \oint_{-\infty}^{\infty} \frac{q(t)dt}{t-z} = u_2^+(x) + iv_2^+(x). \tag{4.3.12}$$

The displacement derivatives u', v' are expressed in (4.2.20), (4.2.21) in terms of the real and imaginary parts of $\phi(z_1)$, $\psi(z_2)$. Equations (4.3.8), (4.3.9) give

$$k_1 K\,\mathrm{Re}\{\phi^+(x)\} = -(1+k_2^2)v_1^+(x) + 2k_2 u_2^+(x) \tag{4.3.13}$$

$$k_1 K\,\mathrm{Im}\{\phi^+(x)\} = (1+k_2^2)u_1^+(x) + 2k_2 v_2^+(x) \tag{4.3.14}$$

$$K\,\mathrm{Re}\{\psi^+(x)\} = 2k_1 v_1^+(x) - (1+k_2^2)u_2^+(x) \tag{4.3.15}$$

$$K\,\mathrm{Im}\{\psi^+(x)\} = -2k_1 u_1^+(x) - (1+k_2^2)v_2^+(x) \tag{4.3.16}$$

which when substituted into (4.2.20), (4.2.21) yield

$$Ku'(x) = 2(1+k_2^2 - 2k_1k_2)u_1^+(x) + 2k_2(1-k_2^2)v_2^+(x). \tag{4.3.17}$$

$$Kv'(x) = 2k_1(1-k_2^2)v_1^+(x) - 2(1+k_2^2 - 2k_1k_2)u_2^+(x). \tag{4.3.18}$$

We need to check that equations (4.3.17), (4.3.18) reduce to (2.2.34), (2.2.35) respectively, when $c \to 0$. For small c,

$$K \simeq -2\mu c^2 \left(\frac{c_1^2 - c_2^2}{c_1^2 c_2^2}\right), \quad 2(1+k_2^2 - 2k_1k_2) \simeq \frac{2c^2}{c_1^2}, \quad 1-k_2^2 \simeq \frac{c^2}{c_2^2}$$

so that, on dividing through by $2c^2/c_2^2$ we find

$$\mu\left(\frac{c_1^2 - c_2^2}{c_1^2}\right)u'(x) = -\frac{c_2^2}{c_1^2}u_1^+(x) - v_2^+(x)$$

$$\mu\left(\frac{c_1^2-c_2^2}{c_1^2}\right)v'(x)=-v_1^+(x)+\frac{c_2^2}{c_1^2}u_2^+(x).$$

This is equivalent to (2.2.34), because

$$\beta=\frac{\kappa-1}{\kappa+1}=\frac{\mu}{\lambda+2\mu}=\frac{c_2^2}{c_1^2},\quad \frac{1}{2\vartheta}=\mu\left(\frac{\lambda+\mu}{\lambda+2\mu}\right)=\mu\left(\frac{c_1^2-c_2^2}{c_1^2}\right).$$

Since $u_1^+, v_1^+, u_2^+, v_2^+$ are given by (2.2.32), (2.2.33), we conclude that on the boundary,

$$Ku'(x)=(1+k_2^2-2k_1k_2)p(x)-\frac{k_2(1-k_2^2)}{\pi}\oint_{-\infty}^{\infty}\frac{q(t)dt}{t-x}, \tag{4.3.19}$$

$$Kv'(x)=\frac{-k_1(1-k_2^2)}{\pi}\oint_{-\infty}^{\infty}\frac{p(t)dt}{t-x}-(1+k_2^2-2k_1k_2)q(x). \tag{4.3.20}$$

4.4 Boundary Value Problems for a Moving Punch

We now apply the results obtained in Section 4.3 to the problem of a punch moving with speed c along the boundary of the upper half-plane, when there is limiting friction over the contact region $(-a, a)$ in the moving coordinates. As before, the boundary conditions are

$$\sigma_{yy}=0=\sigma_{xy} \text{ for } |x|>a \tag{4.4.1}$$

$$v'=f'(x),\quad \sigma_{xy}+\rho\sigma_{yy}=0 \text{ for } |x|<a. \tag{4.4.2}$$

These lead to the equations

$$u_1^+(x)=0=u_2^+(x) \text{ for } |x|>a \tag{4.4.3}$$

$$\left.\begin{aligned}2k_1(1-k_2^2)v_1^+(x)-2(1+k_2^2-2k_1k_2)u_2^+(x)&=Kh'(x)\\ u_2^+(x)+\rho u_1^+(x)&=0\end{aligned}\right\} \text{ for } |x|<a. \tag{4.4.4}$$

These may be combined to give a boundary value problem for $w_1(z)$:

$$u_1^+(x)=0 \text{ for } |x|>a \tag{4.4.5}$$

$$2k_1(1-k_2^2)v_1^+(x)+2(1+k_2^2-2k_1k_2)\rho u_1^+(x)=Kh'(x),\ |x|<a. \tag{4.4.6}$$

In the notation introduced for the Riemann–Hilbert problem in Section 3.2,

$$c(x)=1,\quad d(x)=0 \text{ for } |x|>a \tag{4.4.7}$$

$$c(x) = (1 + k_2^2 - 2k_1k_2)\rho/[k_1(1 - k_2^2)] = \gamma, \quad d(x) = 1, \text{ for } |x| < a \tag{4.4.8}$$

$$f(x) = Kh'(x)/[2k_1(1 - k_2^2)], \text{ for } |x| < a. \tag{4.4.9}$$

With this new notation, the problem in the moving coordinates is exactly the same as that for the stationary punch in Section 3.7; it is therefore unnecessary to take the analysis any further; we merely use the solutions obtained for a stationary punch and reword them in terms of the moving coordinates. We note that, for small c,

$$\gamma = \rho\beta\left\{1 + \left[\frac{1}{2} + \frac{1}{4}\left(\frac{c_1^2 - c_2^2}{c_2^2}\right)^2\right]\frac{c^2}{c_1^2} + \cdots\right\}, \tag{4.4.10}$$

$$\alpha = \frac{K}{2k_1(1 - k_2^2)} = \frac{-1}{2\vartheta}\left\{1 + \left[\frac{1}{2} - \frac{1}{4}\left(\frac{c_1^2 - c_2^2}{c_2^2}\right)\right]\frac{c^2}{c_1^2} + \cdots\right\}. \tag{4.4.11}$$

This shows that if c is 10% of the speed of propagation c_1, then the relative differences between the values of α, γ at $c/c_1 = 0.1$ and at $c = 0$ are of the order of 1%, and may thus be neglected.

It is somewhat curious that Galin abruptly terminates the study of a moving punch at this point – by saying that if $c/c_1 = 0.1$, then the effect of movement can be ignored.

4.5 Complex Variable Formulation for a Plane Anisotropic Elastic Body

We follow the lines laid down in Chapter 2 for an isotropic body; of the three sets of governing equations: the strain-displacement equations, the stress-strain equations, and the equilibrium equations, only the second set is changed.

As for isotropic plane problems, one may envisage two kinds of ideal problems: *plain strain* and *plane stress*. In the former, $u = u(x, y)$, $v = v(x, y)$, $w = 0$ so that $\varepsilon_{13} = 0 = \varepsilon_{23} = \varepsilon_{33}$. In the latter, $\sigma_{13} = 0 = \sigma_{23} = \sigma_{33}$. To find the relations between the strains $\varepsilon_{11}, \varepsilon_{22}, \varepsilon_{12}$ and the stresses $\sigma_{11}, \sigma_{22}, \sigma_{12}$, we start from the general equations in one of the forms

$$\sigma_{ij} = c_{ijkl}\varepsilon_{kl} \text{ or } \varepsilon_{ij} = s_{ijkl}\sigma_{kl}, \tag{4.5.1}$$

using the former for plane strain, the latter for plane stress. The coefficients c_{ijkl} and s_{ijkl} satisfy the *symmetry relations*

$$\left.\begin{aligned} c_{ijkl} &= c_{jikl} = c_{ijlk} = c_{klij} \\ s_{ijkl} &= s_{jikl} = s_{ijlk} = s_{klij} \end{aligned}\right\} \tag{4.5.2}$$

For plane stress, we obtain

$$\varepsilon_{11} = s_{1111}\sigma_{11} + s_{1122}\sigma_{22} + 2s_{1112}\sigma_{12}$$
$$\varepsilon_{22} = s_{1122}\sigma_{11} + s_{2222}\sigma_{22} + 2s_{2212}\sigma_{12}$$
$$\varepsilon_{12} = s_{1211}\sigma_{11} + s_{1222}\sigma_{22} + 2s_{1212}\sigma_{12}$$

On multiplying the last equation throughout by 2 we find

$$\begin{bmatrix} \varepsilon_{11} \\ \varepsilon_{22} \\ 2\varepsilon_{12} \end{bmatrix} = \begin{bmatrix} \beta_{11} & \beta_{12} & \beta_{13} \\ \beta_{21} & \beta_{22} & \beta_{23} \\ \beta_{31} & \beta_{32} & \beta_{33} \end{bmatrix} \begin{bmatrix} \sigma_{11} \\ \sigma_{22} \\ \sigma_{12} \end{bmatrix} \tag{4.5.3}$$

where the matrix (β_{ij}) is symmetric. For plane strain, we start from (4.5.1a), and then invert to give an equation similar to (4.5.3), with different coefficients β_{ij}.

Since the equilibrium equations are unchanged, i.e., they are still (2.1.1), we may still use the *Airy stress function* solution (2.1.16)

$$\sigma_{xx} = \frac{\partial^2 U}{\partial y^2}, \quad \sigma_{xy} = -\frac{\partial^2 U}{\partial x \partial y}, \quad \sigma_{yy} = \frac{\partial^2 U}{\partial x^2}. \tag{4.5.4}$$

We substitute these into the strain-stress relation (4.5.3), and then substitute the strains into the compatibility equation (2.1.3); the result is

$$\beta_{22}U_{,1111} - 2\beta_{23}U_{,1112} + (2\beta_{12} + \beta_{33})U_{,1122} - 2\beta_{13}U_{,1222} + \beta_{11}U_{,2222} = 0 \tag{4.5.5}$$

where $_{,1} = \partial/\partial x, \quad _{,2} = \partial/\partial y$.

In the isotropic plane stress case

$$2\mu\beta_{11} = 1 - \nu = 2\mu\beta_{22}, \quad 2\mu\beta_{12} = -\nu, \quad 2\mu\beta_{33} = 1, \quad \beta_{13} = 0 = \beta_{23},$$

so that (4.5.5) reduces to the biharmonic equation

$$U_{,1111} + 2U_{,1122} + U_{,2222} = 0. \tag{4.5.6}$$

Just as $U(x, y) = \phi(x \pm iy) = \phi(z), \phi(\bar{z})$ are solutions of (4.5.6), so we may seek solutions of (4.5.5) in the form $U(x + \mu y)$, where μ is a root of

$$\beta_{22} - 2\beta_{23}\mu + (2\beta_{12} + \beta_{33})\mu^2 - 2\beta_{13}\mu^3 + \beta_{11}\mu^4 = 0. \tag{4.5.7}$$

An examination of the signs and relative magnitudes of the β_{ij} shows that the roots of (4.5.7) appear in complex-conjugate pairs $\mu_1, \bar{\mu}_1; \mu_2, \bar{\mu}_2$. In the isotropic case, the roots μ_1, μ_2 are equal; that is why there are solutions of (4.5.6) of the form $z\phi(\bar{z})$ and $\bar{z}\phi(z)$ also.

The general (real) solution for U is thus

$$U = \phi_1(z_1) + \bar{\phi}_1(\bar{z}_1) + \phi_2(z_2) + \bar{\phi}_2(\bar{z}_2) \tag{4.5.8}$$

where $z_1 = x + \mu_1 y$, $z_2 = x + \mu_2 y$, and $\mu_1, \bar{\mu}_1; \mu_2, \bar{\mu}_2$ are the roots of (4.5.7). Thus

$$\sigma_{xx} = \mu_1^2\Phi_1(z_1) + \bar{\mu}_1^2\bar{\Phi}_1(\bar{z}_1) + \mu_2^2\Phi_2(z_2) + \bar{\mu}_2^2\bar{\Phi}_2(\bar{z}_2), \tag{4.5.9}$$

$$\sigma_{yy} = \Phi_1(z_1) + \bar{\Phi}_1(\bar{z}_1) + \Phi_2(z_2) + \bar{\Phi}_2(\bar{z}_2), \tag{4.5.10}$$

$$\sigma_{xy} = -\mu_1\Phi_1(z_1) - \bar{\mu}_1\bar{\Phi}_1(\bar{z}_1) - \mu_2\Phi_2(z_2) - \bar{\mu}_2\bar{\Phi}_2(\bar{z}_2), \tag{4.5.11}$$

from which, using the strain-stress relations (4.5.3), we obtain

$$\partial u/\partial x = p_1\Phi_1(z_1) + \bar{p}_1\bar{\Phi}_1(\bar{z}_1) + p_2\Phi_2(z_2) + \bar{p}_2\bar{\Phi}_2(\bar{z}_2), \tag{4.5.12}$$

$$\partial v/\partial x = q_1\Phi_1(z_1) + \bar{q}_1\bar{\Phi}_1(\bar{z}_1) + q_2\Phi_2(z_2) + \bar{q}_2\bar{\Phi}_2(\bar{z}_2), \tag{4.5.13}$$

where

$$\Phi_1(z_1) = \phi_1''(z_1), \quad \Phi_2(z_2) = \phi_2''(z_2), \tag{4.5.14}$$

$$p_i = \beta_{11}\mu_i^2 + \beta_{12} - \beta_{13}\mu_i, \quad q_i = (\beta_{12}\mu_i^2 + \beta_{22} - \beta_{23}\mu_i)/\mu_i \tag{4.5.15}$$

for $i = 1, 2$.

Now we proceed as before: take combinations of σ_{yy} and σ_{xy} that do not involve one or other of $\Phi_2(z_2)$ and $\Phi_1(z_1)$. We have

$$\mu_1\sigma_{yy} + \sigma_{xy} = (\mu_1 - \bar{\mu}_1)\bar{\Phi}_1(\bar{z}_1) + (\mu_1 - \mu_2)\Phi_2(z_2) + (\mu_1 - \bar{\mu}_2)\bar{\Phi}_2(\bar{z}_2),$$

$$\mu_2\sigma_{yy} + \sigma_{xy} = (\mu_2 - \mu_1)\bar{\Phi}_1(z_1) + (\mu_2 - \bar{\mu}_1)\Phi_1(\bar{z}_1) + (\mu_2 - \bar{\mu}_2)\bar{\Phi}_2(\bar{z}_2);$$

if

$$\sigma_{yy}(x,0) = -p(x), \quad \sigma_{xy}(x,0) = -q(x) \tag{4.5.16}$$

then on integrating and using (2.2.2) we find

$$(\mu_2 - \mu_1)\Phi_1(z_1) = -\frac{1}{2\pi i}\int_{-\infty}^{\infty}\frac{(\mu_2 p(x) + q(x))}{x - z_1}dx = -\mu_2 w_1(z_1) - w_2(z_1), \tag{4.5.17}$$

$$(\mu_2 - \mu_1)\Phi_2(z_2) = \frac{1}{2\pi i}\int_{-\infty}^{\infty}\frac{(\mu_1 p(x) + q(x))}{x - z_2}dx = \mu_1 w_1(z_2) + w_2(z_2). \tag{4.5.18}$$

The values of $w_1(z)$ and $w_2(z)$ on the upper side of the x-axis are given by (2.2.30), (2.2.31); these give the values of $u'(x)$, $v'(x)$ on the axis as

$$u'(x) = 2A_1u_1^+(x) - 2B_1v_1^+(x) + 2A_2u_2^+(x) - 2B_2v_2^+(x), \tag{4.5.19}$$

$$v'(x) = 2C_1u_1^+(x) - 2D_1v_1^+(x) + 2C_2u_2^+(x) - 2D_2v_2^+(x), \tag{4.5.20}$$

where

$$\frac{p_2\mu_1 - p_1\mu_2}{\mu_2 - \mu_1} = \beta_{11}(\mu_1\mu_2) - \beta_{12} = A_1 + iB_1, \tag{4.5.21}$$

$$\frac{p_2 - p_1}{\mu_2 - \mu_1} = \beta_{11}(\mu_1 + \mu_2) - \beta_{13} = A_2 + iB_2, \tag{4.5.22}$$

$$\frac{q_2\mu_1 - q_1\mu_2}{\mu_2 - \mu_1} = \beta_{23} - \beta_{22}\frac{(\mu_1 + \mu_2)}{\mu_1\mu_2} = C_1 + iD_1, \tag{4.5.23}$$

$$\frac{q_2 - q_1}{\mu_2 - \mu_1} = \beta_{12} - \frac{\beta_{22}}{\mu_1\mu_2} = C_2 + iD_2. \tag{4.5.24}$$

When the medium is *orthotropic*, $\beta_{13} = 0 = \beta_{23}$ and the equation (4.5.7) is biquadratic with purely imaginary roots $\mu_1 = i\nu_1, \mu_2 = i\nu_2$. Now $\mu_1\mu_2 = -\nu_1\nu_2$ and $\mu_1 + \mu_2 = i(\nu_1 + \nu_2)$, so that

$$B_1 = 0 = A_2 = C_1 = D_2 \tag{4.5.25}$$

and

$$A_1 = -\beta_{11}\nu_1\nu_2 - \beta_{12}, \; B_2 = \nu_1 + \nu_2, \; D_1 = \beta_{22}(\nu_1 + \nu_2)/\nu_1\nu_2, \; C_2 = \beta_{22}/\nu_1\nu_2. \tag{4.5.26}$$

Now, on the x-axis,

$$u'(x) = 2A_1u_1^+(x) - 2B_2v_2^+(x) \tag{4.5.27}$$

$$v'(x) = -2D_1v_1^+(x) + 2C_2u_2^+(x). \tag{4.5.28}$$

In particular, when the medium is isotropic and in plane strain, equation (2.1.7) shows that

$$\beta_{11} = \beta_{22} = (1-\nu)/2\mu, \quad \beta_{12} = -\nu/2\mu, \quad \beta_{13} = 0 = \beta_{23}$$

and $\mu_1 = i = \mu_2$, so that $\nu_1 = 1 = \nu_2$, and

$$A_1 = -\beta_{11} - \beta_{22} = -(1-\nu)/2\mu, \; B_2 = 2\beta_{11} = 2(1-\nu)/2\mu = D_1, \; C_2 = -A_1$$

and equations (4.5.27), (4.5.28) reduce to equations (2.2.22) and (2.2.23) respectively. Savin (1939) investigated contact problems for a frictionless anisotropic body.

4.6 Contact Problems for an Anisotropic Half-Plane

First consider the analogue of the problem discussed in Section 3.6: a single frictional punch indenting the half-plane. The boundary conditions are of Type I outside the punch, and Type IV under the punch:

$$\sigma_{yy} = 0 = \sigma_{xy} \text{ for } x \in (-\infty, -a) \cup (b, \infty)$$

$$v'(x) = h'(x), \quad \sigma_{xy} + \rho\sigma_{yy} = 0 \text{ for } x \in (-a, b).$$

In terms of the potentials $w_1(z_1)$ and $w_2(z_2)$, these are

$$u_1^+(x) = 0 = u_2^+(x) \text{ for } x \in (-\infty, -a) \cup (b, \infty)$$

$$\left.\begin{aligned} 2C_1u_1^+(x) - 2D_1v_1^+(x) + 2C_2u_2^+(x) - 2D_2v_2^+(x) &= h'(x) \\ u_2^+(x) + \rho u_1^+(x) &= 0 \end{aligned}\right\} \text{for } x \in (-a, b).$$

Now on $(-a, b)$

$$v_2^+(x) = -\frac{1}{2\pi}\int_{-a}^{b} \frac{q(t)dt}{t-x} = \frac{\rho}{2\pi}\int_{-a}^{b} \frac{p(t)dt}{t-x} = -\rho v_1^+(x),$$

so that the condition on $(-a, b)$ may be written

$$2(C_1 - \rho C_2)u_1^+(x) - 2(D_1 - \rho D_2)v_1^+(x) = h'(x). \tag{4.6.1}$$

This may be written as

$$v_1^+(x) + \gamma u_1^+(x) = f(x) \text{ for } x \in (-a, b), \tag{4.6.2}$$

where

$$\gamma = -(C_1 - \rho C_2)/(D_1 - \rho D_2), \tag{4.6.3}$$

$$f(x) = -h'(x)/[2(D_1 - \rho D_2)]. \tag{4.6.4}$$

Equation (4.6.2) is formally equivalent to equation (3.6.2) derived in the isotropic case. There are, however, some points to notice. In the general anisotropic case, if $\rho = 0$ (no friction), then

$$\gamma = -C_1/D_1$$

and this is not zero. In other words, for general anisotropy, the boundary value problem for the frictionless case is similar to that for the isotropic frictional case. If the material is orthotropic, then equation (4.5.25) shows that

$$\gamma = \rho C_2/D_1$$

so that $\gamma = 0$ when $\rho = 0$.

We conclude that the boundary value problem in the general anisotropic case with friction is equivalent to that studied in Section 3.6, and thus the analysis derived there may be applied without change to the general problem.

As a second boundary value problem, we suppose that the upper half-plane is impressed by n punches, as in Figure 3.3.2. Thus, the boundary conditions are of Type I outside the punches and of Type III under the punches:

$$\sigma_{yy} = 0 = \sigma_{xy} \text{ outside}$$

$$u'(x) = g'_{1,i}(x), \quad v'(x) = g'_{2,i}(x) \text{ under the } i\text{th punch.}$$

For convenience, we call the two regions $\mathcal{F}$ (free) and $\mathcal{P}$ (punch), and write $P = \cup P_i$. In terms of the functions w_1 and w_2 we have

$$u_1^+(x) = 0 = u_2^+(x), \quad x \in \mathcal{F} \tag{4.6.5}$$

$$\left.\begin{aligned} 2A_1u_1^+(x) - 2B_1v_1^+(x) + 2A_2u_2^+(x) - 2B_2v_2^+(x) = g'_{1,i} \\ 2C_1u_1^+(x) - 2D_1v_1^+(x) + 2C_2u_2^+(x) - 2D_2v_2^+(x) = g'_{2,i} \end{aligned}\right\} \text{on } P_i. \tag{4.6.6}$$

Multiply the second equation by α and add it to the first:

$$\begin{aligned} &2(A_1 + \alpha C_1)u_1^+(x) - 2(B_1 + \alpha D_1)v_1^+(x) + 2(A_2 + \alpha C_2)u_2^+(x) \\ &-2(B_2 + \alpha D_2)v_2^+(x) = g'_{1,i} + \alpha g'_{2,i} \text{ on } P_i. \end{aligned} \tag{4.6.7}$$

Now choose α so that

$$\frac{A_1 + \alpha C_1}{A_2 + \alpha C_2} = \frac{B_1 + \alpha D_1}{B_2 + \alpha D_2} = \delta.$$

This is a quadratic equation for α:

$$(C_1D_2 - C_2D_1)\alpha^2 + (B_2C_1 + A_1D_2 - B_1C_2 - A_2D_1)\alpha + (A_1B_2 - A_2B_1) = 0. \tag{4.6.8}$$

When the medium is orthotropic, this equation reduces to

$$-C_2D_1\alpha^2 + A_1B_2 = 0$$

and since $-A_1B_2/(C_2D_1) > 0$ the roots are purely imaginary. In general, the roots of (4.6.8) are a complex conjugate pair, $\alpha, \bar{\alpha}$; the corresponding values of δ will form a complex conjugate pair also.

With α chosen to satisfy (4.6.8), we divide equation (4.6.7) through by $A_2 + \alpha C_2$ and find

$$u_2^+(x) + \delta u_1^+(x) - \sigma(v_2^+(x) + \delta v_1^+(x)) = g_i(x) \text{ on } P_i \tag{4.6.9}$$

where

$$\sigma = (B_2 + \alpha D_2)/(A_2 + \alpha C_2),\ g_i(x) = \{g'_{1,i}(x) + \alpha g'_{2,i}(x)\}/2(A_2 + \alpha C_2). \tag{4.6.10}$$

Equation (4.6.9) appears at first sight to be of the same form as (3.6.2), an equation linking the real and imaginary parts of the function. However, because δ is not real, but complex, the quantities $u_2^+(x) + \delta u_1^+(x)$ and $v_2^+(x) + \delta v_1^+(x)$ are not real. To solve (4.6.9), we must rewrite it in the form used by Muskhelishvili, as a relation between the values of a function on either side of the x-axis. Introduce two functions

$$w_3(z) = \frac{1}{2\pi i} \oint_P \frac{q(x) + \delta p(x)}{x - z} dz = w_2(z) + \delta w_1(z) \tag{4.6.11}$$

$$w_4(z) = \frac{1}{2\pi i} \oint_P \frac{q(x) + \bar{\delta} p(x)}{x - z} dz = w_2(z) + \bar{\delta} w_1(z). \tag{4.6.12}$$

Then, using the Plemelj formulae, (2.2.11), (2.2.12), we find

$$u_2^+(x) + \delta u_1^+(x) = \frac{1}{2}(q(x) + \delta p(x)$$
$$= \frac{1}{2}(w_3^+(x) - w_3^-(x))$$

$$v_2^+(x) + \delta v_1^+(x) = \frac{1}{2\pi i}\oint_P \frac{(q(t) + \delta p(t)\}}{t - x}dt = \frac{1}{2}(w_3^+(x) + w_3^-(x))$$

so that equations (4.6.5), (4.6.9) become

$$w_3^+(x) - w_3^-(x) = 0 \text{ on } \mathcal{F}, \tag{4.6.13}$$

$$w_3^+(x) - \kappa w_3^-(x) = f_i(x) \text{ on } P_i, \tag{4.6.14}$$

where

$$\kappa = \left(\frac{1+\sigma}{1-\sigma}\right), \quad f_i(x) = \frac{2g_i(x)}{1-\sigma}. \tag{4.6.15}$$

The equations for $w_4(z)$ are exactly the same, except that κ, σ are replaced by $\bar{\kappa}, \bar{\sigma}$.

Let us examine equations (4.6.13), (4.6.14) for the special case of one punch occupying $(-a, a)$. As we showed in Section 3.2, we must first find a solution $X(z)$ of the homogeneous problem

$$X^+(x) - X^-(x) = 0 \text{ for } |x| > a, \tag{4.6.16}$$

$$X^+(x) - \kappa X^-(x) = 0 \text{ for } |x| < a. \tag{4.6.17}$$

We choose $X(z)$ of the form

$$X(z) = (z + a)^{-\theta}(z - a)^{-1+\theta}. \tag{4.6.18}$$

For $|x| < a$,

$$X^+(x) = \exp[-i\pi(1-\theta)](x + a)^{-\theta}(a - x)^{-1+\theta}$$
$$X^-(x) = \exp[i\pi(1-\theta)](x + a)^{-\theta}(a - x)^{-1+\theta}.$$

Thus, we must choose θ so that

$$\exp[-2i\pi(1-\theta) = \exp[2i\pi\theta] = \kappa. \tag{4.6.19}$$

With this homogeneous solution, we divide equations (4.6.13), (4.6.14) through by $X^+(x)$, and put

$$F(z) = w_3(z)/X(z). \tag{4.6.20}$$

Equations (4.6.13), (4.6.14) become

$$F^+(x) - F^-(x) = 0 \text{ for } |x| > a \tag{4.6.21}$$

$$F^+(x) - F^-(x) = f(x)/X^+(x) \text{ for } |x| < a \tag{4.6.22}$$

which have the solution

$$F(z) = \frac{1}{2\pi i}\int_{-a}^{a} \frac{f(x)dx}{X^{+}(x)(x-z)} + \frac{N(z)}{D(z)}, \tag{4.6.23}$$

where $N(z)$, $D(z)$ are polynomials.

As an example, consider the problem of a single punch with a plane base. The punch, which is rigidly linked to the elastic anisotropic half-plane, is acted upon by a normal force P and tangential force Q. The potential $w_3(z)$ satisfies the equations (4.6.13), (4.6.14), while $w_4(z)$ satisfies the equations with κ replaced by $\bar{\kappa}$. We must choose θ according to (4.6.19), and such that $w_3(z)$, given by

$$w_3(z) = C_3(z+a)^{-\theta}(z-a)^{-1+\theta} \tag{4.6.24}$$

has integrable singularities at $\pm a$. Thus, if $\theta = \xi + i\eta$, we must have $\xi < 1$ and $\xi > 0$, i.e., $0 < \xi < 1$.

$$w_4(z) = C_4(z+a)^{-\phi}(z-a)^{-1+\phi} \tag{4.6.25}$$

where $\exp(2i\phi) = \bar{\kappa}$, and $\phi = \xi' + i\eta'$, $0 < \xi' < 1$. We may find C_3 and C_4 by examining the behaviour of $w_3(z)$, $w_4(z)$ as $z \to \infty$. Equations (4.6.11), (4.6.12) show that, for large $|z|$,

$$w_3(z) = -\frac{1}{2\pi i z}(Q + \delta P), \quad w_4(z) = -\frac{1}{2\pi i z}(\bar{Q} + \bar{\delta} P)$$

so that

$$C_3 = -\frac{1}{2\pi i}(Q + \delta P), \quad C_4 = -\frac{1}{2\pi i}(Q + \bar{\delta} P). \tag{4.6.26}$$

To complete the solution of the problem, we retrace the steps in the analysis:

$w_3(z)$, $w_4(z)$ are given by (4.6.24), (4.6.25);

$w_1(z)$, $w_2(z)$ are related to $w_3(z)$, $w_4(z)$ by (4.6.11), (4.6.12), $\Phi_1(z_1)$, $\Phi_2(z_2)$ are related to w_1, w_2 by (4.5.17), (4.5.18). The stresses in the half-plane are given by (4.5.9)–(4.5.11), and the strains are given by (4.5.3).

4.7 Stick-Slip Contact

In Section 3.7, we considered a punch indenting an elastic half-plane when there was limiting friction equilibrium all over the region of contact. In this section we suppose that the contact region is split into three parts, a central part in which the indenter sticks, and parts on either side where there is slip in the presence of limiting friction. We suppose that the face of the indenter is flat, and the elastic medium occupies the upper half-plane.

To formulate the problem, we go back to the conventions for the stresses at the beginning of Section 2.1. The elastic half-plane occupies the region $y > 0$, the + side of the x-axis. The half-plane thus exerts a force per unit area $\sigma_{xy}\hat{\imath} + \sigma_{yy}\hat{\jmath}$ on the punch; the punch exerts $-\sigma_{xy}\hat{\imath} - \sigma_{yy}\hat{\jmath} = q\hat{\imath} + p\hat{\jmath}$ on the half-plane. As the punch is pressed into the half-plane, the material on the surface tends to move *away* from the centre, the origin. Thus, in the right-hand slip zone $(b, 1)$ the displacement $u(x)$ will be positive; since the friction force always opposes the motion, $q(x)$ will be negative; it will have its critical value $q(x) = -\rho p(x)$. Similarly, in the left-hand slip zone $(-1, -a)$, $q(x)$ will be positive: $q(x) = \rho p(x)$.

If there is no net sideways force on the punch, then there will be symmetry: $a = b$, $p(-x) = p(x)$, $q(-x) = -q(x)$; $Q = \int_{-1}^{1} q(x)dx = 0$.

If there is a net sideways force on the punch, $Q \neq 0$, then symmetry will be lost. Particularly if ρ is small, there will be only a small range of values of Q/P such that the whole punch does not slip sideways: $-a < b$. Now we proceed to the analysis.

In the notation of Section 3.1, the boundary conditions are of Type I outside the punch, Type III in the stick zone, Type V in the slip zones. We assume that the indenter has width 2, and is acted on by a normal force P and shear force Q as in Figure 4.7.1.

The stresses on the boundary of the half-plane are

$$\sigma_{yy}(x, 0) = -p(x), \quad \sigma_{xy}(x, 0) = -q(x). \tag{4.7.1}$$

In the stick zone BC, $-a < x < b$, the ratio of shear stress to normal stress is insufficient to cause slipping:

$$\left|\frac{q(x)}{p(x)}\right| < \rho. \tag{4.7.2}$$

The normal pressure under the punch should be positive everywhere: $p(x) > 0$. Outside the punch there are no surface stresses:

$$p(x) = 0 = q(x) \text{ for } |x| > 1.$$

The displacements under the punch are

$$v(x) = 0 \text{ on } AD, \quad u(x) = 0 \text{ on } BC.$$

Thus the boundary conditions are

$$p(x) = 0 = q(x) \text{ for } |x| > 1 \tag{4.7.3}$$

$$q(x) + \rho p(x) = 0, \quad v = 0 \text{ for } b < x < 1 \tag{4.7.4}$$

$$u = 0 = v \quad \text{for} \quad -a < x < b \tag{4.7.5}$$

$$q(x) - \rho p(x) = 0, \quad v = 0 \text{ for } -1 < x < -a \tag{4.7.6}$$

In terms of the potentials w_1, w_2, the boundary conditions are

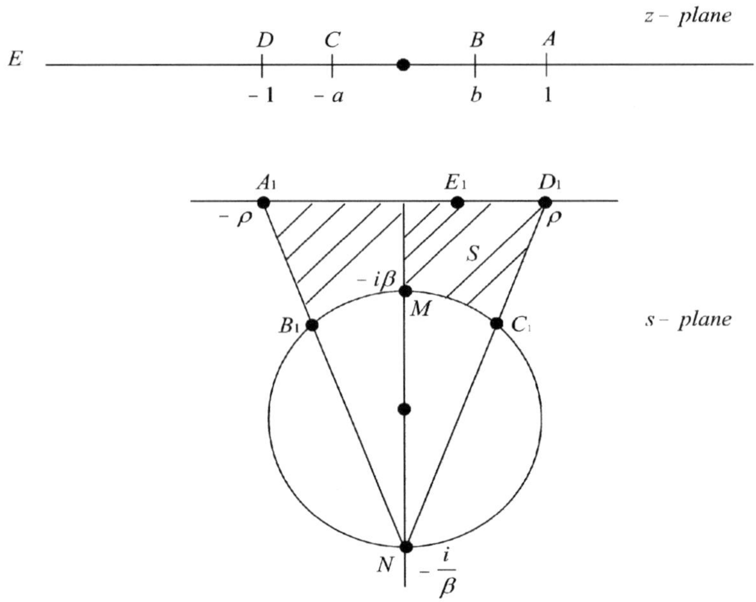

Fig. 4.7.1 $s(z)$ maps the upper z-plane onto S.

$$u_1^+(x) = 0 = u_2^+(x) \text{ for } |x| > 1 \tag{4.7.7}$$

$$u_2^+(x) + \rho u_1^+(x) = 0, \quad -v_1^+(x) + \beta u_2^+(x) = 0 \text{ for } b < x < 1 \tag{4.7.8}$$

$$\beta u_1^+(x) + v_2^+(x) = 0 = -v_1^+(x) + \beta u_2^+(x) \text{ for } -a < x < b \tag{4.7.9}$$

$$u_2^+(x) - \rho u_1^+(x) = 0, \quad -v_1^+(x) + \beta u_2^+(x) = 0 \text{ for } -1 < x < -a \tag{4.7.10}$$

and in addition

$$\left|\frac{q(x)}{p(x)}\right| < \rho \text{ for } -a < x < b. \tag{4.7.11}$$

We introduce the function

$$s(z) = \frac{w_2(z)}{w_1(z)}. \tag{4.7.12}$$

The boundary condition (4.7.7) implies

$$\text{Im}(s(z)) = 0 \text{ for } |x| > 1. \tag{4.7.13}$$

Now consider the condition (4.7.8). First, we note that $-v_1 + \beta u_2 = \text{Im}(-w_1 + i\beta w_2) = 0$. Secondly, $u_2 + \rho u_1 = 0$ and $-v_1 + \beta u_2 = 0$ imply $v_1 + \beta\rho u_1 = \text{Im}(w_1 + i\beta\rho w_1) = 0$. Dividing one by the other, we have

$$\text{Im}\left(\frac{iw_1 + \beta w_2}{(i - \beta\rho)w_1}\right) = 0 \quad \text{on} \quad b < x < 1$$

and this we may write as

$$\text{Im}\left(\frac{i + \beta s(z)}{(i - \beta\rho)}\right) = 0 \quad \text{on} \quad b < x < 1. \tag{4.7.14}$$

Treating the conditions (4.7.9), (4.7.10) in the same way, we find

$$\text{Im}\left(\frac{i + \beta s(z)}{is(z) - \beta}\right) = 0 \quad \text{on} \quad -a < x < b \tag{4.7.15}$$

and

$$\text{Im}\left(\frac{i + \beta s(z)}{i + \beta\rho}\right) = 0 \quad \text{on} \quad -1 < x < -a. \tag{4.7.16}$$

We note that all these conditions, (4.7.13)–(4.7.16) apply on the + side of the x-axis.

Each of these conditions has the form

$$\text{Im}\left(\frac{c + ds(z)}{e + fs(z)}\right) = 0.$$

But this equation may be written

$$\frac{c + ds}{e + fs} = \frac{\bar{c} + \bar{d}\bar{s}}{\bar{e} + \bar{f}\bar{s}}$$

which, when reduced, is

$$(d\bar{f} - \bar{d}f)s\bar{s} + (\bar{e}d - \bar{c}f)s + (c\bar{f} - e\bar{d})\bar{s} + (c\bar{e} - \bar{c}e) = 0. \tag{4.7.17}$$

This is the equation of a circle, or a straight line if

$$d\bar{f} - \bar{d}f = 0.$$

We can consider $s(z)$ as a mapping of the upper half-plane $\text{Im}(z) > 0$ onto a region S in the s-plane. The boundary of S is divided into four parts: the part $(-\rho, \rho)$ of the real axis where (4.7.13) holds, corresponding to $|x| > 1$; the straight line where (4.7.14) holds, corresponding to $(b, 1)$; an arc of the circle where (4.7.15) holds, corresponding to $(-a, b)$; the straight line where (4.7.16) holds, corresponding to $(-1, -a)$.

The circle corresponding to (4.7.15) is given by (4.7.17) as

$$2\beta i(s\bar{s} + 1) + (\beta^2 + 1)(s - \bar{s}) = 0; \tag{4.7.18}$$

it has centre $\left(0, -\frac{1}{2}\left(\beta + \frac{1}{\beta}\right)\right)$, radius $\frac{1}{2}\left(\frac{1}{\beta} - \beta\right)$, and meets the imaginary axis at $-i\beta$ and $-i/\beta$.

Figure 4.7.1 shows the region S and the points A_1, B_1, C_1, D_1 corresponding to A, B, C, D.

It may easily be verified that the coordinates of B_1 and C_1 are

$$\left(\frac{-(1-\beta^2)\rho}{1+\beta^2\rho^2}, \frac{-\beta(1+\rho^2)}{1+\beta^2\rho^2}\right), \left(\frac{(1-\beta^2)\rho}{1+\beta^2\rho^2}, \frac{\beta(1+\rho^2)}{1+\beta^2\rho^2}\right),$$

and that the four interior angles of the region S are all equal:

$$\alpha = \arctan\left(\frac{1}{\beta\rho}\right) = \frac{\pi}{2} - \arctan(\beta\rho). \tag{4.7.19}$$

The point E_1 in the s-plane corresponding to the point at infinity in the z-plane is given by

$$\lim_{z\to\infty} s(z) = \lim_{z\to\infty} \frac{w_2(z)}{w_1(z)} = \frac{Q}{P} \tag{4.7.20}$$

by equation (2.2.36). Note that this shows that Q cannot be greater in magnitude than ρP, since otherwise E_1 would lie outside S. When $|Q| > \rho P$, the whole punch shifts, and we have a different problem.

Now consider the boundary condition on B_1C_1. On B_1C_1 equation (4.7.15) implies

$$\text{Im}\left(\frac{is(z)-\beta}{i+\beta s(z)}\right) = 0$$

so that

$$\frac{is(z)-\beta}{i+\beta s(z)} = \text{Re}\left\{\frac{is(z)-\beta}{i+\beta s(z)}\right\} = \text{Re}\left\{\frac{iw_2-\beta w_1}{iw_1+\beta w_2}\right\}$$
$$= \text{Re}\left\{\frac{-v_2-\beta u_1+i(u_2-\beta v_1)}{-v_1+\beta u_2+i(u_1+\beta v_2)}\right\}.$$

But, on BC, $v_2 = -\beta u_1$ and $v_1 = \beta u_2$, so that

$$\frac{is(z)-\beta)}{i+\beta s(z)} = \frac{(1-\beta^2)u_2}{(1-\beta^2)u_1} = \frac{u_2}{u_1} = \frac{q(x)}{p(x)}.$$

Suppose that there were two points x_1, x_2 on BC such that

$$\frac{q(x_1)}{p(x_1)} = \frac{q(x_2)}{p(x_2)}$$

then

$$\frac{is(x_1)-\beta}{i+\beta s(x_1)} = \frac{is(x_2)-\beta}{i+\beta s(x_2)}$$

which implies $s(x_1) = s(x_2)$. This imples that the mapping is not one-to-one. We deduce that if the mapping is one-to-one then the ratio $q(x)/p(x)$ must vary monotonically from its value ρ at $x = -a$ to $-\rho$ at $x = b$. This means that the condition (4.7.11) is fulfilled, $p(x) > 0$, and the arc B_1C_1 is the upper part of the circle as shown in Figure 4.7.1, including the point $s(z) = -i\beta$ where $q(x) = 0$, and not $s(z) = -i/\beta$ where $p(x) = 0$.

The crucial part of the solution to the problem is the determination of the function $s(z)$ that maps the upper half-plane on the region S. Assuming for the moment that we have found $s(z)$, we return to the boundary conditions and show how we can find $w_1(z)$.

Consider the two boundary conditions

$$u_1^+(x) = 0 \quad \text{on } |x| > 1 \tag{4.7.21}$$

$$-v_1^+(x) + \beta u_2^+(x) = 0 \quad \text{on } |x| < 1. \tag{4.7.22}$$

Write $s(x) = s_1 + is_2$, then $w_2^+ = (s_1 + is_2)w_1^+$ gives $u_2^+ = s_1u_1^+ - s_2v_1^+$, so that

$$\beta s_1 u_1^+ - (1 + \beta s_2)v_1^+ = 0 \quad \text{on } |x| < 1 \tag{4.7.23}$$

$$u_1^+ = 0 \quad \text{on } |x| > 1 \tag{4.7.24}$$

Equation (4.7.23) has the form (3.2.3). Let

$$\frac{i + \beta s}{-i + \beta \bar{s}} = \exp(2i\phi) \tag{4.7.25}$$

then $(i + \beta s) = r\exp(i\phi)$ for some real r. Now equation (4.7.23) may be written

$$\text{Re}\{(i + \beta s)w_1^+(x)\} = 0 \quad \text{on } |x| < 1$$

so that

$$\text{Re}\{\exp(i\phi)w_1^+(x)\} = 0 \quad \text{on } |x| < 1 \tag{4.7.26}$$

$$\text{Re}\{w_1^+(x)\} = 0 \quad \text{on } |x| > 1. \tag{4.7.27}$$

Equation (4.7.25) gives

$$\phi(x) = \frac{1}{2i}\ell n\left(\frac{i + \beta s(x)}{-i + \beta\bar{s}(x)}\right). \tag{4.7.28}$$

According to Section 3.2, the function $\Omega(z)$ holomorphic in the upper half-plane which is such that $\Omega_1^+(x) = \phi(x)$, is

$$\Omega(z) = \frac{2}{2\pi i}\int_{-1}^{1}\frac{\phi(t)dt}{t - z}. \tag{4.7.29}$$

We can now write (4.7.26) as

$$\mathrm{Re}\{\exp(i\Omega_1^+(x))w_1^+(x)\}\exp(-\Omega_2^+(x)) = 0$$

i.e.,

$$\mathrm{Re}\{\exp(i\Omega^+(x))w_1^+(x)\} = 0, \quad |x| < 1. \tag{4.7.30}$$

On $|x| > 1$, $\Omega_1^+(x) = 0$, so that (4.7.27) imples that (4.7.30) holds on $|x| > 1$ also. Thus, the function $\exp(i\Omega(z)w_1(z))$ has its real part zero on the whole of the x-axis; the function must be a rational function with imaginary coefficients such that its poles are situated on the x-axis – and they can be only $z = \pm 1$;

$$w_1(z) = \frac{i\exp(-i\Omega(z))N(z)}{(z-1)(z+1)} \tag{4.7.31}$$

where $N(z)$ is a polynomial with real coefficients, and

$$w_2(z) = s(z)w_1(z). \tag{4.7.32}$$

Before proceeding further, we investigate the purported solution we have obtained, and see that it fulfills all the stated conditions (4.7.13)–(4.7.16). Consider the conditions on BC. We have satisfied the conditions

$$\mathrm{Im}\left(\frac{i+\beta s(z)}{is(z)-\beta}\right) = 0, \quad \mathrm{Re}(iw_1 + \beta w_2) = 0. \tag{4.7.33}$$

The first of these states that

$$\mathrm{Im}\left(\frac{iw_1+\beta w_2}{iw_2-\beta w_1}\right) = 0$$

which, together with the second of (4.7.33), states that

$$\mathrm{Re}(iw_2 - \beta w_1) = -v_2 - \beta u_1 = 0$$

which is the other condition to be satisfied on BC. It can be verified that all the other conditions in (4.7.7)–(4.7.10) are satisfied similarly.

The problem of finding the mapping $s(z)$ that maps the upper half-plane $\mathrm{Im}(z) > 0$ onto the region S, with interior angles α, and with the point at infinity being mapped to E_1, is a difficult problem that requires the solution of a differential equation of Fuchsian type:

$$s'' + \left(\frac{1-\alpha/\pi}{z} + \frac{1-\alpha/\pi}{z-1} + \frac{1-\alpha/\pi}{z-a}\right)s' + \frac{(1-\alpha/\pi)(1-2\alpha/\pi)(z-\lambda)}{z(z-1)(z-a)}s = 0.$$

The difficulties of this problem consist in the determination of the parameters α and λ. Only when $\rho = 0$ and consequently $\alpha = \pi/2$, do we obtain a particular case of a Lamé differential equation, and the problem becomes elementary.

Instead, we consider a problem that is near, in some sense, to the actual problem. We replace the shaded region S by the region S_1 in Figure 4.7.2. Now the

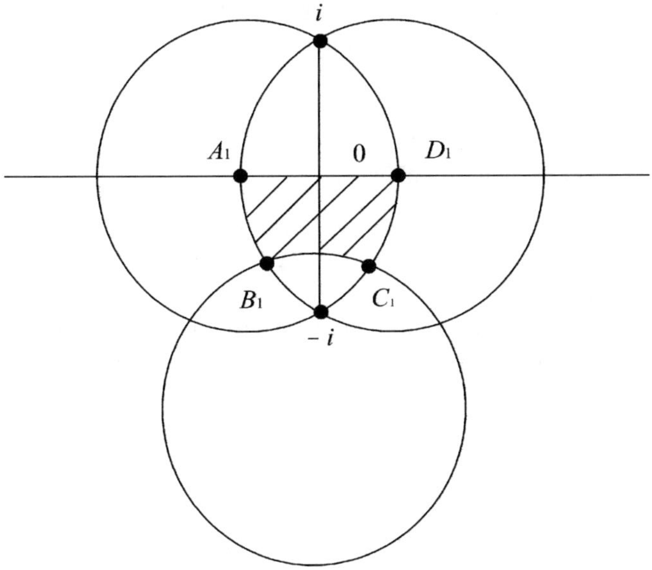

Fig. 4.7.2 The shaded region S_1 in the s_1-plane is bounded by arcs of circles, and the real axis.

straight lines A_1B_1 and D_1C_1 have been replaced by arcs of circles. We first seek the mapping that maps the upper half-plane on S_1, and maps the point at infinity z_1-plane onto the *origin* in the s_1-plane; according to equation (4.7.20), this is the case $Q = 0$.

The circle through A_1, B_1 has centre $(1-\rho^2)/(2\rho)$ and radius $(1+\rho^2)/(2\rho)$ so that its equation is

$$\left(s_1 - \frac{(1-\rho^2)}{2\rho}\right)\left(\bar{s}_1 - \frac{(1-\rho^2)}{2\rho}\right) = \left(\frac{1+\rho^2}{2\rho}\right)^2.$$

The corresponding circle through C_1, D_1 is

$$\left(s_1 + \frac{(1-\rho^2)}{2\rho}\right)\left(\bar{s}_1 + \frac{(1-\rho^2)}{2\rho}\right) = \left(\frac{1+\rho^2}{2\rho}\right)^2$$

and it is seen immediately that these circles intersect at $\pm i$, as shown in Figure 4.7.2. Now the interior angles of the region S_1 are all $\pi/2$. (Galin's Figure 29 incorrectly shows them as α.)

Instead of finding the mapping of the upper half-plane on S_1, we seek the inverse mapping, of S_1 onto the upper half z_1-plane.

First, we apply the bilinear mapping

$$\xi = \frac{s_1 + i}{is_1 + 1}. \tag{4.7.34}$$

This will map $s_1 = +i$ to $\xi = \infty$, $s_1 = -i$ to $\xi = 0$, and hence will map the circles through A_1, B_1 and C_1, D_1 into straight lines through the origin.

Since

$$s_1 = \frac{i - \xi}{i\xi - 1},$$

the map of the real axis in the s_1-plane is

$$\frac{i - \xi}{i\xi - 1} = \frac{-i - \bar{\xi}}{-i\bar{\xi} - 1}$$

which simplifies to $\xi\bar{\xi} = 1$, the unit circle, centre the origin. After some routine calculation, we find similarly that the map of the circle through B_1 and C_1 is the concentric circle

$$\xi\bar{\xi} = \left(\frac{1 - \beta}{1 + \beta}\right)^2.$$

The region S_2, the map of S_1, is shown in Figure 4.7.3. To simplify the notation, we note equation (2.2.21) and put

$$\frac{1 + \beta}{1 - \beta} = \kappa, \qquad \rho = \tan\gamma \tag{4.7.35}$$

then A_2, B_2, C_2, D_2 are given by

$$\exp\left\{i\left(\frac{\pi}{2} + 2\gamma\right)\right\}, \quad \frac{1}{\kappa}\exp\left\{i\left(\frac{\pi}{2} + 2\gamma\right)\right\},$$
$$\frac{1}{\kappa}\exp\left\{i\left(\frac{\pi}{2} - 2\gamma\right)\right\}, \quad \exp\left\{i\left(\frac{\pi}{2} - 2\gamma\right)\right\}$$

respectively, as shown in Figure 4.7.3.

The next step is simple: map S_2 onto a rectangle, by means of

$$\zeta = \ell n\xi. \tag{4.7.36}$$

Now A_3, B_3, C_3, D_3 are given by

$$i\left(\frac{\pi}{2} + 2\gamma\right), -\ell n\kappa + i\left(\frac{\pi}{2} + 2\gamma\right), -\ell n\kappa + i\left(\frac{\pi}{2} - 2\gamma\right), i\left(\frac{\pi}{2} - 2\gamma\right)$$

as shown in Figure 4.7.4

Now we map this region, by a magnification, rotation and displacement, onto the so-called *fundamental* rectangle with sides $2K$ and $2K'$ of elliptic functions (Nehari, 1952, p. 280), shown in Figure 4.7.4. Take

$$\eta = c\left(i\zeta + \frac{\pi}{2} + i\ell n\kappa\right) \tag{4.7.37}$$

then A_4, B_4, C_4, D_4 are

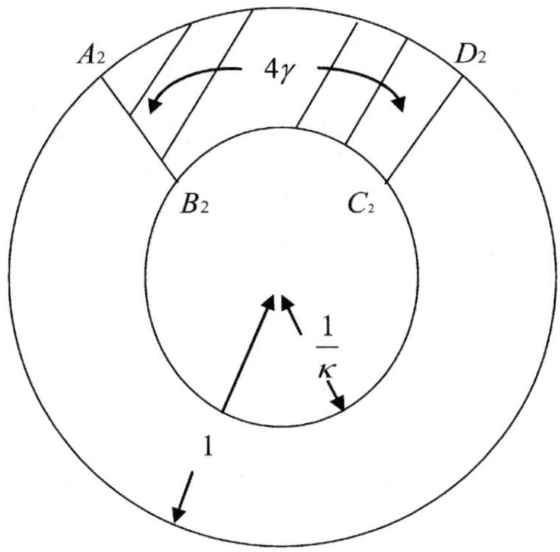

Fig. 4.7.3 The region S_2 in the ξ-plane is bounded by concentric circles and radii.

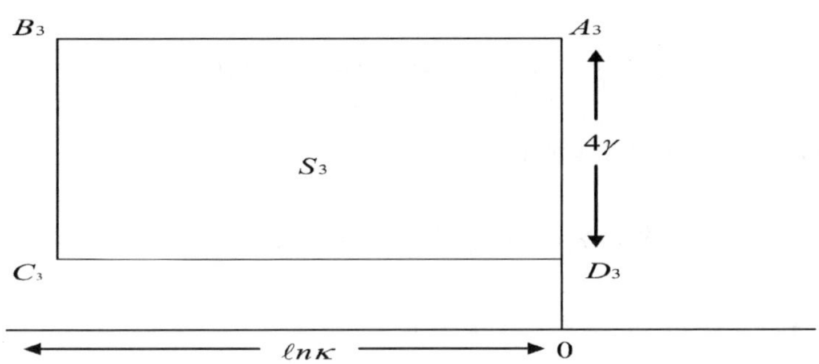

Fig. 4.7.4 The region S_3 in the ζ-plane is a rectangle.

$$c(-2\gamma + i\ell n\kappa),\ -2c\gamma,\ 2c\gamma,\ c(2\gamma + i\ell n\kappa).$$

These are to be $-K + iK'$, $-K$, K, $K + iK'$, so that we must choose

$$2c\gamma = K, \quad c\ell n\kappa = K' \tag{4.7.38}$$

and thus

$$\frac{2\gamma}{\ell n\kappa} = \frac{K}{K'}, \quad c = \frac{K}{2\gamma}. \tag{4.7.39}$$

The function that maps the fundamental rectangle onto the upper half-plane, $\text{Im}(\tau) > 0$ is the elliptic function

$$\tau = sn\eta. \tag{4.7.40}$$

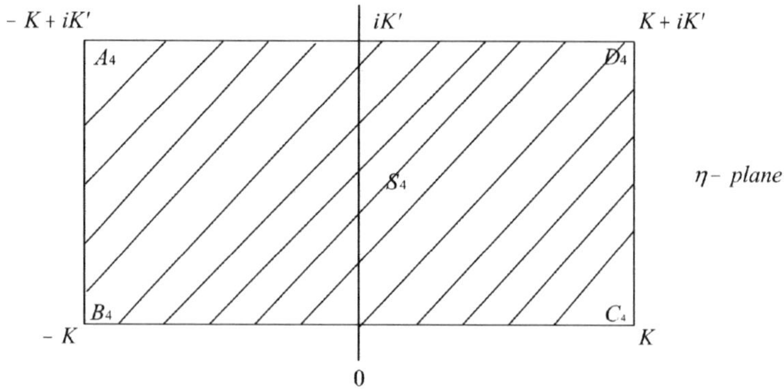

Fig. 4.7.5 The fundamental rectangle in the η-plane.

This maps the points A_4, B_4, C_4, D_4 into $-1/k$, -1, 1, $1/k$ respectively, and maps $\eta = iK'$ into $\tau = \infty$. The inverse of the elliptic function *sn* is the elliptic integral

$$\eta = \int_0^\tau \frac{dt}{\sqrt{(1-t^2)(1-k^2t^2)}} = F(\tau, k). \tag{4.7.41}$$

As a check, we note that this maps $\tau = 1$ to

$$K = K(k) = \int_0^1 \frac{dt}{\sqrt{(1-t^2)(1-k^2t^2)}} \tag{4.7.42}$$

and maps $\tau = 1/k$ to $\eta = K(k) + iK'(k)$, where

$$K'(k) = \int_0^1 \frac{dt}{\sqrt{(1-t^2)(1-k'^2t^2)}} = K(k') \tag{4.7.43}$$

and $k' = \sqrt{1-k^2}$.

The next transformation

$$z_1 = -k\tau \tag{4.7.44}$$

maps $-1/k$, -1, 1, $1/k$ in the τ-plane onto 1, k, $-k$, -1 in the z_1-plane.

After all this, we have mapped S_1 onto the upper half of the z_1-plane: $s_1(z_1)$ is the inverse of this mapping. The steps from z_1 to s_1 are as follows:

$$\tau = -z_1/k \tag{4.7.45}$$

$$\eta = F(\tau, k) \tag{4.7.46}$$

$$\zeta = \frac{i\pi}{2} - \ell n\kappa - \frac{i\eta}{c} \tag{4.7.47}$$

Table 4.7.1 The variation of k with ρ.

ρ	0.1	0.3	0.5	0.7	0.9
k	0.039	0.701	0.945	0.989	0.997

$$\xi = \exp \zeta \tag{4.7.48}$$

$$s_1 = \frac{i - \xi}{i\xi - 1} \tag{4.7.49}$$

As another check, we note that E, the point at infinity in the z_1-plane is mapped onto 0 in the s_1-plane: $\tau = \infty$ implies $\eta = iK'$, $\zeta = \frac{i\pi}{2}$, $\xi = i$, $s_1 = 0$.

This is thus the case $Q = 0$; $a = b = k$. The values of k are given in Table 4.7.1.

In general, when $Q \neq 0$, according to equation (4.7.20), the point at infinity in the z-plane should be mapped into Q/P in the s_1-plane.

Under the mapping we have obtained, $s_1^* = Q/P$ corresponds to the point $z_1 = z^* = z_1(Q/P)$. This means that we should find a mapping that maps the upper z_1-plane onto the upper z-plane, with the points $1, -1, z_1^*$ in the z_1-plane being mapped onto $1, -1, \infty$ in the z-plane. This mapping is

$$z = \frac{z_1 z_1^* - 1}{z_1^* - z_1}. \tag{4.7.50}$$

This maps $k, -k$ in the z_1-plane onto

$$z = \frac{kz_1^* - 1}{z_1^* - k} = a, \qquad z = \frac{-kz_1^* - 1}{z_1^* + k} = -b$$

in the z-plane.

Thus the final mapping maps the upper z-plane onto S_1, with the point at infinity in the z-plane being mapped to $s_1 = Q/P$.

Note that the function $s_1(z)$ maps the upper z-half plane onto S_1 (shown in Figure 4.7.2) in the s_1-plane. The inverse map of S (shown in Figure 4.7.1) in the z_1-plane is not quite the half-plane, but one with shallow notches taken out at $z_1 = \pm 1$, $\pm k$. When $\rho = 0 \cdot 6$, $\nu = 0 \cdot 3$, the maximal depth of these notches is $0 \cdot 0010$. These notches will transform into small notches around ± 1, a, $-b$ in the z-plane.

The parameter k is given by equations (4.7.39), (4.7.42). Equations 16.38.5 and 16.38.7 of Abramowitz and Stegun (1972) give

$$\left(2\frac{K}{\pi}\right)^{\frac{1}{2}} = 1 + 2q + 2q^4 + 2q^9 + \cdots = \vartheta_3(0, q) \tag{4.7.51}$$

$$\left(2\frac{kK}{\pi}\right)^{\frac{1}{2}} = 2q^{\frac{1}{4}}(1 + q^2 + q^6 + q^{12} + \cdots = \vartheta_2(0, q) \tag{4.7.52}$$

so that

$$k = \vartheta_2^2(0,q)/\vartheta_3^2(0,q) \tag{4.7.53}$$

where

$$q = \exp(-\pi K'/K). \tag{4.7.54}$$

We proceed as follows: $\kappa = 3 - 4\nu$, $\gamma = \tan^{-1}\rho$; the first of equations (4.7.39) gives K/K' and then equation (4.7.54) gives q, and equation (4.7.53) gives k.

Figure 4.7.6 shows the values of $-a$ and b for different values of $Q/P = \alpha$ and ρ when $\nu = 0.3$, i.e., $\kappa = 1.8$. We note that the steps in the calculation:

$$s_1^* = Q/P = \alpha = \tan\theta; \quad \xi = i\exp(-2i\theta) = \exp\left(i\frac{\pi}{2} - 2i\theta\right);$$

$$\zeta = i\left(\frac{\pi}{2} - 2\theta\right); \quad \eta = 2c\theta + ic\ell n\kappa = K\theta/\gamma + iK' = u + iK';$$

By entry 16.8.1 of Abramowitz and Stegun (1972),

$$\tau = sn(\eta) = sn(u + iK') = 1/(ksnu); \quad z = -1/snu = -1/v;$$

$$a = (k+v)/(1+kv), b = (k-v)/(1-kv).$$

We note that when $Q = 0, u = 0 = v, a = k = b$; the stick region is symmetrical about the mid point of the punch. The first two columns of Figure 4.7.6 shows the values of $-a$ and b for $\rho = 0.1$ to 0.9; for $\rho = 0.1$ there is a very small stick zone, but when $\rho \geq 0.5$ almost all of the contact region is a stick zone. The next pairs of columns show $-a$ and b for α/ρ ranging from 0.2 to 0.8. Note that for $\alpha = 0$, $\rho = 0.1$ the stick zone is symmetrical and of length 0.08; for a small change in α, to $(0.1)(0.2) = 0.02$, the stick zone shifts markedly: 0.3 to the left; but its length remains roughly the same. When $\alpha = 0.2$ and $\rho \geq 0.5$ almost all the contact region is a stick zone; we can analyse the problem as if there were adhesive contact. On the other hand, when $\rho = 0.1$,.the stick zone is small for all values of α, and one may assume that there is no stick zone. We note that a and b are functions of k and v, and that v is a function of $u = K\theta/\gamma$. For a given value of k, the length of the stick zone, $a + b$, decreases as v increases:

$$a + b = \frac{2k(1-v^2)}{1-k^2v^2} = 2k - \frac{2k(1-v^2)v^2}{1-k^2v^2}.$$

Galin mentions the mapping (4.7.50), but does not identify a and b. Of course, finding k, or a and b, is just one step in the solution of the problem. Once $s(z)$, or its approximation $s_1(z)$ have been found, it is necessary to find $w_1(z)$, $w_2(z)$ and the contact stresses. Galin's solution, while being a significant step forward at the time (the 1950s) is incomplete.

$\rho \backslash \alpha$	0		0.2		0.4		0.6		0.8	
	$-a$	b	$-a$	b	$-a$	b	$-a$	b	$-a$	b
0.1	–0.04	0.04	–0.34	–0.27	–0.61	–0.56	–0.82	–0.82	–0.96	–0.95
0.3	–0.70	0.70	–0.85	0.45	–0.93	0.07	–0.97	–0.41	–0.99	–0.83
0.5	–0.95	0.95	–0.98	0.84	–0.99	0.59	–1.0	0.05	–1.0	–0.65
0.7	–0.99	0.99	–1.0	0.95	–1.0	0.79	–1.0	0.32	–1.0	–0.52
0.9	–1.0	1.00	–1.0	0.98	–1.0	0.88	–1.0	0.45	–1.0	–0.45

Fig. 4.7.6 The limits of the stick zone.

4.8 Contact of Two Elastic Bodies

In all the contact problems we have considered so far, we have assumed that one of the bodies in contact is absolutely rigid – a rigid punch. We shall now show that in a number of cases, a contact between *two elastic bodies*, we can reduce the problem to that of determining holomorphic functions subject to mixed boundary conditions of the same type as those we considered earlier.

We assume that the radii of curvature of both bodies are large compared to the dimension of the contact region, and replace each body by a semi-infinite plane. Thus, body I occupies the semi-infinite plane $y \geq 0$, and body II occupies $y \leq 0$.

Let the equations of the surfaces of the bodies before deformation ocurrs be

$$y = f_1(x) \text{ and } y = -f_2(x)$$

respectively. Place the origin 0 at the point of initial contact between the bodies. Under the action of the contact forces, body I is displaced down by the amount δ_1, and body II is displaced up by δ_2. Suppose that, after contact, point A of body I comes in contact with point B of body II. Point A, which originally was at $(x, y_1) = (x, f_1(x))$ is displaced δ_1 down and $v_1(x, y)$ up: it is at $(x, f_1(x) + v_1(x, y_1) - \delta_1)$. Point B, which originally was at $(x, y_2) = (x, -f_2(x))$ is now at $(x, -f_2(x) - v_2(x, y_2) + \delta_2)$. After contact, points A and B coincide, so that

$$f_1(x) + v_1(x, y_1) - \delta_1 = -f_2(x) - v_2(x, y_2) + \delta_2.$$

We assume, as is usual in linear elasticity theory, that we can apply this equation *on* the x-axis, so that

$$v_1(x, y_1) = v_1(x, 0), \quad v_2(x, y_2) = v_2(x, 0), \text{ and}$$

$$f_1(x) + v_1(x, 0) - \delta_1 = -f_2(x) - v_2(x, 0) + \delta_2$$

and, on differentiating w.r.t x, we have

$$v_1'(x, 0) + v_2'(x, 0) = -f_1'(x) - f_2'(x). \tag{4.8.1}$$

This equation applies in the contact region $(-a, b)$. Now apply equation (2.2.23) to each body, assuming that

$$\sigma_{yy}(x,0+) = -p(x), \quad \sigma_{xy}(x,0+) = -q(x).$$

For body I, we have

$$\frac{v_1'(x,0)}{\vartheta_1} = \frac{1}{\pi}\int_{-a}^{b}\frac{p(t)dt}{t-x} + \beta_1 q(x) \tag{4.8.2}$$

and for body II,

$$\sigma_{yy}(x,0-) = -p(x), \quad \sigma_{xy}(x,0-) = -q(x)$$

so that

$$\frac{v_2'(x,0)}{\vartheta_2} = \frac{1}{\pi}\int_{-a}^{b}\frac{p(t)dt}{t-x} - \beta_2 q_2(x). \tag{4.8.3}$$

The negative sign appears because the body II occupies the lower half-plane, not the upper one.

Combining equations (4.8.2), (4.8.3) and substituting them in (4.8.1), we find

$$(\vartheta_1+\vartheta_2)\,\frac{1}{\pi}\int_{-a}^{b}\frac{p(t)dt}{t-x} + \left(\vartheta_1\beta_1 - \vartheta_2\beta_2\right)q(x) = -f'(x)$$

which may be written

$$\frac{1}{\pi}\int_{-a}^{b}\frac{p(t)dt}{t-x} + \beta q(x) = \frac{-f'(x)}{\vartheta} \tag{4.8.4}$$

where

$$\vartheta = \vartheta_1 + \vartheta_2, \quad \beta = \left(\vartheta_1\beta_1 - \vartheta_2\beta_2\right)/\left(\vartheta_1+\vartheta_2\right) \tag{4.8.5}$$

and

$$f(x) = f_1(x) + f_2(x). \tag{4.8.6}$$

We note that, for plane stree, ν is replaced by $\nu^* = \nu/(1+\nu)$, so that

$$\frac{1}{\vartheta_i^*} = \frac{4\mu}{\kappa_i^*+1} = \frac{E_i}{2}, \quad \beta_i^* = \frac{1-\nu_i}{2} \tag{4.8.7}$$

and

$$\vartheta^* = \vartheta_1^* + \vartheta_2^*, \quad \beta^* = (\vartheta_1^*\beta_1^* - \vartheta_2^*\beta_2^*)/(\vartheta_1^*+\vartheta_2^*) \tag{4.8.8}$$

[Note that β is one of *Dundurs' mismatch parameters*, see Dundurs and Stippes (1970).] We note that when $\mu_2/\mu_1 \to \infty$, $\beta = \beta_1$, and when $\mu_2/\mu_1 \to 0$, $\beta = -\beta_2$; when $\mu_2 = \mu_1$ and $\kappa_2 = \kappa_1$, then $\beta = 0$.

We conclude that problems related to the contact of two elastic bodies reduce to an equation exactly similar to that relating to the contact of a rigid body (a punch) indenting an elastic body occupying the upper half-space. The analysis developed earlier for that problem may therefore be applied to the two-body problem (see Poritsky, 1950).

Galin concludes Part I of the book with a Section 14 entitled, *An approximate estimate of plastic deformations arising under a punch*. This is based on work due to Sokolovsky (1950), and a method of analogy described by Galin (1948a). We omit this section.

Chapter 5
Contact Problems in Three Dimensions

5.1 The Papkovich–Neuber Solution

The fundamental equations governing static linear elasticity in three dimensions, in the absence of body forces, are the strain-displacement equations (2.1.2), (2.1.5); the stress-strain equations (2.1.10) and the equilibrium equations

$$\left.\begin{aligned}\frac{\partial\sigma_{xx}}{\partial x}+\frac{\partial\sigma_{xy}}{\partial y}+\frac{\partial\sigma_{xz}}{\partial z}&=0\\ \frac{\partial\sigma_{yx}}{\partial x}+\frac{\partial\sigma_{yy}}{\partial y}+\frac{\partial\sigma_{yz}}{\partial z}&=0\\ \frac{\partial\sigma_{zx}}{\partial x}+\frac{\partial\sigma_{zy}}{\partial y}+\frac{\partial\sigma_{zz}}{\partial z}&=0\end{aligned}\right\} \tag{5.1.1}$$

Expressing the stresses in terms of the strains, and the strains in terms of the displacements, we find that the three equilibrium equations may be written as one vector equation for the displacement vector $\mathbf{d} = u\mathbf{i} + v\mathbf{j} + w\mathbf{k}$, namely

$$\nabla^2\mathbf{d} + \frac{1}{1-2\nu}\mathbf{grad}\,\mathrm{div}\,\mathbf{d} = 0, \quad \mathbf{grad} \equiv \nabla. \tag{5.1.2}$$

This is *Navier's equation* for the displacements. The *Papkovich–Neuber solution* (Papkovich, 1932a, 1932b; Neuber, 1934l Lur'e, 1939), of this equation expresses **d** in terms of a vector function $\boldsymbol{\psi}$ and a scalar function ϕ, both of which are harmonic:

$$\nabla^2\boldsymbol{\psi} = \mathbf{0}, \quad \nabla^2\phi = 0. \tag{5.1.3}$$

The solution is

$$2\mu\mathbf{d} = 4(1-\nu)\boldsymbol{\psi} - \nabla\{(\mathbf{r}\cdot\boldsymbol{\psi}) + \phi\}. \tag{5.1.4}$$

It may be shown that the displacement field **d** due to one pair $\boldsymbol{\psi}, \phi$ is identical to that due to the pair $\boldsymbol{\psi}^*, \phi^*$, where

$$\boldsymbol{\psi}^* = \boldsymbol{\psi} + \nabla\xi, \quad \phi^* = \phi + 4(1-\nu)\xi - \mathbf{r}\cdot\nabla\xi, \quad \nabla^2\xi = 0. \tag{5.1.5}$$

This means that is always possible (though not always desirable) to choose one of the components of $\boldsymbol{\psi}$ to be zero. See Gladwell (1980, section 1.10) for further details and historical references.

The stresses σ_{xz} and σ_{yz} corresponding to the displacements (5.1.4) are

$$\sigma_{xz} = 2(1-\nu)\left\{\frac{\partial\psi_1}{\partial z} + \frac{\partial\psi_3}{\partial x}\right\} - \frac{\partial^2}{\partial x\partial z}\left(x\psi_1 + y\psi_2 + z\psi_3 + \phi\right), \quad (5.1.6)$$

$$\sigma_{yz} = 2(1-\nu)\left\{\frac{\partial\psi_2}{\partial z} + \frac{\partial\psi_3}{\partial y}\right\} - \frac{\partial^2}{\partial y\partial z}\left(x\psi_1 + y\psi_2 + z\psi_3 + \phi\right), \quad (5.1.7)$$

We now seek a solution which has the property that the plane $z = 0$ is free of shear stress:

$$\sigma_{xz} = 0 = \sigma_{yz} \text{ on } z = 0. \quad (5.1.8)$$

To effect this, we first note that ψ_3 being harmonic implies that

$$\chi = x\frac{\partial\psi_3}{\partial x} + y\frac{\partial\psi_3}{\partial y} + z\frac{\partial\psi_3}{\partial z} \quad (5.1.9)$$

is harmonic. Now choose $\boldsymbol{\psi}$, ϕ, so that

$$\frac{\partial\psi_1}{\partial z} = -\alpha\frac{\partial\psi_3}{\partial x}, \quad \frac{\partial\psi_2}{\partial z} = -\alpha\frac{\partial\psi_3}{\partial y}, \quad \frac{\partial\phi}{\partial z} = \alpha\chi, \quad (5.1.10)$$

for some constant α, then

$$\frac{\partial}{\partial z}(x\psi_1 + y\psi_2 + z\psi_3 + \phi) = \psi_3 + (\alpha+1)z\frac{\partial\psi_3}{\partial z}. \quad (5.1.11)$$

Substituting this expression into (5.1.6), (5.1.7), we find

$$\sigma_{xz} = \{-2(1-\nu)\alpha + (1-2\nu)\}\frac{\partial\psi_3}{\partial x} - (\alpha+1)z\frac{\partial^2\psi_3}{\partial x\partial z}, \quad (5.1.12)$$

$$\sigma_{yz} = \{-2(1-\nu)\alpha + (1-2\nu)\}\frac{\partial\psi_3}{\partial y} - (\alpha+1)z\frac{\partial^2\psi_3}{\partial y\partial z}. \quad (5.1.13)$$

This will satisfy the condition (5.1.8) if α is chosen to be

$$\alpha = (1-2\nu)/(2-2\nu). \quad (5.1.14)$$

Now

$$x\psi_1 + y\psi_2 + z\psi_3 - \phi = (\alpha+1)z\psi_3 + \alpha\int_z^\infty \psi_3 dz. \quad (5.1.15)$$

The displacements are given by

$$2\mu u = (3-4\nu)\alpha\int_z^\infty \frac{\partial\psi_3}{\partial x}dz - (\alpha+1)z\frac{\partial\psi_3}{\partial x}, \quad (5.1.16)$$

$$2\mu v = (3 - 4\nu)\alpha \int_z^\infty \frac{\partial \psi_3}{\partial y} dz - (\alpha + 1) z \frac{\partial \psi_3}{\partial y}, \tag{5.1.17}$$

$$2\mu w = (3 - 4\nu)\psi_3 - (\alpha + 1) z \frac{\partial \psi_3}{\partial z}, \tag{5.1.18}$$

and the stresses are

$$\sigma_{xz} = -(\alpha + 1) z \frac{\partial^2 \psi_3}{\partial x \partial z}, \quad \sigma_{yz} = -(\alpha + 1) z \frac{\partial^2 \psi_3}{\partial y \partial z}, \tag{5.1.19}$$

$$\sigma_{zz} = (\alpha + 1) \frac{\partial \psi_3}{\partial z} - (\alpha + 1) z \frac{\partial^2 \psi}{\partial z^2}. \tag{5.1.20}$$

Noting that

$$\alpha + 1 = (3 - 4\nu)/(2 - 2\nu) \tag{5.1.21}$$

we introduce

$$\psi = \left(\frac{3 - 4\nu}{2 - 2\nu} \right) \psi_3 \tag{5.1.22}$$

so that, in particular,

$$2\mu w = (2 - 2\nu)\psi - z \frac{\partial \psi}{\partial z}, \tag{5.1.23}$$

$$\sigma_{zz} = \frac{\partial \psi}{\partial z} - z \frac{\partial^2 \psi}{\partial z^2}. \tag{5.1.24}$$

We now assume that a normal pressure $p(x, y)$ is applied to the half space $z > 0$ over a finite region Ω of the plane $z = 0$, so that we have the boundary condition on $z = 0$:

$$\sigma_{zz} = \begin{cases} -p(x, y) \text{ if } (x, y) \in \Omega \\ 0 \qquad\qquad \text{otherwise.} \end{cases} \tag{5.1.25}$$

Thus, we must seek ψ so that, on $z = 0$,

$$\frac{\partial \psi}{\partial z} = \begin{cases} -p(x, y) \text{ if } (x, y) \in \Omega \\ 0 \qquad\qquad \text{otherwise.} \end{cases} \tag{5.1.26}$$

Consider the harmonic function

$$\psi = \frac{1}{2\pi} \int_\Omega \int \frac{p(\xi, \eta) d\xi d\eta}{\{(x - \xi)^2 + (y - \eta)^2 + z^2\}^{\frac{1}{2}}} \tag{5.1.27}$$

a so-called *single layer potential*, for which

$$\frac{\partial \psi}{\partial z} = \frac{-z}{2\pi} \int_\Omega \int \frac{p(\xi, \eta) d\xi d\eta}{\{(z - \xi)^2 + (y - \eta)^2 + z^2\}^{\frac{3}{2}}}. \tag{5.1.28}$$

If $(x, y) \notin \Omega$, the integrand is everywhere finite, and

$$\lim_{z\to 0} \frac{\partial \psi}{\partial z} = 0. \tag{5.1.29}$$

If $(x, y) \in \Omega$, we divide Ω into two parts, a circle of radius ε around (x, y), and the remainder. In the limit $z \to 0$, the contribution from the latter will be zero. For the former,

$$\begin{aligned}\frac{\partial \psi}{\partial z} &= \frac{-z}{2\pi} \int_0^{\varepsilon} \int_0^{2\pi} \frac{p(x, y) r dr d\theta}{(r^2 + z^2)^{\frac{3}{2}}} \\ &= p(x, y) \left[\frac{z}{(r^2 + z^2)^{\frac{1}{2}}} \right]_0^{\varepsilon} \\ &= p(x, y) \left\{ \frac{z}{(\varepsilon^2 + z^2)-^{\frac{1}{2}}} - 1 \right\}\end{aligned} \tag{5.1.30}$$

so that, for $(x, y) \in \Omega$,

$$\lim_{z\to 0} \frac{\partial \psi}{\partial z}(x, y, z) = -p(x, y). \tag{5.1.31}$$

Thus, (5.1.27) gives the required ψ satisfying (5.1.26); the displacements and stresses may then be computed using (5.1.16)–(5.1.20). In particular, the displacement w on $z = 0$ is given by

$$w(x, y, 0) = \frac{\vartheta}{2\pi} \int_{\Omega}\int \frac{p(\xi, \eta) d\xi d\eta}{\{(x - \xi)^2 + (y - \eta)^2\}^{\frac{1}{2}}}. \tag{5.1.32}$$

where

$$\vartheta = \frac{1 - \nu}{\mu}. \tag{5.1.33}$$

It follows from this equation that if the displacement of a punch in the region Ω is known, and if there is no tangential load, i.e., shear stresses σ_{xz}, σ_{yz}, are zero, then the value of a harmonic function $\psi(x, y, z)$ is known on the side $z = 0+$, and because the potential ψ in (5.1.27) is even in z, it is known on the side $z = 0-$ also. This is a particular case of Dirichlet's problem (see e.g., Courant and Hilbert, 1953, p. 261): the values of a harmonic function vanishing at infinity are known on the boundary of a degenerate volume, namely the two sides of the region Ω.

5.2 Solutions of Laplace's Equation in Certain Curvilinear Coordinates

We showed in Section 5.1 that the problem of determining the pressure of a rigid punch on a semi-infinite elastic solid, when there is no friction, can be reduced to a particular case of Dirichlet's problem. In this problem, the values of a certain harmonic function are prescribed on the two sides of a region in the plane $z = 0$. It is natural to try to establish in what cases this problem has a solution which can be effectively solved.

It is possible to find such a solution under the following conditions:

- there is a system of curvilinear coordinates in which the two-sided plane surface Ω is one of the coordinate surfaces
- Laplace's equation is *separable* in these coordinates i.e., the required harmonic functions can be expressed as a product of three functions, each of which is a function of one coordinate only
- the coordinate system is an orthogonal system

One system with these properties is the *ellipsoidal coordinate* system ρ, σ, τ. *Ellipsoidal coordinates* ρ, σ, τ are related to cartesian coordiantes x, y, z through the equations

$$k^2x^2 = a^2\rho^2\sigma^2\tau^2, \quad k'^2k^2y^2 = a^2(\rho^2 - k^2)(\sigma^2 - k^2)(k^2 - \tau^2), \tag{5.2.1}$$

$$k'^2z^2 = a^2(\rho^2 - 1)(1 - \sigma^2)(1 - \tau^2). \tag{5.2.2}$$

Here $k'^2 = 1 - k^2$, and

$$1 \le \rho < \infty, \quad k \le \sigma \le 1, \quad 0 \le \tau \le k.$$

The coordinate surfaces $\rho =$ constant are ellipsoids

$$\frac{x^2}{a^2\rho^2} + \frac{y^2}{a^2(\rho^2 - k^2)} + \frac{z^2}{a^2(\rho^2 - 1)} = 1. \tag{5.2.3}$$

The coordinate surfaces $\sigma =$ constant are hyperboloids of one sheet:

$$\frac{x^2}{a^2\sigma^2} + \frac{y^2}{a^2(\sigma^2 - k^2)} - \frac{z^2}{a^2(1 - \sigma^2)} = 1, \tag{5.2.4}$$

while the surfaces $\tau =$ constant are hyperboloids of two sheets:

$$\frac{x^2}{a^2\tau^2} - \frac{y^2}{a^2(k^2 - \tau^2)} - \frac{z^2}{a^2(1 - \tau^2)} = 1. \tag{5.2.5}$$

In the limit $\rho \to 1$, the ellipsoid degenerates into the double-sided elliptical disc

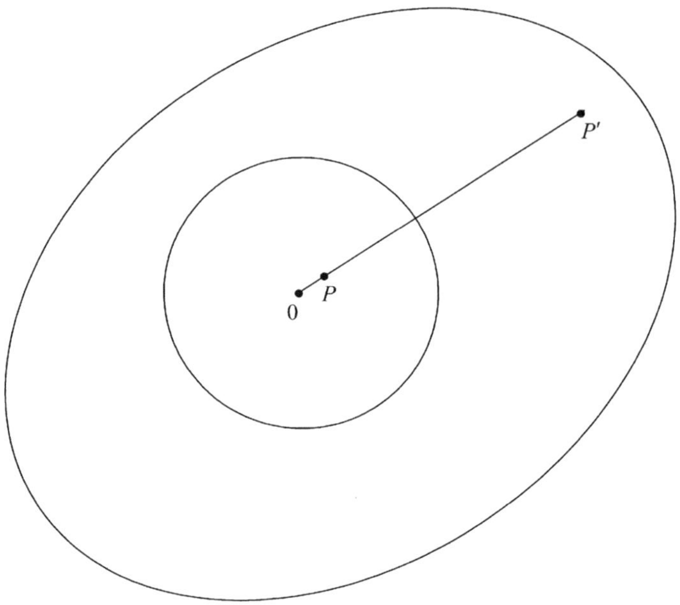

Fig. 5.2.1 $OP \cdot OP' = a^2$.

$$\frac{x^2}{a^2} + \frac{y^2}{k'^2 a^2} \leq 1. \tag{5.2.6}$$

When this system of coordinates is used, the harmonic function ϕ can be represented as a product of three *Lamé functions*

$$\phi(x, y, z) = E_m^n(\rho) E_m^n(\sigma) E_m^n(\tau), \tag{5.2.7}$$

as described in Hobson (1931, chapter 11) or Gladwell (1980, chapter 12).

We may generalise this result by using

Theorem 2 (Kelvin's Theorem). *If $\psi(x, y, z)$ is harmonic, so is $R^{-1}\psi(x', y', z')$ where $x' = a^2x/R^2$, $y' = a^2y/R^2$, $z' = a^2z/R^2$ and $R^2 = x^2 + y^2 + z^2$.*

To apply Kelvin's Theorem, we note that the point $(x', y', 0)$ is the *inverse* of $(x, y, 0)$ in the circle $x^2 + y^2 = a^2$, as shown in Figure 5.2.1.

Thus, if the values of the function $\psi(x, y, 0)$ are known on the inside of a region Ω in the plane $z = 0$, then the values of the function $\psi(x', y', 0)$ are known on the inside of Ω^*, the inverse of Ω in the circle $x^2 + y^2 = a^2$. This means that we can solve Dirichlet's problem for Ω^*, knowing the solution for Ω. Figure 5.2.2 shows the inverse of an ellipse in a circle for various configurations.

If, in (5.2.1), (5.2.2) we write

$$\rho = \cosh \xi, \quad \sigma^2 = k^2 \cos^2 \theta + \sin^2 \theta, \quad \tau = k \cos \phi, \tag{5.2.8}$$

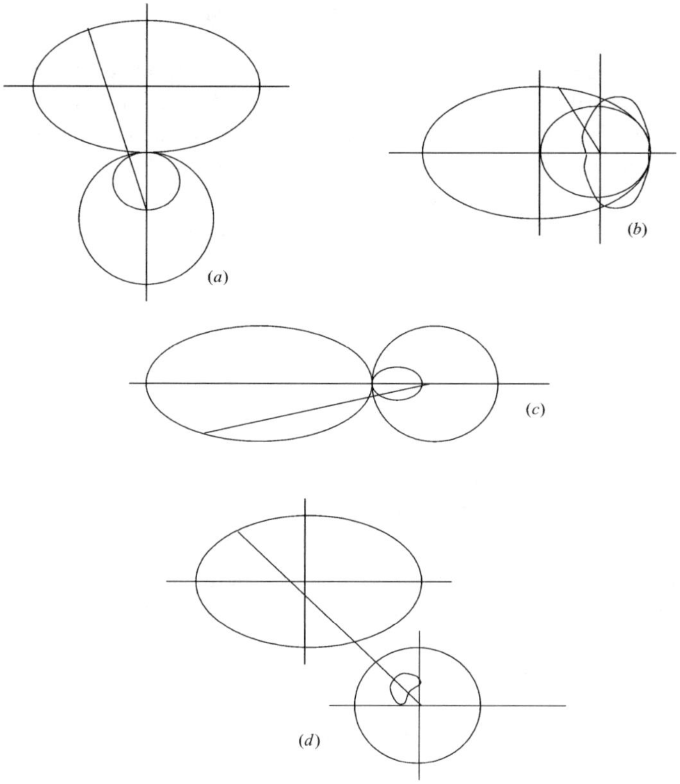

Fig. 5.2.2 Inverses of ellipses in a circle.

we find

$$x^2 = a^2 \cosh^2 \xi \, (k^2 \cos^2 \theta + \sin^2 \theta) \cos^2 \phi, \tag{5.2.9}$$

$$y^2 = a^2 (\cosh^2 \xi - k^2) \sin^2 \theta \sin^2 \phi, \tag{5.2.10}$$

$$z^2 = a^2 \sinh^2 \xi \cos^2 \theta (1 - k^2 \sin^2 \phi), \tag{5.2.11}$$

so that as $k \to 0$ we obtain the *oblate spheroidal* system

$$x = a \cosh \xi \sin \theta \cos \phi, \quad y = a \cosh \xi \sin \theta \sin \phi, \quad z = a \sinh \xi \cos \theta. \tag{5.2.12}$$

This coordinate system can be used when the punch is a solid of resolution. It should be noted that, in this case, as shown in Section 5.3 and later, a more convenient solution of a number of problems can be obtained without making use of the theory of special functions.

The connection between *spherical polar coordinates* and cartesian coordinates is well known. This system, with the *x-axis as the major axis*, is characterised by

$$x = r\cos\phi, \quad y = r\cos\theta\sin\phi, \quad z = r\sin\theta\sin\phi. \tag{5.2.13}$$

This system, although it does not possess all the properties listed above, turns out to be convenient for solving contact problems for a wedge-like punch.

Lebedev (1937) introduced a coordinate system in which one of the families of coordinate surfaces is a family of tori of oval cross-section. Here the tori degenerate into a plane ring. This system of coordinates could be used to solve the problem of a punch with annular cross-section. However, the special functions associated with it have not been tabulated, which makes it difficult to apply them.

5.3 Circular Punches

This section of Galin's book is rather difficult to follow as many results are given without explanations. It is made more difficult by the presence of persistent typographical errors in equations. For these reasons, we depart from his text and introduce explanatory material.

Suppose that a rigid punch is pressed under the action of external forces, against the surface of a semi-infinite elastic solid and, as a result, sets up a state of stress in the solid.

One of the objects of the investigation is the determination of the distribution of pressure over the contact region, and of the relation between the force acting on the punch and the magnitude of the displacement.

We shall assume that the semi-infinite solid occupies the region $z \geq 0$, and that there is no friction. The boundary conditions on $z = 0$ are

$$w = f(x, y), \quad \sigma_{xz} = 0 = \sigma_{yz} \text{ if } (x, y) \in \Omega$$

$$\sigma_{xz} = 0 = \sigma_{yz} = \sigma_{zz} \quad \text{otherwise.}$$

Here w is the displacement in the z-direction, i.e., normal to the original surface of the body.

In terms of the *Papkovich–Neuber solution* ψ, we must find ψ so that

$$\vartheta\psi(x, y, 0) = f(x, y) \quad \text{if } (x, y) \in \Omega \tag{5.3.1}$$

$$\left.\frac{\partial\psi}{\partial z}\right|_{z=0} = 0 \qquad \text{otherwise.} \tag{5.3.2}$$

In Section 5.1 we showed that the so-called *single layer potential* (5.1.27) satisfied equation (5.3.2); see equation (5.1.29). If we use that potential, we must find $p(\xi, \eta)$ so that ψ satisfies equation (5.3.1): we must solve an integral equation. Instead, we return to the *oblate spheroidal coordinates* (5.2.12), putting $a\sinh\xi = t$. One is tempted to put $a\sinh\xi = r$, as Galin did, but this risks confusion with the ordinary radial coordinate of polar coordinates; we can use t because there is no temporal variation in any of the problems we consider. Thus,

$$x = (t^2 + a^2)^{\frac{1}{2}} \sin\theta \cos\phi, \quad y = (t^2 + a^2)^{\frac{1}{2}} \sin\theta \sin\phi, \quad z = t\cos\theta.$$

Note that, time and time again, the text of Galin has $\sqrt{r^2 + a^2 \sin\eta}$ instead of $\sqrt{r^2 + a^2} \sin\eta$; see his equations (2.3) and (3.3) for example.

The coordinate surfaces t = constant are oblate spheroids;

$$\frac{x^2 + y^2}{t^2 + a^2} + \frac{z^2}{t^2} = 1, \tag{5.3.3}$$

the coordinate surfaces θ = constant are hyperboloids of one sheet:

$$\frac{x^2 + y^2}{a^2 \sin^2\theta} - \frac{z^2}{a^2 \cos^2\theta} = 1, \tag{5.3.4}$$

while the coordinate surfaces ϕ = constant are planes:

$$y = x \tan\phi. \tag{5.3.5}$$

As $t \to 0$, the coordinate surface t = constant tends to the double-sided disc Ω : $z = 0, x^2 + y^2 \le a^2$.

Hobson (1900) finds the *Green's function* for the disc, and hence fnds the harmonic function $\psi(t, \theta, \phi)$ satisfying

$$\psi(0, \theta, \phi) = f(\theta, \phi) \text{ on } \Omega. \tag{5.3.6}$$

The function is an even function of z, so that it will satisfy

$$\left.\frac{\partial\psi}{\partial z}\right|_{z=0} . = 0, \quad \text{outside } \Omega.$$

(Hobson finds the Green's function after a long and complicated analysis, and ironically states that 'This formula agrees with one obtained by Heine [in his Kugelfunctionen, Vol. II, p. 132] by a different and somewhat complicated procedure'.)

Hobson finds the potential ψ at the point (x, y, z) with spheroidal coordinates (t, θ, ϕ); it involves the distance R between (x, y, z) and the point $(\xi, \eta, 0)$ with coordinates $(0, \theta_0, \phi_0)$ on the disc:

$$R^2 = (x - \xi)^2 + (y - \eta)^2 + z^2. \tag{5.3.7}$$

The expression for ψ is

$$\psi(t, \theta, \phi) = \frac{az}{\pi^2 \cos\theta} \int_\Omega \int \frac{\sin\theta_0 f(\theta_0, \phi_0)}{R^2} \{1 + M \arctan M\}\, d\theta_0 d\phi_0 \tag{5.3.8}$$

where

$$M = \frac{a \cos\theta \cos\theta_0}{R}, \tag{5.3.9}$$

and the integration is over the disc: $0 \le \theta_0 \le \frac{\pi}{2}, 0 \le \phi_0 \le 2\pi$. [Note that Hobson omits the factor 'a' in front of the integral.]Galin now reverts to Cartesian coordinates. The relation between the elements of area is

$$dS = a^2 \sin\theta_0 \cos\theta_0 d\theta_0 d\phi_0 = d\xi d\eta, \tag{5.3.10}$$

so that

$$\psi = \frac{z}{\pi^2} \int_\Omega \int \frac{f(\xi,\eta)}{R^3} \left\{ \frac{1}{M} + \arctan M \right\} d\xi d\eta. \tag{5.3.11}$$

To express M in Cartesian coordinates, we note that equation (5.3.3) gives

$$a\cos\theta_0 = (a^2 - \xi^2 - \eta^2)^{\frac{1}{2}}, \quad \cos\theta = z/t. \tag{5.3.12}$$

Equation (5.3.4) gives t^2 as the positive root of the quadratic equation

$$t^4 - (x^2 + y^2 + z^2 - a^2)t^2 - a^2 z^2 = 0, \tag{5.3.13}$$

so that

$$t^2 = (x^2 + y^2 + z^2 - a^2 + \rho^2)/2, \tag{5.3.14}$$

where

$$\rho^2 = \{(x^2 + y^2 + z^2 - a^2)^2 + 4a^2 z^2\}^{\frac{1}{2}}. \tag{5.3.15}$$

Thus

$$\cos\theta = \left(\frac{2z^2}{x^2 + y^2 + z^2 - a^2 + \rho^2} \right)^{\frac{1}{2}}, \tag{5.3.16}$$

and

$$M = \left(\frac{2z^2(a^2 - \xi^2 - \eta^2)}{R^2(x^2 + y^2 + z^2 - a^2 + \rho^2)} \right)^{\frac{1}{2}}. \tag{5.3.17}$$

We can use equation (5.3.11) to find the force P which must be applied to the punch with equation $z = f(x, y)$ when it is indenting the surface of the elastic body. It follows from equation (5.1.27) that if $s^2 = x^2 + y^2 + z^2$ then

$$\lim_{s\to\infty} \psi(x, y, z) = \frac{P}{2\pi s} \tag{5.3.18}$$

where

$$P = \int_\Omega \int p(\xi,\eta) d\xi d\eta. \tag{5.3.19}$$

Now equation (5.1.32) shows that, under the punch

$$\vartheta\psi(\xi, \eta, 0) = f(\xi, \eta), \tag{5.3.20}$$

where

$$\vartheta = (1 - \nu)/\mu, \tag{5.3.21}$$

so that equation (5.3.11) gives

$$\vartheta\psi(x, y, z) = \frac{z}{\pi^2}\int_\Omega\int \frac{f(\xi,\eta)}{R^3}\left\{\frac{1}{M} + \arctan M\right\} d\xi d\eta. \tag{5.3.22}$$

As $s \to \infty$, M, given by (5.3.17), tends to zero. Thus

$$\vartheta\psi(x, y, z) \to \frac{z}{\pi^2}\int_\Omega\int \frac{f(\xi,\eta)R(s^2 - a^2 + \rho^2)^{\frac{1}{2}}}{R^3 z\sqrt{2}(a^2 - \xi^2 - \eta^2)^{\frac{1}{2}}} d\xi d\eta. \tag{5.3.23}$$

As $s \to \infty$, M, then $\rho \to s$, $R \to s$, so that

$$\vartheta\psi(x, y, z) \to \frac{1}{\pi^2 s}\int_\Omega\int \frac{f(\xi,\eta)d\xi d\eta}{(a^2 - \xi^2 - \eta^2)^{\frac{1}{2}}}, \tag{5.3.24}$$

which, when expressed in ordinary polar coordinates (r, θ) is

$$\vartheta\psi(x, y, z) \to \frac{1}{\pi^2 s}\int_0^{2\pi}\int_0^a \frac{f(r,\theta)rdrd\theta}{(a^2 - r^2)^{\frac{1}{2}}}. \tag{5.3.25}$$

Therefore,

$$\vartheta P = \frac{2}{\pi}\cdot\int_0^{2\pi}\int_0^a \frac{f(r,\theta)rdrd\theta}{(a^2 - r^2)^{\frac{1}{2}}}. \tag{5.3.26}$$

If the punch is a flat-ended cylinder, and the penetration distance is d, then

$$\vartheta P = 4ad \tag{5.3.27}$$

which is in agreement with the result due to Boussinesq.

5.4 The Green's Function for the Exterior of a Circular Disc

We seek a harmonic function $K(x, y, z, \xi, \eta)$ satisfying the following conditions, in which Ω is the interior of the circle of radius a on the plane $z = 0$, and Ω' is the outside of the circle:

1. $K(x, y, z, \xi, \eta) = 0$ in Ω'
2. $K(x, y, z, \xi, \eta)$ behaves like

$$\frac{1}{R} = \frac{1}{\{(x-\xi)^2 + (y-\eta)^2 + z^2\}^{\frac{1}{2}}}$$

   near $(\xi, \eta, 0) \in \Omega$
3. $K(x, y, z, \xi, \eta)$ is continuous over the whole space
4. $\partial K(x, y, z, \xi, \eta)/\partial z$ is continuous at each point in Ω.

We omit the first part of Part 1, Section 4 of Galin, and proceed straight to the *Legendre function* solution. Recall that in the spheroidal coordinates t, θ, ϕ of equation (5.3.3), the function

$$H(x, y, z) = \frac{2}{\pi} Q_0 \left(\frac{ik}{a}\right) = \frac{2}{\pi} \operatorname{arccot}\left(\frac{t}{a}\right) \tag{5.4.1}$$

is harmonic; it is equal to unity inside the circle $S_1 : x^2+y^2 = a^2$. Since the interior of the circle Ω is given by $t = 0$, H satisfies

$$H(x, y, z) = 1, \quad (x, y, 0) \in \Omega, \tag{5.4.2}$$

$$\frac{\partial H}{\partial z}(x, y, z)\,|_{z=0} = 0, \quad (x, y, 0) \in \Omega'. \tag{5.4.3}$$

The spheroidal coordinate t is given in terms of x, y, z by equations (5.3.14), (5.3.15), so that

$$\begin{aligned} H(x, y, z) &= 1 - \frac{2}{\pi} \arctan\left(\frac{t}{a}\right) \\ &= 1 - \frac{2}{\pi} \arctan\left\{\left(\frac{x^2 + y^2 + z^2 - a^2 + \rho^2}{2a^2}\right)^{\frac{1}{2}}\right\}. \end{aligned} \tag{5.4.4}$$

We now carry out various coordinate transformations starting from $H(x, y, z)$ with 'a' replaced by 'α'; we relabel this function $H_1(x_1, y_1, z_1, \alpha)$; if we introduce new variables (x_2, y_2, z_2)

$$x_2 = x_1 + \beta, \quad y_2 = y_1, \quad z_2 = z_1, \quad \beta < \alpha \tag{5.4.5}$$

we obtain the harmonic function

$$H_2(x_2, y_2, z_2) = H_1(x_2 - \beta, y_2, z_2). \tag{5.4.6}$$

This is equal to unity inside the circle S_2: $(x_2 - \beta)^2 + y_2^2 \le \alpha^2$, as shown in Figure 5.4.1. Now apply *Kelvin's Theorem* 3.2.1. The coordinate transformation

$$(x_3, y_3, z_3) = (x_2, y_2, z_2)/(x_2^2 + y_2^2 + z_2^2) \tag{5.4.7}$$

maps the interior of the circle S_2 onto the outside of the circle S_3, shown in Figure 5.4.1, given by

$$\left(x_3 + \frac{\beta}{\alpha^2 - \beta^2}\right)^2 + y_3^2 = \left(\frac{\alpha}{\alpha^2 - \beta^2}\right)^2. \tag{5.4.8}$$

Kelvin's Theorem states that the function

$$\frac{1}{(x_3^2+y_3^2+z_3^2)^{\frac{1}{2}}} H_2\{(x_3, y_3, z_3)\,/(x_3^2+y_3^2+z_3^2)\}$$

is harmonic in the coordinate system (x_3, y_3, z_3); it is equal to $(x_3^2+y_3^2+z_3^2)^{-\frac{1}{2}}$ outside S_3. Thus

$$H_3(x_3, y_3, z_3) = (x_3^2+y_3^2+z_3^2)^{-\frac{1}{2}}[1 - H_2\{(x_3, y_3, z_3)/(x_3^2+y_3^2+z_3^2)\}] \quad (5.4.9)$$

is harmonic,and equal to zero outside S_3.

Now we shift the axes again, and take

$$x_4 = x_3 + \frac{\beta}{\alpha^2-\beta^2} = x_3 + \gamma, \quad y_4 = y_3, \quad z_4 = z_3, \qquad (5.4.10)$$

and put

$$H_4(x_4, y_4, z_4) = H_3(x_4 - \gamma, y_4, z_4). \qquad (5.4.11)$$

Finally, we rotate the axes to obtain new axes (x, y, z), where

$$x = x_4 \cos\theta - y_4 \sin\theta, \quad y = x_4 \sin\theta + y_4 \cos\theta, \quad z = z_4, \qquad (5.4.12)$$

and put $\xi = \gamma\cos\theta$, $\eta = \gamma\sin\theta$. If we write $H_4(x_4, y_4, z_4)$ in terms of x, y, z we find it is

$$K(x, y, z, \xi, \eta) = \frac{2}{\pi R}\arctan\left\{\left(\frac{(a^2-\xi^2-\eta^2)(a^2-x^2-y^2-z^2+\rho^2}{2a^2R^2}\right)^{\frac{1}{2}}\right\} \qquad (5.4.13)$$

where $R^2 = (x-\xi)^2 + (y-\eta)^2 + z^2$, and ρ^2 is given by equaion (5.3.15).

We note that as (x, y, z) approaches $(\xi, \eta, , 0)$ the argument of arctan behaves like $\frac{1}{R}$, so that

$$K \to \frac{2}{\pi R}\cdot\frac{\pi}{2} = \frac{1}{R}. \qquad (5.4.14)$$

On the other hand, for $(x, y, 0)$ outside the circle Ω, $\rho^2 = x^2 + y\rho^2 - a^2$, so that the argument of arctan is zero: $K = 0$.

If now we form the function

$$\Psi(x, y, z) = \frac{-1}{2\pi}\int_\Omega\int K(x, y, z, \xi, \eta)g(\xi, \eta)d\xi d\eta. \qquad (5.4.15)$$

then we find that $\Psi(x, y, z)$ is harmonic; condition 1) implies that $\Psi(x, y, 0) = 0$ in Ω'; and equation (5.4.3) implies that

$$\frac{\partial K}{\partial z}(x, y, z)|_{z=0} = 0 \text{ in } \Omega, \text{ except near } (\xi, \eta, 0). \qquad (5.4.16)$$

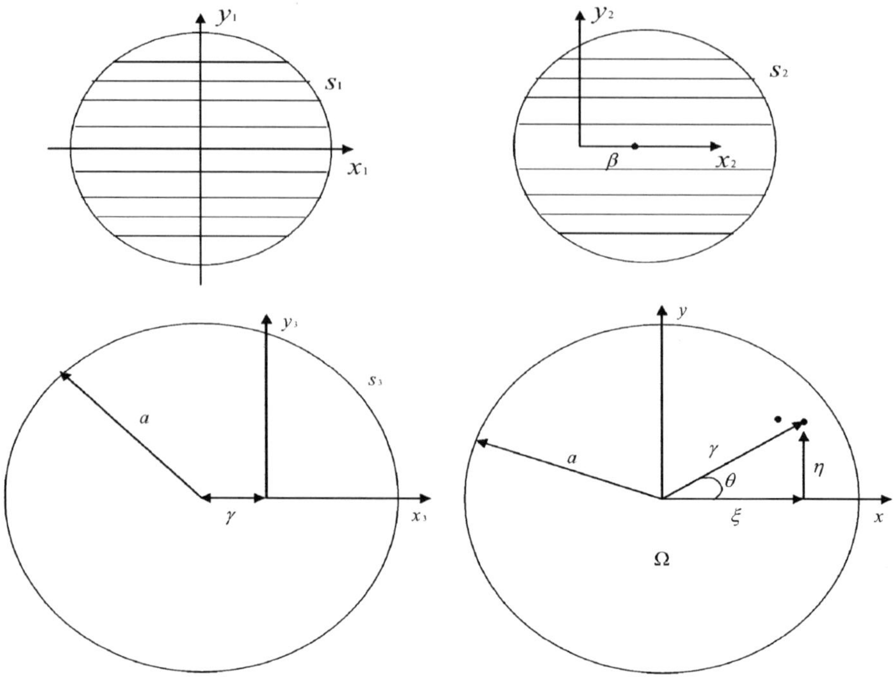

Fig. 5.4.1 The four circles used in computing K.

Thus, to form $\partial\Psi/\partial z$ at a point $(x, y, 0)$ in Ω we integrate over a circle of radius ε around $(x, y, 0)$, using the fact that for $(\xi, \eta, 0)$ near (x, y, z), $\partial K/\partial z$ behaves like $-z/R^3$. Thus

$$\begin{aligned}\frac{\partial\Psi}{\partial z}(x, y, z)|_{z=0} &= \frac{1}{2\pi}\lim_{z\to 0}\int\int \frac{zg(\xi, \eta)d\xi d\eta}{\{(x-\xi)^2 + (y-\eta)^2 + z^2\}^{\frac{3}{2}}} \\ &= g(x, y)\lim_{z\to 0}\int_0^{\varepsilon}\frac{zrdr}{(r^2+z^2)^{\frac{3}{2}}} = g(x, y). \qquad (5.4.17)\end{aligned}$$

It is amusing that Galin echoes Hobson in his comment on the derivation of K: 'This (Green's) function was determined by Kochin (1941) on the basis of results obtained by Sommerfeld (1897). Sommerfeld's method, which makes use of a potential in Riemannian space, is quite complicated.'

5.5 Axisymmetric Frictionless Contact Problems

This section deals with the solution of contact problems in the theory of elasticity for an *axisymmetric punch* pressed into an elastic solid occupying the half-space $z \geq 0$. A force P, directed along the axis of rotation, acts on the punch. We shall suppose that the function describing the bounding surface of the punch possesses derivatives at points on the edge of the area of contact. Here the pressure is zero on the circumference of the contact area.

If the punch is of such a form, that when the force P is increased above a certain value, new parts of the bounding surface *do not* come into contact with the surface of the solid, then the dimensions of the region of contact are given. In this case the pressure is generally infinite at the edge of the contact region. The pressure at these points is, generally, infinite. The solution of this problem is given in Section 5.6.

In cylindrical polar coordinates r, θ, z, the boundary conditions for the determination of the stress distribution are as follows:

$$w = f(r) + c, \quad \sigma_{z\theta} = 0 = \sigma_{zr} \text{ on } \Omega, \tag{5.5.1}$$

$$\sigma_{zz} = 0 = \sigma_{z\theta} = \sigma_{zr} \text{ on } \Omega', \tag{5.5.2}$$

where Ω is the disc $r \leq a, z = 0$.

The equation of the surface bounding the punch is $z = f(r)$, with $f(0) = 0$. The plane $z = 0$ coincides with the surface of the undeformed semi-infinite solid, so that c is the displacement of the apex of the punch.

In Cartesian coordinates, the conditions are as follows:

$$w = f((x^2 + y^2)^{\frac{1}{2}}) + c, \quad \sigma_{xz} = 0 = \sigma_{yz} \quad \text{on } \Omega, \tag{5.5.3}$$

$$\sigma_{xz} = 0 = \sigma_{yz} = \sigma_{zz} \quad \text{on } \Omega'. \tag{5.5.4}$$

In terms of the *Papkovich–Neuber* potential ψ of Section 5.1, the conditions are as follows

$$\vartheta\psi(x, y, 0) = f((x^2 + y^2)^{\frac{1}{2}}) + 0 \quad \text{on } \Omega, \tag{5.5.6}$$

$$\frac{\partial \psi}{\partial z}(x, y, z)|_{z=0} = 0 \quad \text{on } \Omega'. \tag{5.5.7}$$

Since $\psi(x, y, z)$ is harmonic, we have

$$\frac{\partial^2 \psi}{\partial z^2} = -\left(\frac{\partial^2 \psi}{\partial x^2} + \frac{\partial^2 \psi}{\partial y^2}\right) = -\Delta\psi = -\frac{1}{\vartheta}\Delta f \quad \text{on } \Omega, \tag{5.5.8}$$

where Δ denotes the operator $\frac{\partial^2}{\partial x^2} + \frac{\partial^2}{\partial y^2}$, and ϑ is given in (5.3.21).

We introduce the harmonic function

$$\Psi(x, y, z) = \frac{\partial \psi}{\partial z}, \tag{5.5.9}$$

so that, according to equations (5.1.24), (5.1.25)

$$p(x, y) = -\frac{\partial \psi}{\partial z}|_{z=0} = -\Psi(x, y, 0). \tag{5.5.10}$$

Write $s^2 = x^2 + y^2 + z^2$; (5.1.28) shows that, as $s \to \infty$

$$\Psi(x, y, z) \to \frac{-Pz}{2\pi s^3}. \tag{5.5.11}$$

For the determination of Ψ we have the following equations

$$\vartheta \frac{\partial \Psi}{\partial z}(x, y, z)|_{z=0} = -\triangle f \quad \text{on } \Omega, \tag{5.5.12}$$

$$\Psi(x, y, 0) = 0 \qquad \text{on } \Omega'. \tag{5.5.13}$$

We may use the *Green's function* $K(x, y, z, \xi, \eta)$ of Section 5.4 to find

$$\vartheta \Psi(x, y, z) = \frac{1}{2\pi} \int_\Omega \int \triangle f(\xi, \eta) K(x, y, z, \xi, \eta) d\xi d\eta. \tag{5.5.14}$$

Consider $\Psi(x, y, z)$ as $s \to \infty$. We note that, as $s \to \infty$,

$$K(x, y, z, \xi, \eta) \to \frac{2}{\pi s} \arctan \left\{ \left(\frac{(a^2 - \xi^2 - \eta^2)}{s^4} z^2 \right)^{\frac{1}{2}} \right\} \to \frac{2z}{\pi s^3}(a^2 - \xi^2 - \eta^2)^{\frac{1}{2}} \tag{5.5.15}$$

so that

$$\vartheta \Psi(x, y, z) \to \frac{z}{\pi^2 s^3} \int_\Omega \int \triangle f(\xi, \eta) \left(a^2 - \xi^2 - \eta^2\right)^{\frac{1}{2}} d\xi d\eta. \tag{5.5.16}$$

Since $f(\xi, \eta)$ depends only on $\left(\xi^2 + \eta^2\right)^{\frac{1}{2}} = \gamma$, we have

$$\vartheta \Psi(x, y, z) \to \frac{2}{\pi} \cdot \frac{z}{s^3} \int_0^a \gamma \triangle f(\gamma)(a^2 - \gamma^2)^{\frac{1}{2}} d\gamma. \tag{5.5.17}$$

Comparing this with (5.5.11), we deduce that

$$\vartheta P = -4 \int_0^a \gamma \triangle f(\gamma)(a^2 - \gamma^2)^{\frac{1}{2}} d\gamma. \tag{5.5.18}$$

The pressure on the surface is given by

$$\vartheta p(x, y) = \frac{-1}{2\pi} \int_\Omega \int \triangle f(\xi, \eta) K(x, y, 0, \xi, \eta) d\xi d\eta. \tag{5.5.19}$$

Equation (5.4.13) gives

$$K(x, y, 0, \xi, \eta) = \frac{2}{\pi R} \arctan \left\{ \left(\frac{(a^2 - \xi^2 - \eta^2)(a^2 - x^2 - y^2 + \rho^2)}{2a^2 R^2} \right)^{\frac{1}{2}} \right\} \tag{5.5.20}$$

where now, for $x^2 + y^2 \leq a^2$, equation (5.3.15) gives $\rho^2 = a^2 - x^2 - y^2$, so that

$$K(x, y, 0, \xi, \eta) = \frac{2}{\pi R} \arctan \left\{ \left(\frac{(a^2 - \xi^2 - \eta^2)(a^2 - x^2 - y^2)}{a^2 R^2} \right)^{\frac{1}{2}} \right\}. \tag{5.5.21}$$

Now introduce polar coordinates

$$\xi = r_0 \cos\theta_0, \quad \eta = r_0 \sin\theta_0, \quad x = r\cos\theta, \quad y = r\sin\theta, \tag{5.5.22}$$

note that

$$R^2 = (x - \xi)^2 + (y - \eta)^2 = r^2 + r_0^2 - 2rr_0 \cos(\theta - \theta_0). \tag{5.5.23}$$

and write $\Delta f(\xi, \eta) = \Delta f(r_0)$, $p(x, y) = p(r)$; we find

$$\vartheta p(r) = -\frac{1}{2\pi} \cdot \frac{2}{\pi} \cdot \int_0^a r_0 \Delta f(r_0) H(r, r_0) dr_0, \tag{5.5.24}$$

where

$$H(r, r_0) = \int_0^{2\pi} \frac{1}{R} \arctan \left\{ \left(\frac{(a^2 - r_0^2)(a^2 - r^2)}{a^2 R^2} \right)^{\frac{1}{2}} \right\} d\theta_0. \tag{5.5.25}$$

Since R is periodic with period 2π in θ_0, we may integrate over $(\theta, \theta + 2\pi)$ instead of $(0, 2\pi)$, or, what is equivalent, may replace the term $\cos(\theta - \theta_0)$ in R^2 by $\cos\theta_0$.

Having found an expression for $\Psi(x, y, z)$, we revert to $\psi(x, y, z)$. Equation (5.5.9) implies

$$\psi(x, y, z) = \int^z \Psi(x, y, z) dz,$$

and since $\psi(x, y, z) \to 0$ as $z \to \infty$, we have

$$0 = \int^\infty \Psi(x, y, z) dz,$$

so that

$$\Psi(x, y, z) = \int_\infty^z \Psi(x, y, z) dz = -\int_z^\infty \Psi(x, y, z) dz \tag{5.5.26}$$

and hence

$$\vartheta\psi(x,y,0) = -\vartheta\int_0^\infty \Psi(x,y,z)dz,$$
$$= \frac{-1}{2\pi}\int_\Omega\int \triangle f(\xi,\eta)d\xi d\eta\int_0^\infty K(x,y,z,\xi,\eta)dz. \quad (5.5.27)$$

Now the integral over z is a function of R^2, $\xi^2+\eta^2$ and x^2+y^2, so that in the polar coordinates (5.5.22), it is a function of R^2, r and r_0. Again, since R^2 is periodic with period 2π in θ_0, $\Psi(x,y,0)$ is a function of r only. Now the integral over z is a function of R^2, $\xi^2+\eta^2$ and x^2+y^2, so that in the polar coordinates (5.5.22), it is a function of R^2, r and r_0. Again, since R^2 is periodic with period 2π in θ_0, $\Psi\ (x,y,0)$ is a function of r only

Now, according to equation (5.5.8),

$$\vartheta\triangle\psi = \triangle f \quad \text{on } \Omega. \quad (5.5.28)$$

If, therefore, we extend f to the whole x, y plane so that it is zero as $x^2+y^2\to\infty$, we can write

$$\vartheta\psi\ (x,y,0) = \{f(x,y)+\psi^*(x,y)\}, \quad (5.5.29)$$

where $\Psi^*(x,y)$ is a function of r, $\Psi^*\to 0$ as $r\to\infty$ and $\triangle\Psi^* = 0$. But the only bounded such function of r only is a constant, c. Since $f(x,y)=0$ when $r=0$, we have

$$\vartheta\Psi\ (0,0,0) = c. \quad (5.5.30)$$

Combining this equation with (5.5.27), we deduce

$$c = \frac{-1}{2\pi}\int_\Omega\int \triangle f(\xi,\eta)d\xi d\eta\int_0^\infty K(0,0,z,\xi,\eta)dz. \quad (5.5.31)$$

Equation (5.4.13) gives

$$K(0,0,z,\xi,\eta) = \frac{2}{\pi R}\arctan\left\{\left(\frac{(a^2-\xi^2-\eta^2)}{R^2}\right)^{\frac{1}{2}}\right\}, \quad (5.5.32)$$

where now $R^2=\xi^2+\eta^2+z^2$. Thus, in polar coordinates

$$c = \frac{-2}{\pi}\int_0^a r\triangle f(r)dr\int_0^\infty \frac{1}{(r^2+z^2)^{\frac{1}{2}}}\arctan\left(\frac{(a^2-r^2)}{r^2+z^2}\right)^{\frac{1}{2}}dz. \quad (5.5.33)$$

This we write as

$$c = -\int_0^a r\triangle f(r)G(r)dr, \quad (5.5.34)$$

where

$$G(r) = \frac{2}{\pi} \int_0^\infty \frac{1}{(r^2+z^2)^{\frac{1}{2}}} \arctan\left\{ \left(\frac{(a^2-r^2)}{r^2+z^2} \right)^{\frac{1}{2}} \right\} dz. \tag{5.5.35}$$

Formula 4.4.42 of Abramowitz and Stegun (1972) gives

$$x \arctan x = \sum_{k=1}^{\infty} \frac{2^{2k}(k!)^2}{(2k+1)!} \left(\frac{x^2}{1+x^2} \right)^{k+1}. \tag{5.5.36}$$

If $x = \left\{(a^2-r^2)/(r^2+z^2)\right\}^{\frac{1}{2}}$, then

$$\frac{x^2}{1+x^2} = \frac{a^2-r^2}{a^2+z^2}$$

so that

$$G(r) = \frac{2}{\pi} \cdot \int_0^\infty \frac{1}{(a^2-r^2)^{\frac{1}{2}}} \sum_{k=1}^{\infty} \frac{2^{2k}(k!)^2}{(2k+1)!} \left(\frac{a^2-r^2}{a^2+z^2} \right)^{k+1} dz. \tag{5.5.37}$$

Now, formula 3.2.491 of Gradshteyn and Ryzhik (1965) gives

$$\int_0^\infty \frac{dz}{(a^2+z^2)^{k+1}} = \frac{\pi(2k)!}{(2a)^{2k+1}(k!)^2}, \tag{5.5.38}$$

so that

$$G(r) = \sum_{k=0}^{\infty} \frac{1}{2k+1} \left(\frac{a^2-r^2}{a^2} \right)^{k+\frac{1}{2}}, \tag{5.5.39}$$

and formula 1.6432 of Gradshteyn and Ryzhik (1965) gives

$$G(r) = \tanh^{-1}\left\{ \left(\frac{a^2-r^2}{a^2} \right)^{\frac{1}{2}} \right\} \tag{5.5.40}$$

so that

$$c = -\int_0^a r \Delta f(r) \tanh^{-1}\left\{ \left(\frac{a^2-r^2}{a^2} \right)^{\frac{1}{2}} \right\} dr. \tag{5.5.41}$$

Lur'e (1941) and Shtaerman (1949) consider an axisymmetric punch in the form of a *paraboloid of revolution* with equation $z = -A\rho^\lambda$. With the help of the results we have just derived, we can establish the relation between the force and the displacement.

Suppose that

$$f(r) = -Ar^\lambda, \tag{5.5.42}$$

then

$$\Delta f(r) = f''(r) + r^{-1} f'(r) = -\lambda^2 A r^{\lambda-2}, \tag{5.5.43}$$

so that equation (5.5.18) gives

$$\vartheta P = 4A\lambda^2 \cdot \int_0^a r^{\lambda-1}(a^2 - r^2)^{\frac{1}{2}} dr. \tag{5.5.44}$$

The substitution $r = a \sin\theta$ gives

$$\begin{aligned}
\int_0^a r^{\lambda-1}(a^2 - r^2)^{\frac{1}{2}} dr &= a^{\lambda+1} \int_0^{\frac{\pi}{2}} (\sin\theta)^{\lambda-1} \cos^2\theta d\theta, \\
&= a^{\lambda+1} \int_0^{\frac{\pi}{2}} \left\{ (\sin\theta)^{\lambda-1} - (\sin\theta)^{\lambda+1} \right\} d\theta \quad (5.5.45) \\
&= a^{\lambda+1} I.
\end{aligned}$$

Formula 3.6211 of Gradshteyn and Ryzhik (1965) gives

$$\int_0^{\frac{\pi}{2}} (\sin\theta)^{\mu-1} d\theta = 2^{\mu-2} B\left(\frac{\mu}{2}, \frac{\mu}{2}\right), \tag{5.5.46}$$

$$= 2^{\mu-2}\Gamma^2\left(\frac{\mu}{2}\right) / \Gamma(\mu). \tag{5.5.47}$$

so that

$$I = \frac{2^{\lambda-2}\Gamma^2\left(\frac{\lambda}{2}\right)}{(\lambda+1)\Gamma(\lambda)}, \tag{5.5.48}$$

and hence

$$\vartheta P = 4A\lambda^2 \cdot \frac{a^{\lambda+1} 2^{\lambda-2}\Gamma^2\left(\frac{\lambda}{2}\right)}{(\lambda+1)\Gamma(\lambda)}. \tag{5.5.49}$$

The displacement c is given by equation (5.5.41):

$$\begin{aligned}
c &= A\lambda^2 \int_0^a r^{\lambda-1} \tanh^{-1}\left(\frac{(a^2 - r^2)^{\frac{1}{2}}}{a}\right) dr, \\
&= A\lambda^2 a^\lambda \int_0^{\frac{\pi}{2}} (\sin\theta)^{\lambda-1} \tanh^{-1}(\cos\theta) \cos\theta d\theta, \\
&= A\lambda^2 a^\lambda J.
\end{aligned} \tag{5.5.50}$$

Since $\tanh^{-1}(\cos\theta) = -\ell n \tan\left(\frac{\theta}{2}\right)$, routine integration by parts gives

$$J = \frac{2^{\lambda-2}\Gamma^2\left(\frac{\lambda}{2}\right)}{\lambda\Gamma(\lambda)}, \tag{5.5.51}$$

so that

$$c = A a^\lambda \lambda \frac{2^{\lambda-2}\Gamma^2\left(\frac{\lambda}{2}\right)}{\Gamma(\lambda)}. \tag{5.5.52}$$

Equations (5.5.49), (5.5.50) show that P may be written as

$$\vartheta P = 2 \cdot A^{-\frac{1}{\lambda}} \cdot c^{\frac{\lambda+1}{\lambda}} \cdot \chi(\lambda), \tag{5.5.53}$$

where

$$\chi(\lambda) = \frac{\lambda^{\frac{\lambda-1}{\lambda}}}{\lambda+1} \cdot 2^{\frac{2}{\lambda}} \cdot \left[\Gamma(\lambda)/\Gamma^2\left(\frac{\lambda}{2}\right)\right]^{\frac{1}{\lambda}} \tag{5.5.54}$$

has the following values:

λ	1	2	3	4
$\chi(\lambda)$	0.636	0.942	1.127	1.253

Hertz (1882a, 1882b) considered the case $\lambda = 2$ for which

$$\vartheta P = 4 \cdot A \cdot 4 \cdot \frac{a^3}{3}, \qquad c = Aa^2 \cdot 2 \tag{5.5.55}$$

and

$$\vartheta p(r) = \frac{8}{\pi} \cdot A \cdot (a^2 - r^2)^{\frac{1}{2}}. \tag{5.5.56}$$

The relationship between P and c given in (5.5.53) is somewhat misleading because A is not dimensionless; in fact, for $f(r)$ to have the dimension of length, A must have dimension $L^{-\lambda+1}$. If therefore we write $A = Ba^{-\lambda+1}$ where B is dimensionless, then

$$\vartheta P = 4Ba^2 \cdot \frac{\lambda^2 2^{\lambda-2}\Gamma^2\left(\frac{\lambda}{2}\right)}{(\lambda+1)\Gamma(\lambda)}, \tag{5.5.57}$$

$$c = Ba \cdot \frac{\lambda 2^{\lambda-2}\Gamma^2\left(\frac{\lambda}{2}\right)}{\Gamma(\lambda)}, \tag{5.5.58}$$

which yield the simple relationship

$$\frac{\vartheta P}{ac} = \frac{4\lambda}{\lambda+1}. \tag{5.5.59}$$

Galin notes that the derivation of (5.5.56) was carried out by Kochin (1941); it involves the evaluation of a double integral and is extremely complicated. This highlights the fact that the Green's function solution that Galin uses is more laborious than the use of separable solutions in oblate spheroidal coordinates; a historical account of these procedures may be found in Gladwell (1980, section 10.5).

5.6 Axisymmetric Contact with Prescribed Contact Region

Recall the statement made in the second paragraph at the beginning of Section 5.5: there are situations in which the force P increases and new parts of the bounding surface of the punch do not come into contact with the surface of the solid. Now, the contact pressure has a square root singularity $(a^2 - r^2)^{-\frac{1}{2}}$ at the edge, $r = a$, of the contact region. The Papkovich–Neuber potential ψ must satisfy

$$\vartheta\psi(r) = f(r) + c^* \quad \text{on } \Omega, \tag{5.6.1}$$

$$\frac{\partial\psi}{\partial z} = 0 \qquad \text{on } \Omega'. \tag{5.6.2}$$

The potential (5.5.27), which we rename ψ_1, satisfied these equations for the particular value of c given by (5.5.31). We must therefore add to ψ_1 a potential ψ_2 satisfying

$$\vartheta\psi_2 = c^* - c \quad \text{on } \Omega, \tag{5.6.3}$$

$$\frac{\partial\psi_2}{\partial z} = 0 \qquad \text{on } \Omega', \tag{5.6.4}$$

and according to equations (5.4.2), (5.4.3), this is

$$\vartheta\psi_2 = (c^* - c) \cdot \frac{2}{\pi} \cdot \operatorname{arccot}\left(\frac{t}{a}\right) \tag{5.6.5}$$

where, t is given by equations (5.3.14), (5.3.15). [Note that Galin incorrectly has $\arctan\left(\frac{t}{a}\right)$ instead of $\operatorname{arccot}\left(\frac{t}{a}\right)$.] The pressure $p_2(x, y)$ corresponding to ψ_2 is given by

$$-\frac{\vartheta\partial\psi_2}{\partial z}\bigg|_{z=0} = (c^* - c) \cdot \frac{2}{\pi} \cdot \frac{1}{(a^2 - x^2 - y^2)^{\frac{1}{2}}}. \tag{5.6.6}$$

Thus

$$\vartheta p(r) = -\frac{1}{\pi^2}\int_0^a r_0 \Delta f(r_0) H(r, r_0) dr_0 + \frac{2(c^* - c)}{\pi} \cdot \frac{1}{(a^2 - r^2)^{\frac{1}{2}}}. \tag{5.6.7}$$

5.7 Loading Outside a Circular Punch

We consider the case in which a normal pressure $q(x, y)$ is applied to the surface of a semi-infinite elastic body in the region Ω' of the plane $z = 0$ outside the disc Ω of radius a.

The boundary conditions for the Papkovich–Neuber potential ψ on $z = 0$ are

$$\vartheta\psi(x, y, 0) = f(x, y) \quad \text{on } \Omega, \tag{5.7.1}$$

$$\frac{\partial\psi}{\partial z}(x, y, z)|_{z=0} = -q(x, y) \quad \text{on } \Omega'. \tag{5.7.2}$$

The function ψ may be written $\psi = \psi_1 + \psi_2$ where

$$\vartheta\psi_1(x, y, 0) = f(x, y) \quad \text{on } \Omega, \tag{5.7.3}$$

$$\frac{\partial\psi_1}{\partial z}(x, y, z)|_{z=0} = 0 \qquad \text{on } \Omega', \tag{5.7.4}$$

and

$$\psi_2(x, y, z) = 0 \quad \text{on } \Omega, \tag{5.7.5}$$

$$\frac{\partial\psi_2}{\partial z}(x, y, z)|_{z=0} = -q(x, y) \quad \text{on } \Omega'. \tag{5.7.6}$$

The function ψ_1 may be found as in Section 5.5. To find ψ_2, we need to construct the influence function $K_2(x, y, z, \xi, \eta)$ which is harmonic in x, y, z, and behaves like $\frac{1}{R}$ near the point $(\xi, \eta, 0)$ in Ω'. The influence function satisfies

$$K_2(x, y, z, \xi, \eta) = 0 \quad \text{on } \Omega, \tag{5.7.7}$$

$$\frac{\partial}{\partial z}K_2(x, y, z, \xi, \eta)|_{z=0} = 0 \quad \text{on } \Omega', \tag{5.7.8}$$

except near $x = \xi$, $y = \eta$, $z = 0$, where

$$\frac{\partial}{\partial z}K_2(x, y, z, \xi, \eta)|_{z=0} \to \infty. \tag{5.7.9}$$

We shall start from the function $K(x, y, z, \xi, \eta)$ in equation (5.4.13), with a replaced by a^{-1} and apply the *Kelvin* inversion to obtain

$$K'(x, y, z, \xi, \eta) = \frac{1}{(x^2 + y^2 + z^2)^{\frac{1}{2}}}K(x', y', z', \xi', \eta'), \tag{5.7.10}$$

where

$$(x', y', z') = \frac{(x, y, z)}{(x^2 + y^2 + z^2)}, \qquad (\xi', \eta') = \frac{(\xi, \eta)}{\xi^2 + \eta^2}. \tag{5.7.11}$$

Temporarily, let

$$s^2 = x^2 + y^2 + z^2, \quad \sigma^2 = \xi^2 + \eta^2, \tag{5.7.12}$$

then

$$a^{-2} - \xi'^2 - \eta'^2 = (\sigma^2 - a^2)/(a^2\sigma^2), \tag{5.7.13}$$

and

$$R'^2 = (x' - \xi')^2 + (y' - \eta')^2 + z'^2,$$
$$= R^2/(s^2\sigma^2), \tag{5.7.14}$$

$$\rho'^2 = \{(x'^2 + y'^2 + z'^2 - a^{-2})^2 + 4a^{-2}z'^2\}^{\frac{1}{2}},$$
$$= \rho^2/(a^2 s^2), \tag{5.7.15}$$
$$a^{-2} - x'^2 - y'^2 - z'^2 = (s^2 - a^2)/(a^2 s^2),$$

where R^2 is given by equation (5.3.7) and ρ^2 is given by equation (5.3.15). Thus

$$K'(x, y, z, \xi, \eta) = \frac{2\sigma}{\pi R} \arctan\left\{\left(\frac{(\sigma^2 - a^2)(s^2 - a^2 + \rho^2)}{2a^2R^2}\right)^{\frac{1}{2}}\right\}. \tag{5.7.16}$$

Near $x = \xi$, $y = \eta$, $z = 0$ in Ω', this behaves like σ/R so that

$$K_2(x, y, z, \xi, \eta) = \frac{2}{\pi R} \arctan\left\{\left(\frac{(\sigma^2 - a^2)(s^2 - a^2 + \rho^2)}{2a^2R^2}\right)^{\frac{1}{2}}\right\}. \tag{5.7.17}$$

We now determine the derivative of K_2 with respect to z, for $(x, y) \in \Omega$, $(\xi, \eta) \in \Omega'$. We note that, for small z,

$$s^2 - a^2 + \rho^2 = \frac{2a^2z^2}{a^2 - s^2} + \cdots \tag{5.7.18}$$

so that

$$K_2(x, y, z, \xi, \eta) = \frac{2}{\pi R} \arctan\left(\left(\frac{\sigma^2 - a^2}{a^2 - s^2}\right)^{\frac{1}{2}} \frac{z}{R}\right) + \cdots \tag{5.7.19}$$

and

$$\frac{\partial K_2}{\partial z}(x, y, z, \xi, \eta)|_{z=0} \to \frac{2}{\pi R^2}\left(\frac{\sigma^2 - a^2}{a^2 - s^2}\right)^{\frac{1}{2}}. \tag{5.7.20}$$

[Note that Galin's use of Figure 41 to show equation (5.7.20) geometrically, is wrong. The length AD is not $\sqrt{\xi^2 + y^2 - a^2}$ but $\sqrt{\xi^2 + y^2} - a$; similarly BC if not $\sqrt{a^2 - x^2 - y^2}$, but $a - \sqrt{x^2 + y^2}$; for this reason, we omit Figure 41 altogether.]

We now determine the pressure arising over the punch from the action of the normal pressure $q(x, y)$ on the surface of the elastic solid outside the punch.

The pressure is

$$p(x, y) = -\sigma_{zz}(x, y, 0+) = -\frac{\partial \psi_2}{\partial z}(x, y, z)|_{z=0}.$$

Table 5.7.1

$x = 0, y$	$k(0, y)$	$x, y = 0$	$k(x, 0)$
–1.0	∞	–1.0	∞
–0.8	0.646	–0.8	0.354
–0.6	0.536	–0.6	0.318
–0.4	0.508	–0.4	0.339
–0.2	0.500	–0.2	0.396
0.0	0.498	0.0	0.498
0.2	0.500	0.2	0.676
0.4	0.508	0.4	1.011
0.6	0.536	0.6	1.729
0.8	0.646	0.8	3.810
1.0	∞	1.0	∞

For a punch with a plane face,

$$\psi_2(x, y, z) = -\frac{1}{2\pi}\int_{\Omega'}\int q(\xi, \eta)K_2(x, y, z, \xi, \eta)d\xi d\eta.$$

Thus

$$p(x, y) = \frac{1}{2\pi}\int_{\Omega'}\int q(\xi, \eta)\frac{\partial K_2}{\partial z}(x, y, z, \xi, \eta)|_{z=0}d\xi d\eta$$

and, on using (5.7.20), we find

$$p(x, y) = \frac{1}{\pi^2}\int_{\Omega'}\int \left(\frac{\xi^2 + \eta^2 - a^2}{a^2 - x^2 - y^2}\right)^{\frac{1}{2}} \frac{q(\xi, \eta)d\xi d\eta}{(x - \xi)^2 + (y - \eta)^2}.$$

For a concentrated force Q applied at $(x, y) = (\ell, 0)$ the additional pressure is thus

$$p(x, y) = \frac{Q}{\pi^2}\left(\frac{\ell^2 - a^2}{a^2 - x^2 - y^2}\right)^{\frac{1}{2}} \cdot \frac{1}{(x - \ell)^2 + y^2} = \frac{Q}{\pi^2}k(x, y).$$

The values of $k(x, y)$ are given in Table 5.7.1 for points on the x and y axes, for $a = 1, \ell = 1.5$; these results are shown graphically in Figure 5.7.1.

5.8 Axi-Symmetric Contact Problems with Friction

In this section, we consider contact problems in the theory of elasticity in which a punch bounded by a surface of revolution is pressed into a semi-infinite elastic solid, and there is *friction* between the punch and the elastic solid. This problem is of interest because of its application to certain problems in machine construction.

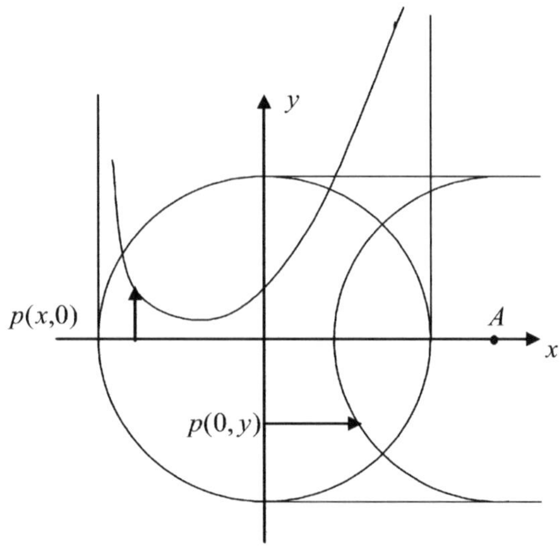

Fig. 5.7.1 The excess pressure on the x- and y-axes.

A force P, directed along the axis of revolution, acts on a punch pressed against a semi-infinite elastic solid. In addition, the punch rotates about the axis of revolution under the action of a moment M. The angular velocity of rotation is ω. Friction occurs between the punch and the elastic solid, and its direction is opposite to that of the motion, i.e., it is perpendicular to the radius of the contact region, which is circular in shape, as in Figure 5.8.1.

The angular speed of a point of the punch is the product of the angular velocity ω and the radial distance r from the axis of rotation: ωr. Since, in general, the coefficient of friction, ρ, between two solid bodies depends on the angular speed, we shall assume that

$$\rho = F(\omega r).$$

Let $O(r, \theta, z)$ be a system of cylindrical polar coordiantes with the z-axis pointing along the axis of rotation. Denote the components of the elastic displacement vector $\mathbf{d}$ by $(u_r, u_\theta, u_z) = (u, v, w)$. The punch applies a pressure $p\mathbf{k}$ to the elastic solid over the contact region Ω; this is applied from $z = 0-$ to $z = 0+$; therefore, following the convention for stresses in Section 2.1,

$$\sigma_{zz} = -p \quad \text{on } \Omega.$$

A point of the punch moves in the $\boldsymbol{\theta}$ direction; the friction force on the punch is thus in the $-\boldsymbol{\theta}$ direction; the friction force on the elastic solid is therefore in the $\boldsymbol{\theta}$ direction. This is the $\boldsymbol{\theta}$ component of the force per unit area on the element with normal $\mathbf{k}$, applied from $z = 0-$ to $z = 0+$. The stress component $\sigma_{\theta z}$ is the $\boldsymbol{\theta}$ component applied from $z = 0+$ to $z = 0-$, so that $\sigma_{\theta z} = -\rho p$. Thus, we have the

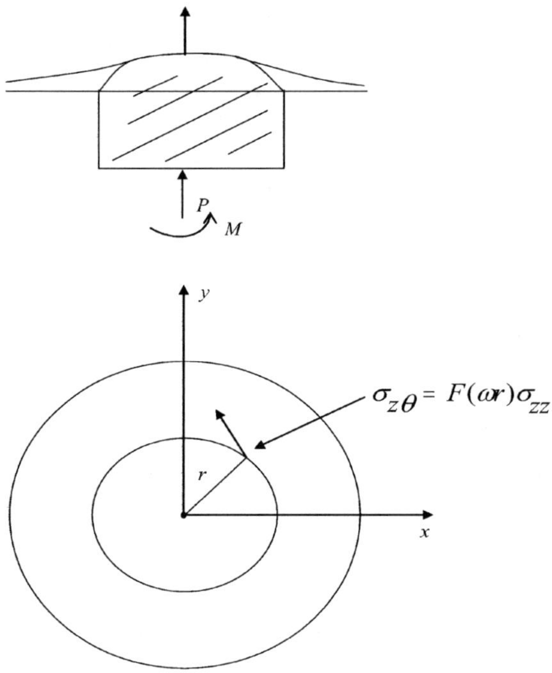

Fig. 5.8.1 A punch is rotated against friction.

boundary conditions

$$w = f(r) + c, \quad \sigma_{z\theta} = F(\omega r)\sigma_{zz}, \quad \sigma_{zr} = 0 \quad \text{on } \Omega \tag{5.8.1}$$

$$\sigma_{zz} = 0 = \sigma_{z\theta} = \sigma_{zr} \quad \text{on } \Omega'. \tag{5.8.2}$$

In cylindrical coordinates, the strains are

$$\varepsilon_{rr} = \frac{\partial u}{\partial r}, \quad \varepsilon_{\theta\theta} = \frac{1}{r}\frac{\partial v}{\partial \theta} + \frac{u}{r}, \quad \varepsilon_{zz} = \frac{\partial w}{\partial z}, \tag{5.8.3}$$

$$\varepsilon_{\theta z} = \frac{1}{2}\left(\frac{1}{r}\frac{\partial w}{\partial \theta} + \frac{\partial v}{\partial z}\right), \quad \varepsilon_{zr} = \frac{1}{2}\left(\frac{\partial u}{\partial z} + \frac{\partial w}{\partial r}\right) \tag{5.8.4}$$

$$\varepsilon_{r\theta} = \frac{1}{2}\left(\frac{\partial v}{\partial r} - \frac{v}{r} + \frac{1}{r}\frac{\partial u}{\partial \theta}\right). \tag{5.8.5}$$

The stress-strain relations are

$$(1-2\nu)\sigma_{rr} = 2\mu\{\nu\varepsilon + (1-2\nu)\varepsilon_{rr}\}, \tag{5.8.6}$$

$$(1-2\nu)\sigma_{\theta\theta} = 2\mu\{\nu\varepsilon + (1-2\nu)\varepsilon_{\theta\theta}\}, \tag{5.8.7}$$

$$(1-2\nu)\sigma_{zz} = 2\mu\{\nu\varepsilon + (1-2\nu)\varepsilon_{zz}\}, \tag{5.8.8}$$

$$\sigma_{\theta z} = 2\mu\varepsilon_{\theta z}, \quad \sigma_{zr} = 2\mu\varepsilon_{zr}, \quad \sigma_{r\theta} = 2\mu\varepsilon_{r\theta}, \tag{5.8.9}$$

where

$$\varepsilon = div\mathbf{d} = \varepsilon_{rr} + \varepsilon_{\theta\theta} + \varepsilon_{zz} \tag{5.8.10}$$

is the dilatation.

The equilibrium equations are

$$\frac{\partial\sigma_{rr}}{\partial r} + \frac{1}{r}\frac{\partial\sigma_{r\theta}}{\partial\theta} + \frac{\partial\sigma_{zr}}{\partial z} + \frac{\sigma_{rr}-\sigma_{\theta\theta}}{r} = 0 \tag{5.8.11}$$

$$\frac{\partial\sigma_{r\theta}}{\partial r} + \frac{1}{r}\frac{\partial\sigma_{\theta\theta}}{\partial\theta} + \frac{\partial\sigma_{\theta z}}{\partial z} + \frac{2\sigma_{r\theta}}{r} = 0 \tag{5.8.12}$$

$$\frac{\partial\sigma_{zr}}{\partial r} + \frac{1}{r}\frac{\partial\sigma_{\theta z}}{\partial\theta} + \frac{\partial\sigma_{zz}}{\partial z} + \frac{\sigma_{zr}}{r} = 0 \tag{5.8.13}$$

In an axisymmetric problem, there is no dependence on θ, so that the strains become

$$\varepsilon_{rr} = \frac{\partial u}{\partial r}, \quad \varepsilon_{\theta\theta} = \frac{u}{r}, \quad \varepsilon_{zz} = \frac{\partial w}{\partial z}, \tag{5.8.14}$$

$$\varepsilon_{\theta z} = \frac{1}{2}\cdot\frac{\partial v}{\partial z}, \quad \varepsilon_{zr} = \frac{1}{2}\left(\frac{\partial u}{\partial z} + \frac{\partial w}{\partial r}\right), \quad \varepsilon_{r\theta} = \frac{1}{2}\left(\frac{\partial v}{\partial r} - \frac{v}{r}\right), \tag{5.8.15}$$

and the equilibrium equations become

$$\frac{\partial\sigma_{rr}}{\partial r} + \frac{\partial\sigma_{zr}}{\partial z} + \frac{\sigma_{rr}-\sigma_{\theta\theta}}{r} = 0, \tag{5.8.16}$$

$$\frac{\partial\sigma_{r\theta}}{\partial r} + \frac{\partial\sigma_{\theta z}}{\partial z} + \frac{2\sigma_{r\theta}}{r} = 0, \tag{5.8.17}$$

$$\frac{\partial\sigma_{zr}}{\partial r} + \frac{\partial\sigma_{zz}}{\partial z} + \frac{\sigma_{zr}}{r} = 0. \tag{5.8.18}$$

Since

$$\varepsilon = \frac{\partial u}{\partial r} + \frac{u}{r} + \frac{\partial w}{\partial z} \tag{5.8.19}$$

now depends only on u and w, the first and third of the equilibrium equations involve only u and w, while the second equation involves only v.

Thus, in the axisymmetric case there are two independent systems of deformation. In the first, $v = 0$, so that $\sigma_{r\theta} = 0 = \sigma_{\theta z}$, and the only stresses are $\sigma_{rr}, \sigma_{\theta\theta}, \sigma_{zz}, \sigma_{rz}$. In the second, $u = 0 = w$ so that $\sigma_{\theta\theta} = 0 = \sigma_{rr} = \sigma_{zz} = \sigma_{rz}$, and only $\sigma_{r\theta}$ and $\sigma_{\theta z}$ are non-zero.

The distribution of stress in our problem, governed by the boundary condition (5.8.1), (5.8.2) can thus be expressed as the superposition of two states * and **, in the first of which $v = 0$, and in the second $u = 0 = w$. Since there is no shear stress σ_{zr} produced by the rotation of the punch we have the boundary condition, for *:

$$w^* = f(r) + c, \quad \sigma^*_{rr} = 0 \quad \text{on } \Omega,$$

$$\sigma^*_{zr} = 0 = \sigma^*_{zz} \qquad \text{on } \Omega',$$

and for **:

$$\sigma^{**}_{z\theta} = F(\omega r)\sigma^*_{zz} \quad \text{on } \Omega,$$

$$\sigma^{**}_{z\theta} = 0 \qquad \text{on } \Omega'.$$

Note that in order to solve the second problem we must first find σ^*_{zz} from the first;

$$\sigma_{zz} = \sigma^*_{zz}, \quad \sigma_{z\theta} = \sigma^{**}_{z\theta}.$$

The moment required to make the punch rotate is

$$M = -\int_\Omega\int r^2\sigma_{z\theta} dr d\theta$$

$$= 2\pi\int_0^a r^2F(\omega r)p(r)dr.$$

For a punch with a plane base

$$p(r) = \frac{P}{2\pi a(a^2 - r^2)^{\frac{1}{2}}}$$

so that in the case that $F(\omega r) = \rho = \text{constant}$,

$$M = \frac{2\pi\rho P}{2\pi a}\int_0^a \frac{r^2 dr}{(a^2 - r^2)^{\frac{1}{2}}} = \frac{\pi}{4}\rho P a.$$

For a punch that is of such a shape that there is a constant pressure in the contact region,

$$p = P/(\pi a^2)$$

the moment is

$$M = \frac{2\pi\rho P}{\pi a^2}\int_0^a r^2 dr = \frac{2}{3}\rho P a.$$

For a punch in the form of a paraboloid in contact over a circle of radius a,

$$p(r) = \frac{3P}{2\pi a^3}(a^2 - r^2)^{\frac{1}{2}}$$

so that

$$M = 2\pi\rho \cdot \frac{3P}{2\pi a^3}\int_0^a r^2(a^2 - r^2)^{\frac{1}{2}}dr$$
$$= \frac{3\pi}{16}\rho P a.$$

Thus, the moment required to turn the punch in the presence of friction depends to a great extent upon its shape.

5.9 A Punch of Elliptic Cross-Section

In this section, we consider problems relating to a punch of elliptic cross-section pressed into an elastic half-space. This leads to a discussion of *Dirichlet's problem* for the exterior of an elliptical disc. This problem can be solved by means of expansions in series of *Lamé functions*. In performing these expansions, a number of difficulties arise so that it is worth establishing some common properties of the solutions obtainable by this method.

The following result will be established: suppose that a punch is pressed against a semi-infinite solid $z \geq 0$, the punch being of elliptical cross-section with semi-axes a and b. Suppose further that there are no frictional forces acting between the punch and the elastic body, and that, outside the contact region, the surface of the elastic body is free from tractions. If the equation of the surface bounding the base of the punch is $z = P_n(x, y)$, where P_n is a polynomial of degree n, then the pressure acting under the punch can be represented in the form

$$p(x, y) = \left(1 - \frac{x^2}{a^2} - \frac{y^2}{b^2}\right)^{-\frac{1}{2}} P_n^*(x, y) \tag{5.9.1}$$

where $P_n^*(x, y)$ is also a polynomial of degree n.

Most of the contact problems we have already considered are special cases of this problem.

The solution (5.9.1) allows us to reduce all problems of this type to the solution of a set of simultaneous linear algebraic equations. Since the form of the punch $P_n(x, y)$ is known beforehand, we can obtain a set of simultaneous equations by means of which we can determine the unknown coefficients in $P_n^*(x, y)$.

To prove the required result, we work with the *ellipsoidal coordinates* ρ, σ, τ defined in equations (5.2.1) and (5.2.2). As we noted in Section 5.2, as $\rho \to 1$ the ellipsoid ρ = constant degenerates into the interior of the double-sided elliptical disc (5.2.6), namely

$$\frac{x^2}{a^2} + \frac{y^2}{b^2} \leq 1$$

where $b^2 = a^2(1 - k^2)$.

The problem considered here can be reduced to the following case of Dirichlet's problem: find a function that is harmonic outside the elliptical disc, and which takes the value $P_n(x, y)$ on the upper and lower sides of the disc, and becomes zero at infinity. We note that the function will be even in z. The presure arising from the punch is proportional to the density of the simple layer whose potential is the harmonic function.

The remainder of this section of Galin's text is written in such a cursory manner that it is impossible to understand it without having at least a rudimentary background in elliptic functions. For this reason, we shall proceed as follows: we first reproduce Galin's account without any major changes, and then give some explanations, taken from Gladwell (1980); that treatment was based on *Jacobian elliptic functions*, following Vol. 3 of Erdelyi (1955) and Arscott (1962). The application to contact problems is based on Shail (1978). It is unfortunate that the basic result contained in (5.9.1), usually known as *Galin's Theorem*, is 'proved' in such a scanty fashion. Galin's theorem has since been definitively proved in a very general context, for certain anisotropic media, as described in Gladwell (1980) and the references given therein. Now we give Galin's account.

It is well-known that the product of three Lamé functions $E_n^m(\rho)$, $E_n^m(\sigma)$, $E_n^m(\tau)$, i.e.,

$$\wedge_n^m(x, y, z) = E_n^m(\rho), E_n^m(\sigma), E_n^m(\tau)$$

is a harmonic function. In future, we shall refer to this product as a *Lamé triple product*. Lamé's functions satisfy Lamé's equation for a definite value of a certain parameter contained in this equation. (See Whittaker and Watson (1952). When expressed in terms of x, y, z, the product $\wedge_n^m(x, y, z)$ is a polynomial in x, y, z.

We now show how we may construct a harmonic function $W_1(x, y, z)$, even in z, which coincides with the polynomial $P_n(x, y)$ when $z = 0$. We shall look for a function $W_1(x, y, z)$ of the form

$$W_1(x, y, z) = P_n(x, y) + z^2 P^{(2)}(x, y) + z^4 P^{(4)}(x, y) + \cdots$$

Applying the Laplace operator

$$\Delta = \frac{\partial^2}{\partial x^2} + \frac{\partial^2}{\partial y^2} + \frac{\partial^2}{\partial z^2}$$

to this function, we obtain

$$\Delta W_1(x, y) = [\Delta_0 P_n + 2P^{(2)}] + z^2[\Delta_0 P^{(2)} + 12P^{(4)}] + \cdots,$$

where

$$\Delta_0 = \frac{\partial^2}{\partial x^2} + \frac{\partial^2}{\partial y^2}.$$

If W_1 is to be harmonic, then

$$P^{(2)} = -\frac{1}{2!}\Delta_0 P_n, \quad P^{(4)} = -\frac{1}{4.3}\Delta_0 P^{(2)} = \frac{1}{4!}\Delta_0^2 P_n, \ldots$$

so that

$$W_1(x, y, z) = \sum_{j=0}^{m} (-)^j \frac{1}{(2j)!}\Delta_0^j P_n(x, y) z^{2j}, \tag{5.9.2}$$

where $m = \left[\frac{n}{2}\right]$, the integral part of $\frac{n}{2}$. [Galin omits to point out that since $P_n(x, y)$ is of degree n, the term $\Delta_0^{m+1} P_n$ is identically zero.] $W_1(x, y, z)$ is a harmonic polynomial of degree n.

Now consider the ellipsoid (5.2.3) corresponding to $\rho = \kappa$. The harmonic function $W_1(x, y, z)$ which equals $P_n(x, y)$ on the part of the plane $z = 0$ inside the ellipsoid, is determined by equation (5.9.2). We now prove that $W_1(x, y, z)$ can be represented as a sum of a finite number of Lamé triple products.

We know that there are $(n+1)^2$ linearly independent harmonic polynomials with degrees less than or equal to n. On the other hand, there are $(n+1)^2$ linearly independent Lamé triple products with degrees less than or equal to n. These conditions are sufficient to allow us to write every harmonic polynomial of degree n as a linear combination of Lamé triple products of degree less than or equal to n.

It is easy to show that these conditions are also necessary. If we enumerate the linearly independent harmonic polynomials $P_1, P_2, P_3, \ldots, P_{n+1}$, and also enmerate the Lamé triple products $\wedge_1, \wedge_2, \ldots, \wedge_{n+1}$ then we can write $r = (n+1)^2$ simultaneous linear equations

$$\sum_{j=1}^{r} B_{ij} P_j = \wedge_i, \quad i = 1, 2, \ldots, r.$$

To be able to represent each Lamé triple product as a linear combination of harmonic polynomials, i.e., to invert the equations, the determinant $\det \mathbf{B}$ must be nonzero. But $\det \mathbf{B}$ is zero iff its rows are linearly dependent. The linear dependence of rows is impossible since it would lead to the linear dependence of the right hand sides, the Lamé triple products, which are known to be linearly independent. Thus, $\det \mathbf{B} \neq 0$, and therefore, every harmonic polynomial can be represented as a linear combination of Lamé triple products. It follows that any harmonic polynomial of degree n is a linear combination of Lamé triple products of degree less than or equal to n:

$$W_1(x, y, z) = \sum_{i=0}^{n} \sum_{j=1}^{2\ell+1} A_{ij} E_i^j(\rho) E_i^j(\sigma) E_i^j(\tau). \tag{5.9.3}$$

We now find the function $W_2(x, y, z)$ which is harmonic in the region outside the ellipsoid $\rho = \kappa$ and is equal to $W_1(x, y, z)$ on the ellipsoid.

We shall make use of *Lamé functions of the second kind* defined by

$$F_n^m(\rho) = (2n+1) E_n^m(\rho) \int_0^\rho \frac{du}{[E_n^m(u)]^2}.$$

The variable u can be found from the following relation

$$\mathcal{P}(u) = u + \frac{1}{3}(A^2 + B^2 + C^2)$$

where A, B, C are the semi-axis of the ellipsoid. The invariants g_2 and g_3 of the Weierstrass elliptic function can be determined from the identity

$$4(A^2 + \rho)(B^2 + \rho)(C^2 + \rho) = 4\mathcal{P}^3(u) + g_2\mathcal{P}(u) - g_3.$$

The harmonic function $W_2(x, y, z)$ which takes the value $W_1(x, y, z)$ on the surface of the ellipsoid $\rho = \kappa$ has the following expression:

$$W_2(x, y, z) = \sum_{i=0}^{n}\sum_{j=1}^{2i+1} A_{ij}\frac{E_i^j(\rho)}{F_i^j(\kappa)}F_i^j(\rho)E_i^j(\sigma)E_i^j(\tau). \tag{5.9.4}$$

Hence the density of the layer distributed over the surface of the ellipsoid is

$$q(x, y, z) = -\frac{1}{4\pi}\left[\left(\frac{\partial W_2}{\partial \rho} - \frac{\partial W_1}{\partial \rho}\right)\frac{\partial \rho}{\partial \nu}\right]_{\rho=\kappa},$$

where $\frac{\partial}{\partial \nu}$ denotes the derivative in the direction of the normal to the ellipsoid. Thus,

$$q(x, y, z) = -\frac{1}{4\pi}\frac{\partial \rho}{\partial \nu}|_{\rho=\kappa}\sum_{i=0}^{n}\sum_{j=1}^{2i+1}$$

$$\times A_{ij}\frac{E_i^j(\kappa)}{F_i^k(\kappa)}E_i^j(\sigma)E_i^j(\tau) * \left[\frac{\partial}{\partial \rho}F_i^j(\rho) - \frac{\partial E_i^j(\rho)}{\partial \rho}\right]_{\rho=\kappa}. \tag{5.9.5}$$

If we let $\kappa \to 1$ in equation (5.9.5), we obtain a harmonic function which is equal to $P_n(x, y)$ on the elliptical disc with semi axes a and b, with $b^2 = a^2(1 - k^2)$. Letting $\kappa \to 1$ in equation (5.9.3) and multiplying by 2, we find the density of a simple layer distributed over the elliptical disc in such a way that the potential due to this simple layer takes the value $P_n(x, y)$ on the disc. We note that

$$\frac{\partial \rho}{\partial \nu}\alpha\left(1 - \frac{x^2}{a^2} - \frac{y^2}{b^2}\right)^{-\frac{1}{2}}$$

on the elliptical disc.

It is not difficult to prove that the double sum in equation (5.9.3) reduces to a polynomial of degree n on the elliptical disc, since each term is a polynomial in x, y, and the orders of the Lamé triple products do not exceed n. Since the pressure under the punch is proportional to the density of the simple layer, we obtain the result

$$p(x, y) = \left(1 - \frac{x^2}{a^2} - \frac{y^2}{b^2}\right)^{-\frac{1}{2}} P_n^*(x, y). \tag{5.9.6}$$

This then is Galin's 'proof'. We now proceed to give some explanations, starting from *potential theory*, as described in Courant and Hilbert (1953, vol. II, section IV, §3).

Let V be a bounded region in $\mathbb{R}^3$ with piecewise smooth boundary S. If u, v have continuous first derivatives in $V + S$, and continuous second derivatives in V, then *Green's Formula* states that

$$\int_V (u\Delta v - v\Delta u)dV = \int_S \left(u\frac{\partial v}{\partial \nu} - v\frac{\partial u}{\partial \nu}\right) dS,$$

where $\frac{\partial}{\partial \nu}$ denotes differentiation in the direction of the outward drawn normal to S.

The function

$$v = \frac{1}{r}, \quad r^2 = (x - \xi)^2 + (y - \eta)^2 + (z - \zeta)^2$$

is harmonic in any region excluding the point $P = (\xi, \eta, \zeta)$. If, therefore, u is harmonic in V, and P lies outside V, then $\Delta u = 0 = \Delta v$ in V and

$$\int_S \left(u\frac{\partial}{\partial \nu}\left(\frac{1}{r}\right) - \frac{1}{r}\frac{\partial u}{\partial \nu}\right) dS = 0, \quad P \text{ outside } V.$$

Now suppose P lies inside V. Let V_ε be the sphere of radius ε around P, and apply Green's Theorem to the region $V - V_\varepsilon$ with boundary $S + S_\varepsilon$:

$$\int_{V-V_\varepsilon} (u\Delta v - v\Delta u)dV = \int_S \left(u\frac{\partial v}{\partial \nu} - v\frac{\partial u}{\partial \nu}\right) dS + \int_{S_\varepsilon} \left(u\frac{\partial v}{\partial \nu} - v\frac{\partial u}{\partial \nu}\right) dS.$$

Both u and v are harmonic in $V - V_\varepsilon$, so that the left-hand side is zero. For the integral over S_ε, $\frac{\partial}{\partial \nu} = -\frac{\partial}{\partial r}$ and

$$\int_{S_\varepsilon} \left(u\frac{\partial v}{\partial \nu} - v\frac{\partial u}{\partial \nu}\right) dS \simeq \left(\frac{u}{\varepsilon^2} + \frac{1}{\varepsilon}\frac{\partial u}{\partial r}\right) 4\pi\varepsilon^2 \to 4\pi u(P).$$

Thus

$$\frac{1}{4\pi}\int_S \left(u\frac{\partial}{\partial \nu}\left(\frac{1}{r}\right) - \frac{1}{r}\frac{\partial u}{\partial \nu}\right) dS = -u, \quad P \text{ inside } V.$$

Now suppose u, is harmonic *outside* the region V. Apply Green's Formula to the bounded region outside V and inside the large sphere V_R of radius R. Provided that $u, v = O(1/R)$, the integral over the surface of the sphere will tend to zero as $R \to \infty$. We obtain as before, that if u is harmonic outside V, then

$$\frac{1}{4\pi}\int_S \left(u\frac{\partial}{\partial \nu}\left(\frac{1}{r}\right) - \frac{1}{r}\frac{\partial u}{\partial \nu}\right) dS = \begin{cases} 0 \; P \text{ inside } V, \\ 1 \; P \text{ outside } V. \end{cases}$$

According to (5.1.23) and (5.1.24), we seek a harmonic $\psi(x, y, z)$ such that

$$\vartheta\psi(x, y, z) = P_n(x, y) \text{ in } \Omega, \tag{5.9.7}$$

$$\frac{\partial\psi}{\partial z}(x, y, z)|_{z=0} = 0 \qquad \text{in } \Omega'. \tag{5.9.8}$$

The pressure $p(x, y)$ is then given by

$$p(x, y) = -\frac{\partial\psi}{\partial z}(x, y, z)|_{z=0} \text{ in } \Omega. \tag{5.9.9}$$

We use the fact that Ω is the limit, as $\kappa \to 1$ of the ellipsoid S_κ given by (5.2.3) with $\rho = \kappa$. We construct $W_1(x, y, z)$ harmonic in V_κ, even in z, and equal to $\vartheta^{-1}P_n(x, y)$ when $z = 0$:

$$\vartheta W_1(x, y, 0) = P_n(x, y), \tag{5.9.10}$$

$$\frac{\partial W_1}{\partial z}(x, y, z)|_{z=0} = 0. \tag{5.9.11}$$

Now we construct $W_2(x, y, z)$ harmonic outside V_κ and equal to $W_1(x, y, z)$ on S_κ. The Green's Formula results we have obtained yield

$$\frac{1}{4\pi}\int_{S_\kappa}\left(W_1\frac{\partial}{\partial\nu}\left(\frac{1}{r}\right) - \frac{1}{r}\frac{\partial W_1}{\partial\nu}\right)dS = \begin{cases} -W_1 \text{ inside } V_\kappa, \\ 0 \text{ outside } V_\kappa, \end{cases} \tag{5.9.12}$$

$$\frac{1}{4\pi}\int_{S_\kappa}\left(W_2\frac{\partial}{\partial\nu}\left(\frac{1}{r}\right) - \frac{1}{r}\frac{\partial W_2}{\partial\nu}\right)dS = \begin{cases} 0 \text{ inside } V_\kappa, \\ W_2 \text{ outside } V_\kappa. \end{cases} \tag{5.9.13}$$

Now subtract equation (5.9.12) from equation (5.9.13), and use the fact that $W_1 = W_2$ on S_κ, the result is

$$\psi \equiv \frac{1}{4\pi}\int_{S_\kappa}\frac{1}{r}\frac{\partial}{\partial\nu}(W_1 - W_2)dS = \begin{cases} W_1, & P \text{ inside } V_\kappa, \\ W_2, & P \text{ outside } V_\kappa. \end{cases} \tag{5.9.14}$$

Now let $\kappa \to 1$. Since W_1 satisfies (5.9.10), (5.9.11) and W_2 is even in z, ψ will satisfy the boundary conditions (5.9.7), (5.9.8). To find $p(x, y)$, we need to find the z-derivative of ψ on S_κ. Write ψ as

$$\psi \equiv \frac{1}{4\pi}\int_{S_\kappa}\frac{\phi(\xi, \eta, \zeta)dS}{\{(x-\xi)^2 + (y-\eta)^2 + (z-\zeta)^2\}^{\frac{1}{2}}}, \tag{5.9.15}$$

so that

$$\frac{\partial\psi}{\partial z} = -\frac{1}{4\pi}\int_{S_\kappa}\frac{(z-\zeta)\phi(\xi, \eta, \zeta)dS}{\{(x-\xi)^2 + (y-\eta)^2 + (z-\zeta)^2\}^{\frac{3}{2}}}.$$

As $\kappa \to 1$, the surface S_κ tends to the upper and lower surfaces of the elliptical disc, so that $\zeta \to 0$. Therefore, as $z \to 0$, the contribution to the integral from all but the vicinity of (x, y, z) tends to zero.

The contribution from a small circle of radius ε around x, y, z is

$$
\begin{aligned}
-\frac{1}{4\pi}\int_0^{\varepsilon} \frac{2\pi r z\phi(x, y, z)}{(r^2+z^2)^{\frac{3}{2}}}dr &= \frac{1}{2}\cdot z\phi(x, y, z)\left[\frac{1}{(r^2+z^2)^{\frac{1}{2}}}\right]_0^{\varepsilon},\\
&= \frac{1}{2}\cdot z\phi(x, y, z)\left[\frac{1}{(\varepsilon^2+z^2)^{\frac{1}{2}}}-\frac{1}{z}\right],\\
&\to -\frac{1}{2}\cdot\phi(x, y, z).
\end{aligned}
$$

Remembering that the integral is over the upper and lower surfaces, we find that

$$p(x, y) = -\frac{\partial\psi}{\partial z}(x, y, z)|_{z=0} = \phi(x, y, z),$$

or

$$p(x, y) = \frac{\partial}{\partial\nu}(W_1 - W_2)|_{\kappa\to 1}. \tag{5.9.16}$$

The ellipsoid S_κ is given by equation (5.2.3) with $\rho = \kappa$, so that the tangent plane at (x', y', z') has equation

$$\frac{xx'}{a^2\kappa^2} + \frac{yy'}{a^2(\kappa^2-k^2)} + \frac{zz'}{a^2(\kappa^2-1)} = 1,$$

and the unit normal has direction cosines

$$\frac{1}{D}\left(\frac{x'}{\kappa^2}, \frac{y'}{\kappa^2-k^2}, \frac{z'}{\kappa^2-1}\right) = (\ell, m, n), \tag{5.9.17}$$

where

$$D^2 = \left(\frac{x'}{\kappa^2}\right)^2 + \left(\frac{y'}{\kappa^2-k^2}\right)^2 + \left(\frac{z'}{\kappa^2-1}\right)^2.$$

Thus

$$\frac{\partial}{\partial\nu} = \ell\frac{\partial}{\partial x} + m\frac{\partial}{\partial y} + n\frac{\partial}{\partial z}.$$

Now

$$\frac{\partial}{\partial\rho} = \frac{\partial x}{\partial\rho}\frac{\partial}{\partial x} + \frac{\partial y}{\partial\rho}\frac{\partial}{\partial y} + \frac{\partial z}{\partial\rho}\frac{\partial}{\partial z},$$

and equation (5.2.1) gives

$$k\frac{\partial x}{\partial\rho} = a\sigma\tau = \frac{kx'}{\rho},$$

$$\frac{\partial x}{\partial \rho} = \frac{x'}{\kappa} \text{ when } x = x', \rho = \kappa.$$

Similarly

$$\frac{\partial y}{\partial \rho} = \frac{\kappa y'}{\kappa^2 - k^2}, \quad \frac{\partial z}{\partial \rho} = \frac{\kappa z'}{\kappa^2 - 1},$$

at (x', y', z'). Thus

$$\frac{\partial}{\partial \rho} = \frac{x'}{\kappa}\frac{\partial}{\partial x} + \frac{\kappa y'}{\kappa^2 - k^2}\frac{\partial}{\partial y} + \frac{\kappa z'}{\kappa^2 - 1}\frac{\partial}{\partial z}. \tag{5.9.18}$$

Comparing equations (5.9.17), (5.9.18), we see that

$$\frac{\partial}{\partial \nu} = \frac{1}{\kappa D}\frac{\partial}{\partial \rho}. \tag{5.9.19}$$

Since (x', y', z') lies on the ellipsoid S_κ, we find

$$D^2 = \frac{a^2}{\kappa^2 - 1}\left(1 - \frac{x'^2}{a^2\kappa^4} - \frac{(1-k^2)y'^2}{a^2(\kappa^2 - k^2)^2}\right), \tag{5.9.20}$$

and

$$p(x, y) = \frac{1}{\kappa D}\frac{\partial}{\partial \rho}(W_1 - W_2)|_{\substack{\rho=\kappa \\ \kappa\to 1}}. \tag{5.9.21}$$

Now inserting the expressions for W_1 and W_2, we see that the generic term in the expansion of $p(x, y)$ is the limit, as $\kappa \to 1$, of

$$Q \equiv \frac{1}{\kappa D}\frac{[E_i^j(\kappa)F_i^{j'}(\kappa) - E_i^{j'}(\kappa)F_i^j(\kappa)]}{F_i^j(\kappa)}E_i^j(\sigma)E_i^j(\tau), \tag{5.9.22}$$

where $'$ denotes $\partial/\partial\rho$. [Note that Galin's equation (5.9.5) is incorrect.]

Before going further, we must address ourselves to the derivation of the Lamé functions E_i^j and F_i^j. These functions are related to elliptic functions. The elliptic functions are defined as in basic texts, e.g., Byrd and Friedman (1971). The basic equation is

$$\zeta = \zeta(y, k) = \int_0^y \frac{dt}{[(1-t^2)(1-k^2t^2)]^{\frac{1}{2}}} = \int_0^\phi \frac{d\theta}{[1 - k^2\sin^2\theta]^{\frac{1}{2}}} = F(\phi, k). \tag{5.9.23}$$

For a given value of y, the first integral defines ζ. The elliptic function $sn\zeta$ gives the inverse of this equation: $sn\zeta$ is the value of y for which the integral yields ζ. We write

$$y = \sin\phi = sn(\zeta, k) = sn\zeta, \quad \phi = am(\zeta, k) = am\zeta$$

so that 'sn' may be considered as an abbreviation for $\sin(am\zeta)$. We note that if $k = 0$, then equation (5.9.23) reduces to

$$\zeta = \zeta(y, 0) = \int_0^y \frac{dt}{(1-t^2)^{\frac{1}{2}}} = \int_0^{\phi} d\theta = \phi = \arcsin y,$$

so that $y = \sin\zeta$.

The functions $cn\zeta$, $dn\zeta$ are

$$cn\zeta = \cos\phi = (1 - sn^2\zeta)^{\frac{1}{2}}, \quad dn\zeta = (1 - k^2\sin^2\phi)^{\frac{1}{2}} = (1 - k^2 sn^2\zeta)^{\frac{1}{2}}.$$

The functions $sn\zeta$, $cn\zeta$, $dn\zeta$ are *doubly periodic* with periods $(4K, 2iK')$, $(4K, 2K + 2iK')$, $(2K, 4iK')$ respectively, where

$$K = K(k) = F(\frac{\pi}{2}, k), \quad K' = K(k'), \tag{5.9.24}$$

and $F(\phi, k)$ is defined in (5.9.23). The function $F(\phi, k)$ is called an *incomplete elliptic integral*, while $K(k)$ is called a *complete elliptic integral*, of the first kind.

The derivatives of $sn\zeta$, $cn\zeta$, $dn\zeta$ are

$$\frac{d}{d\zeta}sn\zeta = cn\zeta dn\zeta, \quad \frac{d}{d\zeta}cn\zeta = -sn\zeta dn\zeta, \quad \frac{d}{d\zeta}dn\zeta = -k^2 sn\zeta cn\zeta. \tag{5.9.25}$$

The function $sn\zeta$ satisfies the equations

$$sn(0) = 0, \quad sn(K) = 1, \quad cn(0) = 1, \quad cn(K) = 0, \tag{5.9.26}$$

$$sn(\zeta + iK') = \frac{1}{ksn\zeta}, \quad cn(\zeta + iK') = -\frac{idn\zeta}{ksn\zeta}. \tag{5.9.27}$$

Now return to the *ellipsoidal coordinates* ρ, σ, τ in equation (5.2.1), (5.2.2), and put

$$\rho = ksn\alpha, \quad \sigma = ksn\beta, \quad \tau = ksn\gamma, \tag{5.9.28}$$

then

$$x = ak^2 sn\alpha sn\beta sn\gamma, \quad y = -ak^2k' cn\alpha cn\beta cn\gamma \quad z = ia(k')^{-1} dn\alpha dn\beta dn\gamma. \tag{5.9.29}$$

In α, β, γ coordinates, Laplace's equation reduces to

$$(sn^2\gamma - sn^2\beta)\frac{\partial^2\psi}{\partial\alpha^2} + (sn^2\alpha - sn^2\gamma)\frac{\partial^2\psi}{\partial\beta^2} + (sn^2\beta - sn^2\alpha)\frac{\partial^2\psi}{\partial\gamma^2} = 0, \tag{5.9.30}$$

which has *separable solutions* $Y(\alpha)Y(\beta)Y(\gamma)$, where $Y(\zeta)$ satisfies *Lamé's equation*

$$\frac{d^2Y}{d\zeta^2} + (h - \ell sn^2\zeta)Y = 0. \tag{5.9.31}$$

It is proved in the standard texts that if one solution of this equation is to be doubly periodic, like $sn\zeta$, then $\ell = k^2 n(n+1)$, where n is a non-negative integer, and h must take one of $2n + 1$ eigenvalues; in that case the second solutions of

(5.9.31) is not periodic at all. Thus, for a given value of $n = i$, there are $2i + 1$ doubly periodic solutions $\mathcal{E}_i^j(\zeta)$, $j = 1, 2, \ldots, 2i + 1$. The Lamé function $E_i^j(\rho)$ is defined as

$$E_i^j(\rho) = \mathcal{E}_i^j(\zeta), \quad \rho = ksn\zeta. \tag{5.9.32}$$

[Sometimes $\mathcal{E}$ is replaced by E, but this can cause confusion.]

If $\mathcal{E}_i^j(\zeta)$ and $\mathcal{F}_i^j(\zeta)$ are two solutions of (5.9.31), then

$$\mathcal{E}_i^j(\zeta)\mathcal{F}_i^{j''}(\zeta) - \mathcal{E}_i^{j''}(\zeta)\mathcal{F}_i^j(\zeta) = 0,$$

where $' = \frac{\partial}{\partial\zeta}$. Thus,

$$\mathcal{E}_i^j(\zeta)\mathcal{F}_i^{j'}(\zeta) - \mathcal{E}_i^{j'}(\zeta)\mathcal{F}_i^j(\zeta) = H_i^j, \tag{5.9.33}$$

a constant. Write $\mathcal{F}_i^j(\zeta) = \mathcal{E}_i^j(\zeta)\mathcal{Z}_i^j(\zeta)$, then

$$(\mathcal{E}_i^j(\zeta))^2\mathcal{Z}_i^{j'}(\zeta) = H_i^j$$

so that

$$\mathcal{Z}_i^j(\zeta) = H_i^j \int_{\zeta_0}^{\zeta} \frac{du}{[\mathcal{E}_i^j(u)]^2}.$$

We need to choose the datum ζ_0. The equations (5.9.27) show that $sn(iK') = \infty$, so that in ρ, σ, τ coordinates $\alpha = iK'$ corresponds to the point at infinity. If $F_i^j(\rho)E_i^j(\sigma)E_i^j(\tau)$ is to be harmonic outside S_κ then we must take $\zeta_0 = iK'$. Equations (5.9.27) give $sn(K + iK') = 1/k$, so that $\rho = 1$ corresponds to $\alpha = K + iK'$, and thus

$$\mathcal{F}_i^j(K + iK') = H_i^j \int_{iK'}^{K+iK'} \frac{du}{[\mathcal{E}_i^j(u)]^2}. \tag{5.9.34}$$

Now return to equation (5.9.22). The bracketed term is

$$P = E_i^j(\rho)F_i^{j'}(\rho) - E_i^{j'}(\rho)F_i^j(\rho)|_{\rho=\kappa}$$

where $' = \partial/\partial\rho$. But $\rho = ksn\alpha$ and

$$\frac{d}{d\alpha} = \frac{d}{d\rho} \cdot \frac{d\rho}{d\alpha} = kcn\alpha dn\alpha \frac{d}{d\rho}$$

or

$$\frac{d}{d\alpha} = k\left(1 - \frac{\rho^2}{k^2}\right)^{\frac{1}{2}} (1 - \rho^2)^{\frac{1}{2}} \frac{d}{d\rho}.$$

Thus

$$P = \frac{\mathcal{E}_i^j(\alpha)\mathcal{F}_i^{j\prime}(\alpha) - \mathcal{E}_i^{j\prime}\alpha)\mathcal{F}_i^j(\alpha)}{(k^2-\kappa^2)^{\frac{1}{2}}(1-\kappa^2)^{\frac{1}{2}}}$$

where now $' = \partial/\partial\zeta$; now use equation (5.9.33) to give

$$P = \frac{H_i^j}{(k^2-\kappa^2)^{\frac{1}{2}}(1-\kappa^2)^{\frac{1}{2}}} = \frac{H_i^j}{(\kappa^2-k^2)^{\frac{1}{2}}(\kappa^2-1)^{\frac{1}{2}}}.$$

Now substitute the expressions for P and D into (5.9.22), to give Q; the terms in $(\kappa^2-1)^{1/2}$ in P and D cancel and we can take the limit as $\kappa \to 1$; the result is that if the displacement under the punch is

$$w(x, y) = \sum_{i=0}^{n}\sum_{j=1}^{2i+1} B_{ij}E_i^j(\sigma)E_i^j(\tau) \tag{5.9.35}$$

then

$$\vartheta p(x, y) = \frac{1}{ak'}\cdot\left(1-\frac{x^2}{a^2}-\frac{y^2}{b^2}\right)^{-\frac{1}{2}}\sum_{i=0}^{n}\sum_{j=1}^{2i+1}\frac{B_{ij}}{c_{ij}}E_i^j(\sigma)E_i^j(\tau) \tag{5.9.36}$$

where

$$c_{ij} = [\mathcal{E}_i^j(K+iK')]^2\int_{iK'}^{K+iK'}\frac{du}{[\mathcal{E}_i^j(u)]^2}. \tag{5.9.37}$$

An alternative treatment of these problems may be found in Gladwell (1980, section 12.3), following Shail (1978). This is based on the fact that, in α, β, γ coordinates, $\gamma = K + iK'$ is the region inside the ellipse

$$\frac{x^2}{a^2}+\frac{y^2}{b^2} = 1,$$

described twice; and $\beta = K + iK'$ is the region of the x, y plane outside the ellipse.

The boundary conditions in α, β, γ coordinates are

$$\vartheta\psi(\alpha, \beta, K+iK') = f(\alpha, \beta)$$
$$\frac{\partial\psi}{\partial\beta}(\alpha, K+iK', \gamma) = 0.$$

If $f(\alpha, \beta) = \mathcal{E}_i^j(\alpha)\mathcal{E}_i^j(\beta)$ then

$$\vartheta\psi(\alpha, \beta, \gamma) = \mathcal{E}_i^j(\alpha)\mathcal{E}_i^j(\beta)\mathcal{F}_i^j(\gamma)/\mathcal{F}_i^j(K+iK').$$

5.10 Forces and Moments on an Elliptical Punch

Galin obtains the forces and moments using a long and complicated analysis which involves the orthogonality properties of Lamé functions, and the asymptotic form of ψ for large distance from the origin. We simplify the analysis by using *Betti's Reciprocal Theorem*. To use this, we first consider special cases of the analysis in Section 5.9.

For a flat-ended punch, $n = 0$; there is only one term in the expansion; $\mathcal{E}_0^1(\zeta) = 1$, so that

$$c_{01} = \int_{iK'}^{K+iK'} du = K.$$

This means that if $w(x, y) = d_0$, then

$$p(x, y) = \frac{1}{\vartheta b} \cdot \frac{d_0}{K} \left(1 - \frac{x^2}{a^2} - \frac{y^2}{b^2}\right)^{-\frac{1}{2}} \tag{5.10.1}$$

and the total load on the punch is

$$P = \frac{1}{\vartheta b} \cdot \frac{d_0}{K} \cdot 2\pi ab = \frac{2\pi a}{\vartheta K} \cdot d_0, \tag{5.10.2}$$

which agrees with known results.

For a punch with slanted flat face,

$$w = d_0 + d_1 x + d_2 y. \tag{5.10.3}$$

The relevant Lamé functions are

$$\mathcal{E}_1^1(\zeta) = sn\zeta, \quad \mathcal{E}_1^2(\zeta) = cn\zeta.$$

For the first, $sn(K + iK') = 1/k$ and, from equation (5.9.27),

$$I_{11} = \int_{iK''}^{K+iK'} \frac{du}{(snu)^2} = \int_0^K (ksnv)^2 dv.$$

Put $snv = \sin\phi$, then $cnvdnvdv = \cos\phi d\phi$, the limits become 0 and $\pi/2$, and

$$I_{11} = \int_0^{\frac{\pi}{2}} \frac{k^2 \sin^2\phi d\phi}{(1 - k^2 \sin^2\phi)^{\frac{1}{2}}} = K - E$$

where

$$E = E(k) = \int_0^{\frac{\pi}{2}} (1 - k^2 \sin^2\phi)^{\frac{1}{2}} d\phi$$

is the *complete elliptic integral of the second kind*. Thus

$$c_{11} = \frac{K - E}{k^2} = D(k).$$

For the second term we have, from equation (5.9.27),

$$cn(K + iK') = -\frac{idnK}{snK} = -\frac{ik'}{k}$$

and

$$I_{12} = \int_{iK''}^{K+iK'} \frac{du}{(cnu)^2} = -\int_0^K \frac{(ksnv)^2}{(dnv)^2} dv.$$

Put $snv = \sin\phi$, then

$$I_{12} = -\int_0^{\frac{\pi}{2}} \frac{k^2 \sin^2 \phi d\phi}{(1 - k^2 \sin^2 \phi)^{\frac{3}{2}}} = -\frac{[E - k'^2 K]}{k'^2}$$

so that

$$c_{12} = \frac{E - k'^2 K}{k^2} = B(k).$$

Thus

$$p(x, y) = \frac{1}{\vartheta b}\left(1 - \frac{x^2}{a^2} - \frac{y^2}{b^2}\right)^{-\frac{1}{2}} \left\{\frac{d_0}{K(k)} + \frac{d_1 x}{D(k)} + \frac{d_2 y}{B(k)}\right\}. \tag{5.10.4}$$

The total load P is given by equation (5.10.2), and the moments are

$$M_1 = \int_S xp(x, y)dS = \frac{2\pi a^3 d_1}{3\vartheta D(k)}, \tag{5.10.5}$$

$$M_2 = \int yp(x, y)dS = \frac{2\pi ab^2 d_2}{3\vartheta B(k)}. \tag{5.10.6}$$

Now we consider two configurations for the punch: the system with displacement $w(x, y)$ given by (5.10.3, and the corresponding pressure distribution (5.10.4); and a starred system with displacement $w^*(x, y) = f^*(x, y)$ and corresponding pressure distribution $p^*(x, y)$. Betti's Reciprocal Theorem states that the work done by $p(x, y)$ on the displacement $w^*(x, y)$ is equal to that done by $p^*(x, y)$ on the displacement $w(x, y)$:

$$\int_S p(x, y)w^*(x, y)dS = \int_S p^*(x, y)w(x, y)dS.$$

Thus

$$\frac{1}{\vartheta b}\int_S \left(1-\frac{x^2}{a^2}-\frac{y^2}{b^2}\right)^{-\frac{1}{2}} \left\{\frac{d_0}{K(k)}+\frac{d_1 x}{D(k)}+\frac{d_2 y}{B(k)}\right\} f^*(x,y)dS$$
$$= \int_S p^*(x,y)\{d_0+d_1x+d_2y\}dS.$$

Since d_0, d_1, d_2 are arbitrary, we can equate corresponding terms, to give the total load and moments of the starred system as

$$P^* = \frac{1}{\vartheta b}K(k)\int_S \left(1-\frac{x^2}{a^2}-\frac{y^2}{b^2}\right)^{-\frac{1}{2}} f^*(x,y)dS \tag{5.10.7}$$

$$M_1^* = \frac{1}{\vartheta b}D(k)\int_S \left(1-\frac{x^2}{a^2}-\frac{y^2}{b^2}\right)^{-\frac{1}{2}} xf^*(x,y)dS \tag{5.10.8}$$

$$M_2^* = \frac{1}{\vartheta b}B(k)\int_S \left(1-\frac{x^2}{a^2}-\frac{y^2}{b^2}\right)^{-\frac{1}{2}} yf^*(x,y)dS \tag{5.10.9}$$

5.11 A Punch of Arbitrary Cross-Section

In this section, we find upper and lower bounds for the ratio of the total load P applied to a punch of arbitrary cross-section, and the displacement w, assumed to be constant over the punch: $w = d$ on Ω.

As shown in equation (5.1.23), the harmonic function $\psi(x,y,z)$ takes the value d/ϑ on both sides of the slit in space in the form of the region Ω; also $\psi(x,y,z)$ tends to zero at infinity. The *Maximum Principle* states that a function harmonic in a region $V \in \mathbb{R}^3$, takes its maximum on the boundary of that region. Hence, at points outside Ω, $\psi(x,y,z) < d/\vartheta$. Circumscribe an ellipse $\mathcal{E}$ around Ω, so that the ellipse is divided into two parts, Ω and $\mathcal{E}/\Omega$, as shown in Figure 5.11.1. Then, in the notation of equation (5.10.7),

$$f^*(x,y) = d \text{ on } \Omega, \quad f^*(x,y) < d \text{ on } \mathcal{E}/\Omega.$$

Equation (5.10.7) gives

$$P = \frac{1}{\vartheta b K(k)}\int_{\mathcal{E}} \left(1-\frac{x^2}{a^2}-\frac{y^2}{b^2}\right)^{-\frac{1}{2}} f^*(x,y)dxdy$$

so that

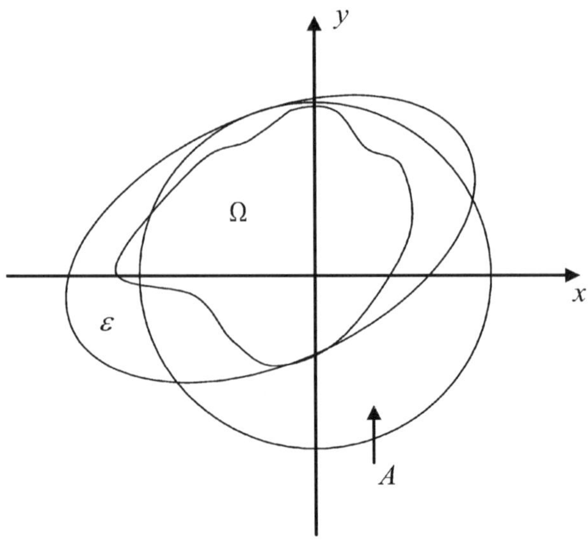

Fig. 5.11.1 The ellipse ε contains Ω; the circular disc A has the same area as Ω.

$$P < \frac{d}{\vartheta b K(k)} \int_{\mathcal{E}} \left(1 - \frac{x^2}{a^2} - \frac{y^2}{b^2}\right)^{-\frac{1}{2}} dS,$$
$$< \frac{2\pi abd}{\vartheta b K(k)} = \frac{2\pi ad}{\vartheta K(k)}. \tag{5.11.1}$$

To obtain a lower bound for P, Galin employs an analogy between the contact problem and an electrical problem, to deduce that a punch with a plane face and arbitrary cross-section has, for a given force, a smaller displacement than a punch with a circular cross-section of the same area. The lower bound for the electrical problem was obtained by Pólya and Szegö (1945), as described in their monograph Pólya and Szegö (1951). Equation (5.3.27) gives

$$P > \frac{4rd}{\vartheta} = 4\sqrt{\frac{A}{\pi} \cdot \frac{d}{\vartheta}}, \tag{5.11.2}$$

where r is the radius of the circle of area A.

This is Galin's analysis. Let us check it by applying the inequalities (5.11.2), to an ellipse. The area of an ellipse is πab, so that Pólya and Szegö's inequality gives

$$\frac{2\pi ad}{\vartheta K(k)} > 4\sqrt{ab \cdot \frac{d}{\vartheta}}$$

and thus

$$K(k) < \frac{\pi}{2}\sqrt{\frac{a}{b}} = \frac{\pi}{2}(1-k^2)^{-\frac{1}{4}}.$$

This is correct: Abramowitz and Stegun (1972, 17.3.11) gives (with $m = k^2$)

$$K(k) = \frac{\pi}{2}\left\{1 + \left(\frac{1}{2}\right)^2 k^2 + \left(\frac{1.3}{2.4}\right)^2 k^4 + \cdots\right\}$$

while

$$\frac{\pi}{2}\left(1-k^2\right)^{-\frac{1}{4}} = \frac{\pi}{2}\left\{1 + \frac{1}{4}k^2 + \frac{1.5}{4.4}\cdot\frac{k^4}{2!} + \cdots\right.$$

and each term in the first series is less than or equal to that in the second.

Now apply both inequalities (5.11.1), (5.11.2) to a rectangular punch of sides $2a$, $2b$, as shown in Figure 5.11.2. The area is $A = 4ab$, so that (5.11.2) gives

$$P > 4\sqrt{\frac{4ab}{\pi}}\cdot\frac{d}{\vartheta}. \tag{5.11.3}$$

Now we must find the circumscribing ellipse. Let its major and minor semi-axes be a', b'.

To find the ellipse of minimum area around the rectangle, we must minimise $\pi a'b'$ subject to $(\pm a, \pm b)$ lying on the ellipse; we must make

$$f(a', b') = \pi a'b' + \lambda\left(\frac{a^2}{a'^2} + \frac{b^2}{b'^2} - 1\right)$$

stationary. We have

$$\frac{\partial f}{\partial a'} = \pi b' - \frac{2\lambda a^2}{a'^3} = 0, \qquad \frac{\partial f}{\partial b'} = \pi a' - \frac{2\lambda b^2}{b'^3} = 0,$$

giving

$$\frac{a^2}{a'^2} = \frac{b^2}{b'^2} = \frac{1}{2}.$$

Putting $b/a = \varepsilon$, we find $b'/a' = \varepsilon = (1-k^2)^{1/2}$, and

$$4\sqrt{\frac{4ab}{\pi}} < \frac{\vartheta P}{d} < \frac{2\pi a'}{K(k)} = \frac{2\pi\sqrt{2}a}{K(k)}.$$

Define

$$\gamma = \frac{\vartheta P}{4rd}, \tag{5.11.4}$$

where r is the radius of the circle of area A, i.e., $r = \sqrt{\frac{A}{\pi}}$, then

$$1 < \gamma < \sqrt{\frac{\pi}{2}} \cdot \frac{\pi}{2(1-k^2)^{\frac{1}{4}} K(k)}. \qquad (5.11.5)$$

For a square, $k = 0$ and

$$1 < \gamma < \sqrt{\frac{\pi}{2}} = 1.253.$$

For any other rectangle, the spread between the bounds is greater, since the second factor in the inequaltiy (5.11.5) is always greater than unity.

5.12 The Pressure of a Wedge-Shaped Punch with Plane Base

Galin's purported solution to this problem is incomplete: it corresponds to the case in which there is an additional load acting on the surface of the elastic body, along a straight line which is the continuation of the mid line of the wedge. For this reason, we have decided that presenting Galin's analysis can serve no useful purpose. The problem has since been solved; the crucial point in the solution is the determination of the singularity in the pressure at the top of the wedge, see Keer and Parihar (1978).

5.13 A Slender Beam on an Elastic Body

In the theory of the bending of a beam on an elastic foundation it is often assumed that the pressure applied by the foundation is proportional to the sag, the beam displacement: the foundation is a so-called *Winkler foundation.*

In some cases, this assumption is not adequate, so that a series of investigations has been carried out, in which the foundation is regarded as a semi-infinite space, and the contact problem is solved approximately. However, this leads to complicated calculations, and therefore, it is useful to establish the limits of application of the assumption.

Here we establish the cases when the assumption approximates to the actual situation, and determine the coefficient of proportionality between the sag and the pressure.

We consider the problem of pressing a rigid beam of finite length into a semi-infinite elastic body. If the beam is elastic, and bends as a result of the action of the applied load, the proportionality of the sag and the applied pressure still holds. Indeed, it is of no consequence whether the beam has the given shape or has acquired it as the result of the action of applied forces.

We suppose that the beam is a slender rectangle in plan, of length $2a$ and width $2b$ as shown in Figure 5.11.2, so that $\varepsilon = b/a$ is small. We assume that the pressure distribution across the width is the same as in plane problems, so that if $p(x)$ is the force per unit length at section x, then

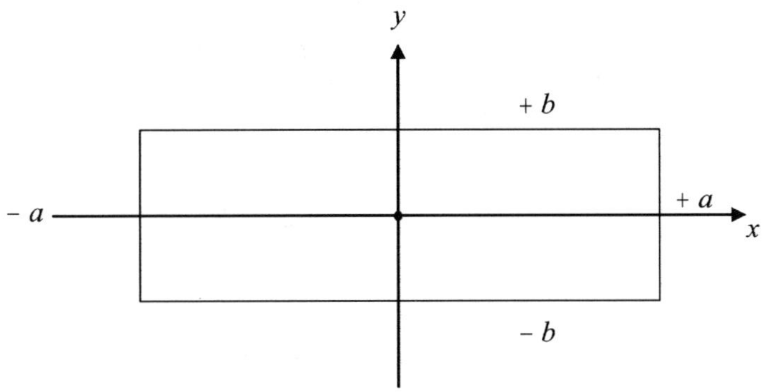

Fig. 5.13.1 The plan of the beam is a rectangle.

$$p(x,y) = \frac{p(x)}{\pi(b^2 - y^2)^{\frac{1}{2}}}, \tag{5.13.1}$$

so that

$$\int_{-b}^{b} p(x,y)dy = p(x). \tag{5.13.2}$$

The function $p(x)$ is found from the condition that the normal displacement of the semi-infinite elastic body, along the x-axis, resulting from this pressure, should be equal to a prescribed function $w(x)$, determined by the shape of the beam in the x, z plane. This assumption is similar to that used in the theory of wings of finite span.

Recalling the theory of Section 5.1, we find that the potential $\psi(x, y, z)$ satisfies

$$\vartheta\psi(x, 0, 0) = w(x) \text{ for } -a < x < a$$

$$\frac{\partial\psi}{\partial z}(x, y, 0) = -p(x, y) = \frac{-p(x)}{\pi(b^2 - y^2)^{\frac{1}{2}}}, \text{ for } -a < x < a, \ -b < y < b.$$

There is no pressure on the free surface, so that

$$\frac{\partial\psi}{\partial z}(x, y, 0) = 0 \text{ outside the rectangle.}$$

Equation (5.1.27) gives

$$w(x) = \frac{\vartheta}{2\pi}\int_{-a}^{a}\int_{-b}^{b} \frac{p(\xi)d\xi d\eta}{\pi(b^2 - \eta^2)^{\frac{1}{2}}\{(x-\xi)^2 + \eta^2\}^{\frac{1}{2}}}.$$

Carrying out the η integration, we find

$$w(x) = \frac{\vartheta}{\pi^2}\int_{-a}^{a} G(x - \xi)p(\xi)d\xi \tag{5.13.3}$$

where

$$G(s) = \frac{1}{2}\int_{-b}^{b} \frac{d\eta}{(b^2-\eta^2)^{\frac{1}{2}}(s^2+\eta^2)^{\frac{1}{2}}}.$$

Putting $\eta = b\sin\theta$ we find

$$G(s) = \int_0^{\frac{\pi}{2}} (s^2 + b^2\sin^2\theta)^{-\frac{1}{2}} d\theta \tag{5.13.4}$$

so that

$$\lim_{b\to 0} G(s) = \frac{\pi}{2}\cdot\frac{1}{|s|} \text{ for } s \neq 0. \tag{5.13.5}$$

Now construct the kernel

$$H_\varepsilon(s) = G(s)/\lambda, \tag{5.13.6}$$

where

$$\lambda = \int_{-a}^{a} G(s)ds = \frac{1}{2}\int_{-a}^{a}\int_{-b}^{b} \frac{d\eta ds}{(b^2-\eta^2)^{\frac{1}{2}}(s^2+\eta^2)^{\frac{1}{2}}}.$$

The s-integral in λ is

$$I = \frac{1}{2}\int_{-a}^{a} \frac{ds}{(s^2+\eta^2)^{\frac{1}{2}}}.$$

Putting $s = \eta\sinh\theta$, $a = \eta\sinh\alpha$, we find

$$I = \alpha = \alpha\sinh^{-1}\left(\frac{a}{\eta}\right),$$

so that

$$\lambda = 2\int_0^b \frac{\sinh^{-1}\left(\frac{a}{\eta}\right)d\eta}{(b^2-\eta^2)^{\frac{1}{2}}} = 2\int_0^1 \frac{\sinh^{-1}\left(\frac{1}{(\varepsilon v)}\right)dv}{(1-v^2)^{\frac{1}{2}}}.$$

Now

$$\begin{aligned}\sinh^{-1}x &= \ell n\{x + (1+x^2)^{\frac{1}{2}}\}\\ &= \ell n x + \ell n\{1 + (1+x^{-2})^{\frac{1}{2}}\}.\end{aligned}$$

But $(1+x^{-2})^{1/2} > 1$, so that $\sinh^{-1}x > \ell n x + \ell n 2$. On the other hand, if $x > 1$ then $(1+x^{-2})^{1/2} < \sqrt{2}$ so that $\sinh^{-1}x < \ell n x + \ell n(1+\sqrt{2})$. Thus, for $\varepsilon < 1$, we have

$$\ell n\left(\frac{1}{\varepsilon}\right) - \ell n v + \ell n 2 < \sinh^{-1}\left(\frac{1}{(\varepsilon v)}\right) < \ell n\left(\frac{1}{\varepsilon}\right) - \ell n v + \ell n(1+\sqrt{2}).$$

For $0 < v < 1$, $-\ell n v > 0$, so that we can neglect the $-\ell n v$ on the left of the inequality. On the other hand, Gradshteyn and Ryzhick (1965, 4.3228) gives

$$-2\int_0^1 \frac{\ell n dv}{(1-v^2)^{\frac{1}{2}}} = -2\int_0^{\frac{\pi}{2}} \ell n(\sin\theta)d\theta < \frac{\pi}{2}.$$

Thus, since

$$2\int_0^1 \frac{dv}{(1-v^2)^{\frac{1}{2}}} = \pi,$$

we have

$$\pi\left\{\ell n\left(\frac{1}{\varepsilon}\right) + \ell n 2\right\} < \lambda < \left\{\ell n\left(\frac{1}{\varepsilon}\right) + \ell n(1+\sqrt{2}) + 1\right\}$$

so that, for small ε,

$$\lambda \sim \pi \ell n\,(1/\varepsilon)\,. \tag{5.13.6}$$

Putting equations (5.13.5), (5.13.6) together, we find that, for small ε,

$$H_\varepsilon(s) \sim \frac{1}{2|s|\ell n\,(1/\varepsilon)}. \tag{5.13.7}$$

Thus $\lim_{\varepsilon\to 0} H_\varepsilon(s)$ behaves like a δ-function:

$$\lim_{\varepsilon\to 0} H_\varepsilon(s) = 0 \text{ if } s \neq 0;\ \lim_{\varepsilon\to 0} H_\varepsilon(0) = \infty$$

and

$$\int_{-a}^{a} H_\varepsilon(s)ds = 1$$

and

$$\lim_{\varepsilon\to 0}\int_{-a}^{a} H_\varepsilon(x-\xi)p(\xi)d\xi = p(x).$$

This means that, for small ε, equation (5.13.3) becomes

$$w(x) = \frac{\vartheta}{\pi^2}\cdot\pi\ell n\,(1/\varepsilon)\cdot p(x).$$

There is proportionality, with

$$k = \frac{p(x)}{w(x)} = \frac{\pi}{\vartheta\ell n\,(1/\varepsilon)}. \tag{5.13.8}$$

We now try to find how fast the pressure on the beam can vary, and proportionality still hold.

Consider the integral in equation (5.13.3). Most of the integral is contributed from the immediate neighbourhood of $\xi = x$. Let

$$I(x_1, x_2) = \int_{x_1}^{x_2} G(x-\xi)p(\xi)d\xi.$$

We seek δ such that, as $\varepsilon \to 0$,

$$f(x,\delta) = \frac{I(-a, x-\delta) + I(x+\delta, a)}{I(x-\delta, x+\delta)} \tag{5.13.9}$$

tends to zero: the contribution from the neighbourhood of x dominates the remaining part.

Equation (5.13.4) shows that $G(s) < \frac{\pi}{2|s|}$. We assume that $p(x)$ is positive, with maximum value p_m, then

$$I(-a, x-\delta) + I(x+\delta, a) < \frac{\pi}{2} \cdot \frac{2}{\delta} \cdot p_m,$$

since the maximum value of $|x - \xi|$ for the integrals is δ. The denominatior of (5.13.9) is

$$I(x-\delta, x+\delta) = \int_{x-\delta}^{x+\delta} G(x-\xi)p(\xi)d\xi.$$

Both factors in the integrand are positive, so the integral form of the Mean Value Theorem gives

$$I(x-\delta, x+\delta) = p(x_0) \int_{-\delta}^{\delta} G(\xi)d\xi,$$

for some $x_0 \in (x-\delta, x+\delta)$. Evaluating the integral as we did for λ, we find

$$\int_{-\delta}^{\delta} G(\xi)d\xi = 2\int_{0}^{\delta} G(\xi)d\xi = 2\int_{0}^{1} \frac{\sinh^{-1}\left(\frac{\delta}{bv}\right)dv}{(1-v^2)^{\frac{1}{2}}}.$$

But $\sinh^{-1}\left(\frac{\delta}{bv}\right) > \ell n\left|\frac{\delta}{b}\right| + \ell n 2 > \ell n\left|\frac{\delta}{b}\right|$, so that

$$I(x-\delta, x+\delta) > \pi \ell n\left(\left|\frac{\delta}{b}\right|\right),$$

and

$$f(x,\delta) < \frac{\pi}{\delta} \cdot p_m \cdot \frac{1}{\pi\, p(x_0) \ell n\left(\left|\frac{\delta}{b}\right|\right)}. \tag{5.13.10}$$

We must now choose δ so that

$$\delta \to 0, \quad \delta \ell n\left(\left|\frac{\delta}{b}\right|\right) \to \infty \text{ as } b \to 0. \tag{5.13.11}$$

One such choice is

$$\delta = \frac{a}{\ell n \ell n\left(\frac{a}{b}\right)} = \frac{a}{\ell n \ell n\left(\frac{1}{\varepsilon}\right)}. \tag{5.13.12}$$

If a is kept constant and $b \to 0$, then $1/\varepsilon \to \infty$ and $\delta \to 0$. On the other hand

$$\ell n\left(\left|\frac{\delta}{b}\right|\right) = \ell n\left[\frac{\frac{1}{\varepsilon}}{\ell n \ell n\left(\frac{1}{\varepsilon}\right)}\right] = \ell n\left(\frac{1}{\varepsilon}\right) - \ell n \ell n \ell n\left(\frac{1}{\varepsilon}\right)$$

so that

$$\delta \ell n\left(\left|\frac{\delta}{b}\right|\right) = \frac{a}{\ell n \ell n\left(\frac{1}{\varepsilon}\right)}\left\{\ell n\left(\frac{1}{\varepsilon}\right) - \ell n \ell n \ell n\left(\frac{1}{\varepsilon}\right)\right\}.$$

Since $\frac{x}{\ell n x} \to \infty$ and $\frac{\ell n x}{x} \to 0$ as $x \to \infty$, we have

$$\delta \ell n\left(\left|\frac{\delta}{b}\right|\right) \to \infty.$$

We conclude that for approximate proportionality to hold, δ given by (5.13.12) should be small, and $p(x)$ should not alter appreciably in an interval of length 2δ.. Then

$$\int_{-a}^{a} H_{\varepsilon}(x-\xi)p(\xi)d\xi \to \int_{x-\delta}^{x+\delta} H_{\varepsilon}(x-\xi)p(\xi)d\xi \to p(x_0)\int_{-\delta}^{\delta} H_{\varepsilon}(\xi)d\xi \to p(x).$$

We now find the pressure distribution $p(x)$ corresponding to constant displacement over the whole of the rectangular contact region: $w(x) = d$. We do so by using Fourier transforms.

Let

$$p^*(\alpha) = \frac{1}{\sqrt{2\pi}}\int_{-\infty}^{\infty} p(x)\exp(i\alpha x)d\alpha, \qquad (5.13.13)$$

then

$$p(x) = \frac{1}{\sqrt{2\pi}}\int_{-\infty}^{\infty} p^*(\alpha)\exp(-i\alpha x)d\alpha. \qquad (5.13.14)$$

Similarly, let

$$G^*(\alpha) = \frac{1}{\sqrt{2\pi}}\int_{-\infty}^{\infty} G(x)\exp(i\alpha x)d\alpha, \qquad (5.13.15)$$

then

$$G(x) = \frac{1}{\sqrt{2\pi}}\int_{-\infty}^{\infty} G^*(\alpha)\exp(-i\alpha x)d\alpha. \qquad (5.13.16)$$

The *Convolution Theorem* for Fourier transforms states that

$$\int_{-\infty}^{\infty} G(x-\xi)p(\xi)d\xi = \int_{-\infty}^{\infty} G^*(\alpha)p^*(\alpha)\exp(-i\alpha x)d\alpha. \qquad (5.13.17)$$

Equation (5.13.15) gives

$$G^*(\alpha) = \frac{1}{\sqrt{2\pi}}\int_{-\infty}^{\infty}\exp(i\alpha s)\left\{\frac{1}{2}\int_{-b}^{b}\frac{d\eta}{(b^2-\eta^2)^{\frac{1}{2}}(s^2+\eta^2)^{\frac{1}{2}}}\right\}ds$$

and

$$I = \int_{-\infty}^{\infty} \frac{\exp(i\alpha s)}{(s^2 + \eta^2)^{\frac{1}{2}}} ds = 2\int_0^{\infty} \frac{\cos(\alpha s)}{(s^2 + \eta^2)^{\frac{1}{2}}} ds.$$

Put $s = \eta \sinh \theta$, then by Abramowitz and Stegun (1972, 9.6.21),

$$I = 2\int_0^{\infty} \cos(\alpha\eta \sinh\theta) d\theta = 2K_0(\alpha\eta)$$

where K_0 is the modified Bessel function of order zero. Now Gradshteyn and Ryzhick (1965, 6.6814), gives

$$G^*(\alpha) = \frac{1}{\sqrt{2\pi}} \int_{-b}^{b} \frac{K_0(\alpha\eta)}{(b^2 - \eta^2)^{\frac{1}{2}}} d\eta = \sqrt{\frac{2}{\pi}} \int_0^{\frac{\pi}{2}} K_0(\alpha b \sin\theta) d\theta$$

$$= \sqrt{\frac{2}{\pi}} \cdot \frac{\pi}{2} \cdot I_0\left(\alpha\frac{b}{2}\right) K_0\left(\alpha\frac{b}{2}\right) \tag{5.13.18}$$

where I_0 is the other modified Bessel function. Thus

$$w(x) = \frac{\vartheta}{\pi^2} \int_{-\infty}^{\infty} G^*(\alpha) p^*(\alpha) \exp(-i\alpha x) d\alpha$$

$$= \frac{2\vartheta}{\pi^2} \int_0^{\infty} G^*(\alpha) p^*(\alpha) \cos(\alpha x) d\alpha$$

$$= \sqrt{\frac{2}{\pi}} \cdot \frac{1}{\pi} \cdot \vartheta \int_0^{\infty} p^*(\alpha) I_0\left(\alpha\frac{b}{2}\right) K_0\left(\alpha\frac{b}{2}\right) \cos\alpha x d\alpha.$$

The pressure $p(x)$ and displacement $w(x)$ are both even functions of x, so that the boundary conditions can be confined to positive x:

$$w(x) = d, \text{ for } 0 \le x < a; \; p(x) = 0 \text{ for } x > a.$$

Introduce the dimensionless quantities

$$t = \frac{x}{a}, \quad \xi = a\alpha, \quad \varepsilon = \frac{b}{a} \tag{5.13.19}$$

and put $p^*(\alpha) = F(\xi)$, then

$$\int_0^{\infty} F(\xi) I_0\left(\varepsilon\frac{\xi}{2}\right) K_0\left(\varepsilon\frac{\xi}{2}\right) \cos t\xi d\xi = c, \quad 0 \le t < 1 \tag{5.13.20}$$

$$\int_0^{\infty} F(\xi) \cos t\xi d\xi = 0, \quad 1 < t < \infty, \tag{5.13.21}$$

where

$$c = \sqrt{\frac{2}{\pi}} \cdot \frac{\pi}{\vartheta} \cdot ad. \tag{5.13.22}$$

We seek the solution of (5.13.20), (5.13.21) in the form

$$F(\xi) = c \sum_{n=0}^{\infty} (-)^n A_{2n} J_{2n}(\xi) \tag{5.13.23}$$

and use Gradshteyn and Ryzhik (1965, 6.6712):

$$\int_0^{\infty} J_{2n}(\xi) \cos t\xi \, d\xi = \begin{cases} (-)^n (1-t^2)^{-\frac{1}{2}} T_{2n}(t) & 0 \le t < 1 \\ 0 & 1 < t < \infty \end{cases}$$

which means that $F(\xi)$ satisfies equation (5.13.21), and also

$$\cos t\xi = J_0(\xi) + 2 \sum_{m=0}^{\infty} (-)^m T_{2m}(t) J_{2m}(\xi), \quad 0 \le t < 1. \tag{5.13.24}$$

Substituting (5.13.23), (5.13.24) into (5.13.20), we find

$$\sum_{n=0}^{\infty} A_{2n} C_{0n} + 2 \sum_{n=0}^{\infty} A_{2n} \sum_{m=0}^{\infty} (-)^m C_{mn} T_{2m}(t) = 1, \quad 0 < t < 1, \tag{5.13.25}$$

where

$$C_{mn} = (-)^n \int_0^{\infty} J_{2m}(\xi) J_{2n}(\xi) I_0\left(\varepsilon \frac{\xi}{2}\right) K_0\left(\varepsilon \frac{\xi}{2}\right) d\xi.$$

Thus we have the equations

$$\sum_{n=0}^{\infty} C_{mn} A_{2n} = \begin{cases} 1 & m = 0 \\ 0 & m = 1, 2, \ldots \end{cases}$$

Once we have solved these equations, the Fourier transform $p^*(\alpha) = F(\xi)$ is given by equation (5.13.23), so that

$$p(x) = \frac{c}{a} \sqrt{\frac{2}{\pi}} \sum_{n=0}^{\infty} (-)^n A_{2n} \int_0^{\infty} J_{2n}(\xi) \cos t\xi \, d\xi$$

$$= \frac{c}{a} \sqrt{\frac{2}{\pi}} \sum_{n=0}^{\infty} (1-t^2)^{-\frac{1}{2}} A_{2n} T_{2n}(t).$$

In particular, the total force exerted by the beam on the half-space is

$$P = \int_{-a}^{a} p(x) dx = \frac{c}{a} \cdot \sqrt{\frac{2}{\pi}} A_0 \cdot \pi a$$

Table 5.13.1 The dimensionless quantities γ and Γ are given in equations (5.11.4), (5.11.5).

ε	0.02	0.05	0.10	0.15	0.2
A_0	0.133	0.161	0.194	0.220	0.243
γ	2.05	1.58	1.34	1.24	1.19
Γ	3.68	2.53	2.22	2.17	2.06

or, with (5.13.22),

$$P = \frac{\pi^2 ad}{\vartheta} A_0.$$

The dimensionless quantity γ is thus

$$\gamma = \sqrt{\frac{\pi}{4ab}} \cdot \frac{\pi^2}{4} a A_0 = \frac{\pi^2}{8} \sqrt{\frac{\pi}{\varepsilon}} A_0. \tag{5.13.27}$$

Galin presents computed values of A_{2n} for $n = 0, \ldots, 10$, but the tabulated values display anomolies. We present just his tabulated values of A_0, the corresponding values of the dimensionless quantity γ defined in (5.13.27), and the upper bound Γ given in equation (5.11.5).

This table shows that γ is nearer the lower bound, 1, given by Pólya and Szegö, than it is to the upper bound Γ provided by the circumscribing ellipse.

5.14 Contact of Two Elastic Bodies

Until now, we have considered contact problems on the assumption that the body exerting the pressure is perfectly rigid. In this section, we show that when both of the bodies in contact are elastic, the problem is reduced to the solution of a mixed problem in potential theory of the same type as we have considered above.

We shall suppose tht the radii of curvature of both bodies are large compared to the linear dimensions of the contact region. We therefore replace each of the bodies by a semi-infinite elastic body. We suppose that in the unstressed state, the bodies are in contact at O, and that, near O, the bodies are geometrically smooth. Taking the common tangent plane at O as the x, y plane, we may express the shapes of the two bodies near O in the form

$$z_1 = f_1(x, y), \quad z_2 = f_2(x, y)$$

where z_1, z_2 are directed into the bodies, as in Figure 5.14.1.

When the bodies are pressed together, there will be a displacement of both. Take the displacements of body 1 to be (u_1, v_1, w_1) relative to $O(x, y, z_1)$, and that of body 2 to be (u_2, v_2, w_2) relative to $O(x, y, z_2)$. Within the area of contact, therefore,

$$z_1 + w_1 = -(z_2 + w_2) + \delta$$

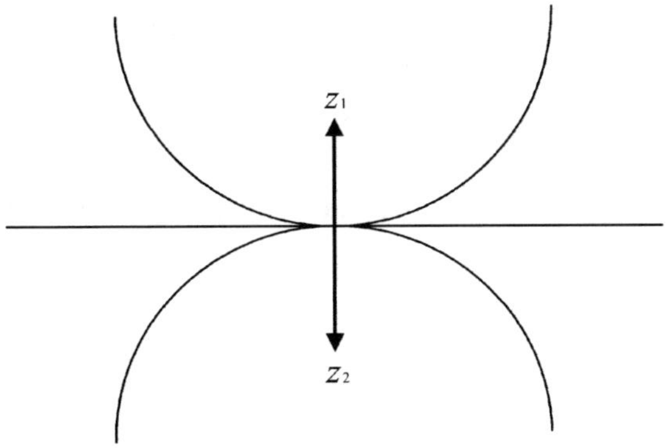

Fig. 5.14.1 Two bodies are initially in contact at 0.

where δ is the approach of the two bodies. Therefore, within the contact region S,

$$w_1 + w_2 = \delta - f_1(x, y) - f_2(x, y).$$

If $p(x, y)$ is the contact pressure, then (5.1.23), (5.1.27) give

$$w_1 = \frac{\vartheta_1}{2\pi} \int_S \frac{p dS}{R}, \quad w_2 = \frac{\vartheta_2}{2\pi} \int_S \frac{p dS}{R}$$

where $R^2 = (x - \xi)^2 + (y - \eta)^2$, and ϑ_i is the elastic constant ϑ for body i. Thus

$$\frac{\vartheta_1 + \vartheta_2}{2\pi} \int_S \frac{p dS}{R} = \delta - f_1(x, y) - f_2(x, y), \quad (x, y) \in S$$

which has precisely the same form as equation (5.1.32).

This means that the analysis developed in this chapter for contact problems involving a rigid punch indenting a half-space may be used for problems relating to the contact of two elastic bodies

Note: Galin, in his equation (14.4) has $\vartheta_1 - \vartheta_2$ not $\vartheta_1 + \vartheta_2$.

5.15 Contact between a Rigid Punch and a Clamped Plate

In the preceding sections of this chapter, we assumed that the elastic body being deformed is of considerable depth and so can be replaced by an elastic half-space. In this section, we consider another extreme case in which the elastic body is a thin clamped plate.

Suppose that the rigid body presses on a thin clamped circular plate of radius a, so that the initial contact occurs at the centre of the plate. Just as in Hertz's theory of contact, we shall retain only the principal terms in the equation of the surface bounding the rigid body: it has the form

$$z = Ax^2 + By^2. \tag{5.15.1}$$

The body is acted on by a force P, directed along the axis of symmetry.

We make the following assumptions:

1. The thickness of the plate is small compared to the dimensions of the contact region.
2. The dimensions of the contact region are small compared with the radius of the plate.
3. The dimensions of the contact region are small compared with the radii of curvature of the punch.

The problem is essentially one of plate theory: determine the normal displacement of the plate, knowing that its shape over the (yet to be determined) contact region is given by (5.15.1).

In thin plate theory, as given for example in Love (1927, section 313), the equation governing the displacement of a plate is the biharmonic equation

$$D\Delta^2 w = p(x, y), \tag{5.15.2}$$

where h is the thickness of the plate, and

$$D = 4h^3/3\vartheta, \quad \vartheta = (1 - \nu)/\mu. \tag{5.15.3}$$

We note on anomoly in the theory. If the punch has the parabolic form given by (5.15.1), and if the punch is in contact with the plate over a region S, then the displacement of the plate has the form

$$w(x, y) = w_1(x, y) = d - Ax^2 - By^2, \quad (x, y) \in S. \tag{5.15.4}$$

But for this $w(x, y)$, $\Delta^2 w \equiv 0$: there is *no* distributed load under the punch! This means that the punch must be supported by a concentrated ring of forces around the boundary, L, of the contact region. The thin plate theory neglects the distributed load as being small compared to this ring load; the ring load has resultant P, balancing the load applied by the punch.

Saint-Venant's principle states that far from the contact region (and note assumption 2), the displacement of the plate will be that due to a concentrated load P applied at the centre of the plate. Love (1927, section 314, (iv)) gives

$$w(x) = w_2(x, y) = C\left\{r^2 \ell n\left(\frac{r^2}{a^2}\right) + a^2 - r^2\right\}, \tag{5.15.5}$$

where $r^2 = x^2 + y^2$, $C = P/(16\pi D)$; this is the form of $w(x, y)$ for large r.

Outside the contact region, there is no applied pressure, so that

$$\Delta^2 w = 0. \tag{5.15.6}$$

Across the boundary L the displacement and its first and second derivatives will be continuous. Thus if $w_3 = w_2 - w_1$ then

$$w_3 = 0, \quad \frac{\partial w_3}{\partial x} = 0 = \frac{\partial w_3}{\partial y}, \quad \frac{\partial^2 w_3}{\partial x^2} = 0 = \frac{\partial^2 w_3}{\partial y^2} = \frac{\partial^2 w_3}{\partial x \partial y} \text{ on } L.$$

There will be a discontinuity in the third derivatives, yielding the required ring load on L. We note that if the second derivatives are identically zero on L, then w_3 must be linear in x, y, and by symmetry, it must be constant.

To find the contour L, we extend the functions w_1 and w_2 to the whole plane. For large r, w is given by (5.15.5) so that

$$\begin{aligned} w_3 &= w_2 - w_1 \\ &= C\left\{r^2 \ell n\left(\frac{r^2}{a^2}\right) + a^2 - r^2\right\} - (d - Ax^2 - By^2) \\ &= Cr^2 \ell n\left(\frac{r^2}{a^2}\right) - (d' - A'x^2 - B'y^2), \end{aligned} \tag{5.15.7}$$

where

$$d' = d - Ca^2, \quad A' = A - C, \quad B' = B - C. \tag{5.15.8}$$

Put $w_4 = w_3 + d'$, then for large r,

$$w_4 = Cr^2 \ell n\left(\frac{r^2}{a^2}\right) + A'x^2 + B'y^2. \tag{5.15.9}$$

We now seek a function $F(x, y)$, biharmonic, with the form (5.15.9) for large r, and satisfying the equations

$$\frac{\partial^2 F}{\partial x^2} = 0 = \frac{\partial^2 F}{\partial y^2} = \frac{\partial^2 F}{\partial x \partial y} \tag{5.15.10}$$

on the contour L. If F is biharmonic, then it may be written

$$F(x, y) = \text{Re}[\bar{z}\phi^*(z) + \psi^*(z)]$$

where ϕ^*, ψ^* are functions of $z = x + iy$, and then

$$\mathcal{L}F(x, y) \equiv \frac{\partial^2 F}{\partial x^2} + \frac{\partial^2 F}{\partial y^2} = 4\,\text{Re}[\Phi^*(z)], \tag{5.15.11}$$

$$\mathcal{M}F(x, y) \equiv \frac{\partial^2 F}{\partial x^2} - \frac{\partial^2 F}{\partial y^2} - 2i\frac{\partial^2 F}{\partial x \partial y} = 2[\bar{z}\Phi^{*\prime}(z) + \Psi^*(z)], \tag{5.15.12}$$

where

$$\Phi^*(z) = \phi^{*\prime}(z), \quad \Psi^*(z) = \psi^{*\prime\prime}(z).$$

For large r, $F(x, y)$ has the form (5.15.9), so that for large r,

$$\mathcal{L}F(x, y) = 8C\left\{\ell n\left(\frac{r}{a}\right) + 1\right\} + 2A' + 2B', \tag{5.15.13}$$

$$\mathcal{M}F(x, y) = 4C\exp(-2i\theta) + 2A' - 2B'. \tag{5.15.14}$$

On the boundary L, we must have

$$\mathcal{L}F(x, y) = 0 = \mathcal{M}F(x, y). \tag{5.15.15}$$

We now transform the region outside L to the exterior of the unit circle Λ in the ζ-plane, i.e., to $|\zeta| > 1$, in such a way that the point at infinity in the z-plane is mapped out the point at infinity in the ζ-plane. The mapping function $\omega(\zeta)$ can be written in the form

$$z = \omega(\zeta) = c\zeta + g(\zeta). \tag{5.15.16}$$

Here $g(\zeta)$ is a function, holomorphic outside Λ, and such that $g(\zeta) \to 0$ as $|\zeta| \to \infty$.

If we write

$$\Phi(\zeta) = \Phi^*[\omega(\zeta)], \quad \Psi(\zeta) = \Psi^*[\omega(\zeta)]$$

then from equations (5.15.11), (5.15.12) we get

$$\mathcal{L}F(x, y) = 4\,\mathrm{Re}[\Phi(\zeta)]$$

$$\mathcal{M}F(x, y) = 2\left\{\frac{\bar{\omega}(\bar{\zeta})}{\omega'(\zeta)}\Phi'(\zeta) + \Psi(\zeta)\right\}$$

so that equation (5.15.15) gives

$$\mathrm{Re}[\Phi(\zeta)] = 0 = \frac{\bar{\omega}(\bar{\zeta})}{\omega'(\zeta)}\Phi'(\zeta) + \Psi(\zeta) \text{ for } |\zeta| = 1 \tag{5.15.17}$$

while equations (5.15.13), (5.15.14) give, for large $|\zeta|$,

$$4\,\mathrm{Re}[\Phi(\zeta)] = 8C\{\ell n|\zeta| + \ell nc - \ell na + 1\} + 2A' + 2B', \tag{5.15.18}$$

$$\frac{\bar{\omega}(\bar{\zeta})}{\omega'(\zeta)}\Phi'(\zeta) + \Psi(\zeta) = 2C\exp(-2i\theta') + A' - B', \tag{5.15.19}$$

where $\zeta = |\zeta|\exp(i\theta')$.

Equation (5.15.18) gives

$$\Phi(\zeta) = 2C\ell n\zeta + h(\zeta),$$

where $h(\zeta)$ is holomorphic outside Λ. On Λ, the circle $|\zeta| = 1$, equation (5.15.17a) gives

$$\mathrm{Re}[\Phi(\zeta)] = \mathrm{Re}(h(\zeta)) = 0.$$

Thus, since $h(\zeta)$ is bounded everywhere outside Λ, including the point at infinity, and has real part zero on Λ, it is identically zero. Thus

$$\Phi(\zeta) = 2C\ell n\zeta,$$

$$8C(\ell nc - \ell na + 1) + 2A' + 2B' = 0 \tag{5.15.21}$$

giving

$$\frac{c}{a} = \exp\{-1 - (A' + B')/4C\} = \exp\left\{-\frac{1}{2} - (A+B)/4C\right\}. \tag{5.15.21}$$

Now consider equation (5.15.19), again for large $|\zeta|$. For large $|\zeta|$, $\omega(\zeta) = c\zeta$, $\omega'(\zeta) = c$, and since $\Phi'(\zeta) = 2C/\zeta$, we have

$$\bar{\zeta} \cdot (2C/\zeta) + \Psi(\zeta) = 2C\exp(-2i\theta) + A' - B'$$

since $\theta' = \theta$ for large $|\zeta|$. Thus

$$\Psi(\zeta) \to A' - B' \text{ for large } |\zeta|. \tag{5.15.22}$$

Finally, we use $\mathcal{M}F(x, y) = 0$ on $|\zeta| = 1$. This gives

$$\Psi(\zeta) = -\frac{\bar{\omega}(\bar{\zeta})}{\omega'(\zeta)} \cdot \frac{2C}{\zeta} \text{ on } |\zeta| = 1.$$

The function $\omega(\zeta)$ is given by equation (5.15.16), where

$$g(\zeta) = \sum_{n=1}^{\infty} a_n \zeta^{-n}.$$

On Λ, $\bar{\zeta} = 1/\zeta$, so that

$$\bar{\omega}(\bar{\zeta}) = c\zeta^{-1} + \sum_{n=1}^{\infty} a_n \zeta^n$$

and equation (5.15.17b) becomes

$$2C\left\{c\zeta^{-1} + \sum_{n=1}^{\infty} a_n \zeta^n\right\} = \zeta\omega'(\zeta)\Psi(\zeta) \text{ on } |\zeta| = 1.$$

Multiply both sides of this with ζ^{-m} and integrate along the contour Λ:

$$2C\int_{\Lambda}\left(c\zeta^{-1} + \sum_{n=1}^{\infty} a_n \zeta^n\right)\zeta^{-m} d\zeta = \int_{\Lambda}\{\zeta\omega'(\zeta)\Psi(\zeta)\}\zeta^{-m} d\zeta.$$

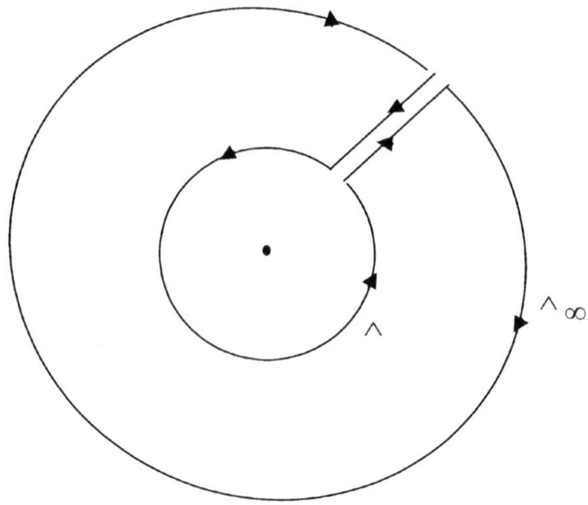

Fig. 5.15.1 The contour $\wedge + \wedge_\infty$.

The function $\zeta\omega'(\zeta)\Psi(\zeta)$ is holomorphic outside the unit circle, except at the point at infinity where it has a simple pole. Therefore, the integral on the right is zero for all $m > 1$, and so is the integral on the left. Therefore, $a_n = 0$ when $n > 1$. To find a_1, we take $m = 1$:

$$2C \int_\Lambda a_1 d\zeta = \int_\Lambda \omega'(\zeta)\Psi(\zeta)d\zeta.$$

But $\omega'(\zeta)\Psi(\zeta)$ is holomorphic outside Λ, so that, by applying Cauchy's Theorem to the contour $\Lambda + \Lambda_\infty$, as in Figure 5.15.1, we find

$$\int_\Lambda \omega'(\zeta)\Psi(\zeta)d\zeta + \int_{\Lambda_\infty} \omega'(\zeta)\Psi(\zeta)d\zeta = 0.$$

Thus, using (5.15.16) and (5.15.22), we find

$$\int_\Lambda \omega'(\zeta)\Psi(\zeta)d\zeta = -\int_{\Lambda_\infty} c(A' - B')d\zeta = 2\pi i c(A' - B')$$

and

$$a_1 = c(A' - B')/2C = c(A - B)/2C$$

so that

$$\omega(\zeta) = c\zeta + a_1\zeta^{-1} \tag{5.15.23}$$

and the contact region is an ellipse with semi-axes $a' = c + a_1$ and $b' = c - a_1$, where c is given by equation (5.15.21).

It is generally recognised that the thin plate theory is valid if the displacements are everywhere small compared to the linear dimensions. For the clamped plate subject to the concentrated load P at the centre, the displacement is given by (5.15.5) so that the maximum displacement is that at the centre, namely Ca^2, so that we require

$$Ca^2 \ll 1.$$

Chapter 6
Viscoelasticity, Wear and Roughness

6.1 Introduction

In this chapter, we consider some contact problems for viscoelastic bodies, in particular, the rolling contact of a viscoelastic cylinder over a viscoelastic base. The problem of rolling contact is formulated in a general form and solved analytically. The friction effects include the retardation of the material in the stick zones and partial slip in the slip zones in the contact region.

In contact problems, there can be relative slip between the contacting bodies. This slip gives rise to wear, and this in turn leads to a redistribution of the contact pressure. We will consider the effects of wear.

Most bodies in contact are not ideally smooth; they have rough surfaces which must be modelled as a surface layer with non-linear behaviour. For such bodies, the contact problem can be reduced to the solution of non-linear integral equations. These results are very important for the analysis of contact compliances.

The results described in this chapter are based on the following papers: Section 6.3 – Galin (1967); Section 6.4 – Mitkevich (1970); Section 6.5 – Goryacheva(1973a); Section 6.6 – Goryacheva (1973b); Section 6.7 – Galin (1976); Section 6.8 – Goryacheva (1977); Sections 6.9 and 6.10 – Goryacheva (1978).

6.2 The Laplace Transform

In this section, we list without proofs, some of the basic properties of the Laplace transform that we will be using in subsequent analysis.

Suppose $f(x)$, defined on $0 \leq x < \infty$, is such that

$$\int_0^\infty \exp(-cx)|f(x)|dx$$

is bounded for some $c > 0$. We define

$$F(s) = \int_0^\infty \exp(-sx) f(x) dx; \qquad (6.2.1)$$

sometimes the notation used is $\tilde{f}(s)$, rather than $F(s)$.

If $F(s)$ is an analytic function of the complex variable s and is of order $0(s^{-k})$ in some half-plane $\mathrm{Re}(s) \geq c$, where c and k are real and $k > 1$, then the integral

$$\frac{1}{2\pi} \lim_{d\to\infty} \int_{\gamma-id}^{\gamma+id} \exp(sx) F(s) ds \qquad (6.2.2)$$

along any line $\mathrm{Re}(s) = \gamma \geq c$ converges to a function $f(x)$ which is independent of γ and whose Laplace transform is $F(s)$. Thus, equation (6.2.2) provides an inverse for the Laplace transform (6.2.1).

One of the most important properties of the Laplace transform is stated in the convolution theorem. Suppose $F(s)$, $G(s)$ are the Laplace transforms of $f(x)$, $g(x)$, then

$$\frac{1}{2\pi i} \int_{\gamma-id}^{\gamma+id} F(s)G(s) \exp(sx) ds = \int_0^x f(\eta) g(x-\eta) d\eta. \qquad (6.2.3)$$

In other words, it states that the Laplace transform of

$$h(x) = \int_0^x f(\eta) g(x-\eta) d\eta \qquad (6.2.4)$$

is

$$H(s) = F(s)G(s). \qquad (6.2.5)$$

We note one application of this result. Suppose $g(x) = 1$, then

$$G(s) = \int_0^\infty \exp(-sx) dx = \frac{1}{s}$$

so that

$$H(s) = F(s)/s \qquad (6.2.6)$$

is the Laplace transform of

$$h(x) = \int_0^x f(\eta) d\eta. \qquad (6.2.7)$$

In other words

$$\int_0^x f(\eta) d\eta = \lim_{d\to\infty} \int_{\gamma-id}^{\gamma+id} \frac{\exp(sx)}{s} F(s) ds. \qquad (6.2.8)$$

6.3 2-D Orthotropic Viscoelastic Bodies

We consider a 2-D formulation for the deformation of glass-plastic bodies. We assume that the glass plastic is *orthotropic* with respect to its elastic and viscoelastic properties. We recall that for plane strain or plane stress problems for an arisotropic body, the strain-stress relations have the form (4.5.3) with $\beta_{13} = 0 = \beta_{23}$, i.e.,

$$\varepsilon_{11} = \beta_{11}\sigma_{11} + \beta_{12}\sigma_{22}, \quad \varepsilon_{22} = \beta_{21}\sigma_{11} + \beta_{22}\sigma_{22}, \tag{6.3.1}$$

$$2\varepsilon_{12} = \beta_{33}\sigma_{12} \tag{6.3.2}$$

where $\beta_{12} = \beta_{21}$.

The fundamental assumption for viscoelasticity is that the strain at time t depends on the *history* of the stress for times τ from some datum time, 0, to time t. If, therefore, the material possesses both elastic and viscoelastic properties, the strain-stress relations have the form

$$\begin{aligned}\varepsilon_{xx}(t) = \beta_{11}\sigma_{xx}(t) &+ \int_0^t K_{11}(t-\tau)\sigma_{xx}(\tau)d\tau \\ &+\beta_{12}\sigma_{yy}(t) + \int_0^t K_{12}(t-\tau)\sigma_{yy}(\tau)d\tau\end{aligned} \tag{6.3.3}$$

$$\begin{aligned}\varepsilon_{yy}(t) = \beta_{21}\sigma_{xx}(t) &+ \int_0^t K_{21}(t-\tau)\sigma_{xx}(\tau)d\tau \\ &+\beta_{22}\sigma_{yy}(t) + \int_0^t K_{22}(t-\tau)\sigma_{yy}(\tau)d\tau\end{aligned} \tag{6.3.4}$$

$$2\varepsilon_{xy}(t) = \beta_{33}\sigma_{xy}(t) + \int_0^t K_{33}(t-\tau)\sigma_{xy}(\tau)d\tau, \tag{6.3.5}$$

where $\beta_{ji} = \beta_{ij}$, $K_{12} = K_{21}$. We assume that these relationships hold in both tension and compression.

Now apply the Laplace transform to equations (6.3.3)–(6.3.5), and use the convolution theorem (6.2.5):

$$\tilde{\varepsilon}_{xx} = \beta^*_{11}\tilde{\sigma}_{xx} + \beta^*_{12}\tilde{\sigma}_{yy}, \quad \tilde{\varepsilon}_{yy} = \beta^*_{21}\sigma_{xx} + \beta^*_{22}\sigma_{yy} \tag{6.3.6}$$

$$2\tilde{\varepsilon}_{xy} = \beta^*_{33}\sigma_{xy} \tag{6.3.7}$$

where

$$\beta^*_{ij} = \beta_{ij} + \tilde{K}_{ij} = \beta^*_{ji}. \tag{6.3.8}$$

Experimental results indicate that for small time intervals the kernels K_{ij} in (6.3.3)–(6.3.5) may be satisfactorily approximated by powers:

$$K_{ij}(t-\tau) = k_{ij}(t-\tau)^{\alpha-1}$$

where, in most situations, α lies in the range (0.74, 0.78). We now use the result

$$\int_0^\infty \exp(-st)t^{\alpha-1}dt = \Gamma(\alpha)s^{-\alpha} \tag{6.3.9}$$

to obtain

$$\beta_{ij}^* = \beta_{ij} + k_{ij}\Gamma(\alpha)s^{-\alpha}. \tag{6.3.10}$$

In the transformed variables, equation (6.3.6), (6.3.7) have precisely the same form as the elastic relations (6.3.1), (6.3.2). This means that formally, the 2-D problem for viscoelastic bodies has been reduced to one that is similar to the elastic problem. However, the determination of the actual, time, deformations and stresses is often difficult because they can be found only by inverting the Laplace transforms, i.e., by reconstructing $\sigma_{ij}(t)$, $\varepsilon_{ij}(t)$ from $\tilde{\sigma}_{ij}(s)$, $\tilde{\varepsilon}_{ij}(s)$. To carry out the inversion, all the singularities of the integrands in the complex s-plane must be analysed.

As an example, we consider a glass plastic plate which can be modelled as an orthotropic viscoelastic half-plane with

$$k_{11} = 0 = k_{12} = k_{22}, \quad k_{33} = k \tag{6.3.11}$$

acted on by a point force at the origin which is zero for time $t < 0$ and has the value P for $t > 0$. First, we need to solve the elastic problem.

Equations (4.5.17), (4.5.18) show that for a point force P at the origin

$$(\mu_2 - \mu_1)\Phi_1(z_1) = \frac{\mu_2 P}{2\pi i z_1}, \quad (\mu_2 - \mu_1)\Phi_2(z_2) = \frac{\mu_1 P}{2\pi i z_2}. \tag{6.3.12}$$

It was noted in Section 4.5 that, for an orthotropic meduim, $\mu_1 = i\nu_1, \mu_2 = i\nu_2$, where ν_1^2, ν_2^2 are the roots of

$$\beta_{22} - (2\beta_{12} + \beta_{33})\nu^2 + \beta_{11}\nu^4 = 0. \tag{6.3.13}$$

Substituting (6.3.12) into equation (4.5.9)–(4.5.11), we find

$$\sigma_{xx} = -\frac{P}{2\pi} \cdot \gamma \cdot x^2 y, \quad \sigma_{xy} = -\frac{P}{2\pi} \cdot \gamma \cdot xy^2, \quad \sigma_{yy} = -\frac{P}{2\pi} \cdot \gamma \cdot y^3 \tag{6.3.14}$$

where

$$\gamma = \frac{2\nu_1\nu_2(\nu_1 + \nu_2)}{|z_1 z_2|^2}. \tag{6.3.15}$$

These Cartesian stresses correspond to a single radial stress

$$\sigma_{rr} = -\frac{P}{2\pi} \cdot \gamma \cdot r^3 \sin\theta \tag{6.3.16}$$

Since $z_1 = x + i\nu_1 y$, $z_2 = x + i\nu_2 y$, we have

$$
\begin{aligned}
|z_1 z_2|^2 &= r^4(\cos^2\theta + \nu_1^2 \sin^2\theta)(\cos^2\theta + \nu_2^2 \sin^2\theta) \\
&= r^4(\cos^4\theta + (\nu_1^2 + \nu_2^2)\cos^2\theta \sin^2\theta + \nu_1^2\nu_2^2 \sin^4\theta).
\end{aligned}
$$

Equation (6.3.13) shows that

$$
\nu_1^2 + \nu_2^2 = \frac{2\beta_{12} + \beta_{33}}{\beta_{11}}, \quad \nu_1^2\nu_2^2 = \frac{\beta_{22}}{\beta_{11}}, \quad \nu_1\nu_2 = \sqrt{\frac{\beta_{22}}{\beta_{11}}}
$$

so that

$$
\begin{aligned}
|z_1 z_2|^2 &= \frac{r^4}{\beta_{11}}\left\{\beta_{11}\cos^4\theta + (2\beta_{12} + \beta_{33})\cos^2\theta \sin^2\theta + \beta_{22}\sin^4\theta\right\} \\
&= \frac{r^4}{\beta_{11}} L(\theta)
\end{aligned}
\tag{6.3.17}
$$

and hence

$$
\sigma_{rr} = -\frac{P}{\pi} \cdot \frac{\sqrt{\beta_{11}\beta_{22}}(\nu_1 + \nu_2)\sin\theta}{rL(\theta)}. \tag{6.3.18}
$$

We can express $\nu_1 + \nu_2$ in terms of the β_{ij}:

$$
(\nu_1 + \nu_2)^2 = \nu_1^2 + \nu_2^2 + 2\nu_1\nu_2 = \frac{2\beta_{12} + \beta_{33} + 2\sqrt{\beta_{11}\beta_{22}}}{\beta_{11}}. \tag{6.3.19}
$$

Now we proceed to the viscoelastic case: we replace β_{33} by $\beta_{33} + k\Gamma(\alpha)s^{-\alpha}$. Then $L(\theta)$ is replaced by

$$
L^*(\theta) = L(\theta) + k\Gamma(\alpha)\cos^2\theta \sin^2\theta s^{-\alpha} = L(\theta) + M(\theta)s^{-\alpha} \tag{6.3.20}
$$

and

$$
\begin{aligned}
\nu_1 + \nu_2 &= \sqrt{(2\beta_{12} + \beta_{33} + 2\sqrt{\beta_{11}\beta_{22}} + k\Gamma(\alpha)s^{-\alpha})/\beta_{11}} \\
&= \sqrt{(A + Bs^{-\alpha})/\beta_{11}}.
\end{aligned}
\tag{6.3.21}
$$

Galin's text again ends abruptly. For more details, see the papers listed at the end of Section 6.2.

6.4 Adhesive Contact for a Viscoelastic Half-Plane

Before formulating the problem for a punch in adhesive contact with an isotropic *viscoelastic* half-plane, we consider the problem for an *elastic* half-plane.

Galin does not solve this problem and, in fact, it is difficult to solve it by using, as he does, just the real and imaginary parts of a holomorphic function of the upper half of the x-axis. Muskhelishvili (1953a) does solve it, by using the values of a holomorphic function on either side of the x-axis. We will use the methods and notation introduced in Section 2.2 and 3.1 to do this.

Suppose a rigid punch with plane base is in adhesive contact with the upper elastic half-plane over the segment $(-a, a)$. According to the classification in Section 3.1, the boundary conditions are of Type III under the punch, and of Type II for the remainder of the x-axis. Thus, in the notation of Section 2.2

$$\beta u_1^+(x) + v_2^+(x) = 0 = v_1^+(x) - \beta u_2^+(x) \text{ on } |x| < a \tag{6.4.1}$$

$$u_1^+(x) = 0 = u_2^+(x) \text{ on } |x| > a. \tag{6.4.2}$$

We will now express these equations in terms of the boundary values $w_1^+(x), w_1^-(x), w_2^+(x), w_2^-(x)$. Combining equations (2.2.28), (2.2.29) with the Plemelj formulae (2.2.9), (2.2.10), we see that

$$w_1^+(x) = u_1^+(x) + iv_1^+(x), \quad w_1^-(x) = -u_1^+(x) + iv_1^+(x)$$
$$w_2^+(x) = u_2^+(x) + iv_2^+(x), \quad w_2^-(x) = -u_2^+(x) + iv_2^+(x)$$

so that

$$2u_j^+(x) = w_j^+(x) - w_j^-(x),$$
$$2iv_j^+(x) = w_j^+(x) + w_j^-(x), \quad j = 1, 2.$$

Now combine the two equations, on $|x| < a$ into one:

$$\beta u_1^+(x) + v_2^+(x) + i(v_1^+(x) - \beta u_2^+(x)) = 0 \quad |x| < a \tag{6.4.3}$$

When expressed in terms of w_1 and w_2, this is

$$(1+\beta)\{w_1^+(x) - iw_2^+(x)\} + (1-\beta)\{w_1^-(x) - iw_2^-(x)\} = 0.$$

Using $-i$ instead of i in equation (6.4.3), we find

$$(1-\beta)\{w_1^+(x) + iw_2^+(x)\} + (1+\beta)\{w_1^-(x) + iw_2^-(x)\} = 0.$$

These may be expressed in terms of two functions

$$\Phi_1(z) = w_1(z) - iw_2(z), \quad \Phi_2(z) = w_1(z) + iw_2(z): \tag{6.4.4}$$

$$(1+\beta)\Phi_1^+(x) + (1-\beta)\Phi_1^-(x) = 0, \quad |x| < a,$$
$$(1-\beta)\Phi_2^+(x) + (1+\beta)\Phi_2^-(x) = 0, \quad |x| < a.$$

Now, writing the two relations on $|x| > a$ as

$$u_1^+(x) + iu_2^+(x) = 0 = u_1^+(x) - iu_2^+(x)$$

we find

$$\Phi_1^+(x) - \Phi_1^-(x) = 0 = \Phi_2^+(x) - \Phi_2^-(x), \quad |x| > a.$$

We now have two separate Riemann–Hilbert problems, one for $\Phi_1(z)$, one for $\Phi_2(z)$:

$$\kappa\Phi_1^+(x) + \Phi_1^-(x) = 0 \text{ on } |x| < a, \tag{6.4.5}$$

$$\Phi_1^+(x) - \Phi_1^-(x) = 0 \text{ on } |x| > a, \tag{6.4.6}$$

and

$$\Phi_2^+(x) + \kappa\Phi_2^-(x) = 0 \text{ on } |x| < a, \tag{6.4.7}$$

$$\Phi_2^+(x) - \Phi_2^-(x) = 0 \text{ on } |x| > a, \tag{6.4.8}$$

where

$$\kappa = \frac{1+\beta}{1-\beta}. \tag{6.4.9}$$

To find the solution of these equations, we return to Section 3.4, and recall the boundary values of

$$\Phi(z) = (z+a)^\theta (z-a)^{1-\theta}.$$

On $|x| < a$, the boundary values are

$$\Phi^+(x) = \exp[i(1-\theta)\pi](a+x)^\theta (a-x)^{1-\theta}$$

$$\Phi^-(x) = \exp[-i(1-\theta)\pi](a+x)^\theta (a-x)^{1-\theta}.$$

On $|x| > a$, the boundary values are equal: $\Phi^+(x) = \Phi^-(x)$. Thus $\Phi(z)$ will satisfy (6.4.5) if

$$\frac{\Phi_1^+(x)}{\Phi_1^-(x)} = \exp[2i(1-\theta)\pi] = -\frac{1}{\kappa}$$

so that

$$\kappa = -\exp[-2i(1-\theta)\pi] = \exp[i\pi - 2i\pi + 2i\theta\pi].$$

Write

$$\kappa = \exp(2\alpha\pi), \tag{6.4.10}$$

then

$$\theta = \frac{1}{2} - i\alpha.$$

Similarly, $\Phi_2(z)$ will satisfy (6.4.7) if $\theta = \frac{1}{2} + i\alpha$.

We recall from Section 2.2 that, for large z,

$$w_1(z) = \frac{iP}{2\pi z}, \quad w_2(z) = \frac{iQ}{2\pi z}$$

where P, Q are the resultant normal and tangential forces applied to the half-plane: in our case, $Q = 0$, so that, from equation (6.3.4), we deduce that

$$\Phi_1(z) = \frac{iP}{2\pi z} = \Phi_2(z) \text{ for large } z.$$

This means that we must take

$$\Phi_1(z) = \frac{iP}{2\pi z}(z+a)^{-\frac{1}{2}-i\alpha}(z-a)^{-\frac{1}{2}+i\alpha} \tag{6.4.11}$$

$$\Phi_2(z) = \frac{iP}{2\pi} \cdot (z+a)^{-\frac{1}{2}+i\alpha}(z-a)^{-\frac{1}{2}-i\alpha}. \tag{6.4.12}$$

Now equations (2.2.30), (2.2.31) give

$$\begin{aligned} p(x) + iq(x) &= (w_1^+(x) - w_1^-(x)) + i(w_2^+(x) - w_2^-(x)) \\ &= \Phi_2^+(x) - \Phi_2^-(x) = \left(\frac{\kappa+1}{\kappa}\right)\Phi_2^+(x) \\ &= \frac{\kappa+1}{\sqrt{\kappa}} \cdot \frac{iP}{2\pi} \cdot \frac{1}{\sqrt{a^2-x^2}} \cdot \left(\frac{a+x}{x-a}\right)^{i\alpha} \end{aligned} \tag{6.4.13}$$

To evaluate this, put

$$\frac{a+x}{a-x} = \exp(2\pi\phi) \tag{6.4.14}$$

then

$$\begin{aligned} \left(\frac{a+x}{x-a}\right)^{i\alpha} &= \exp(2i\alpha\pi\phi) \\ &= \cos(2\alpha\pi\phi) + i\sin(2\alpha\pi\phi) \end{aligned}$$

so that

$$p(x) = \frac{\kappa+1}{\sqrt{\kappa}} \cdot \frac{iP}{2\pi} \cdot \frac{1}{\sqrt{a^2-x^2}}\cos(2\alpha\pi\phi) \tag{6.4.15}$$

$$q(x) = \frac{(\kappa+1)}{\sqrt{\kappa}} \cdot \frac{P}{2\pi} \cdot \frac{1}{\sqrt{a^2-x^2}}\sin(2\alpha\pi\phi). \tag{6.4.16}$$

We note that Muskhelishvili's equations (114.7) have the factor of 2 missing in the denominator. (We can verify (6.4.15) by checking that

$$\int_{-a}^{a} p(x)dx = P.)$$

We now follow Mitkevich (1970); he starts from equation (6.4.15), with the 2 missing in the denominator. Since $2\alpha\pi\phi = \phi\ell n\kappa$, we can write

$$p(x) = \frac{P}{2\pi} \cdot \frac{1}{\sqrt{a^2 - x^2}} \cdot \left\{ \left(\kappa^{\frac{1}{2}} + \kappa^{-\frac{1}{2}} \right) \frac{(\kappa^{i\phi} + \kappa^{-i\phi})}{2} \right\}$$

$$= \frac{P}{4\pi} \cdot \frac{1}{\sqrt{a^2 - x^2}} \left\{ \kappa^{\frac{1}{2}+i\phi} + \kappa^{\frac{1}{2}-i\phi} + \kappa^{-\frac{1}{2}+i\phi} + \kappa^{-\frac{1}{2}-i\phi} \right\} \quad (6.4.17)$$

We recall the stress-strain relations (2.1.11) for plane stress. On inverting these, we find the strain-stress relations

$$E\varepsilon_{xx} = \sigma_{xx} - \nu\sigma_{yy}, \quad E\varepsilon_{yy} = \sigma_{yy} - \nu\sigma_{xx}, \quad E\varepsilon_{xy} = (1+\nu)\sigma_{xy}. \quad (6.4.18)$$

The parameter κ in (6.3.9) is given by (2.1.36) as

$$\kappa = \frac{3-\nu}{1+\nu}. \quad (6.4.19)$$

Now we construct the strain-stress relations for viscoelastic plane stress:

$$E\varepsilon_{xx}(t) = \sigma_{xx}(t) + \int_0^t K_1(t-\tau)\sigma_{xx}(\tau)d\tau - \nu\sigma_{yy}(t) - \nu \int_0^t K_2(t-\tau)\sigma_{yy}(\tau)d\tau, \quad (6.4.20)$$

$$E\varepsilon_{yy(t)} = \sigma_{yy}(t) + \int_0^t K_1(t-\tau)\sigma_{yy}(\tau)d\tau - \nu\sigma_{xx}(t) - \nu \int_0^t K_2(t-\tau)\sigma_{yy}(\tau)d\tau, \quad (6.4.21)$$

$$E\varepsilon_{xy}(t) = (1+\nu)\sigma_{xy}(t) + (1+\nu)\int_0^t K(t-\tau)\sigma_{xy}(\tau)d\tau, \quad (6.4.22)$$

where

$$(1+\nu)K(t) = K_1(t) + \nu K_2(t). \quad (6.4.23)$$

The Laplace transforms of (6.4.20)–(6.4.22) are

$$E\tilde{\varepsilon}_{xx} = (1+\tilde{K}_1)\tilde{\sigma}_{xx} - \nu(1+\tilde{K}_2)\tilde{\sigma}_{yy}, \quad (6.4.24)$$

$$E\tilde{\varepsilon}_{yy} = (1+\tilde{K}_1)\tilde{\sigma}_{yy} - \nu(1+\tilde{K}_2)\tilde{\sigma}_{xx}, \quad (6.4.25)$$

$$E\tilde{\varepsilon}_{xy} = (1+\nu)(1+\tilde{K})\tilde{\sigma}_{xy}. \quad (6.4.26)$$

Experimental data indicate that the creep kernels K_i can be adequately approximated by using exponentials. We shall therefore assume that

$$K_j(t) = k_j \exp(-\delta t), \quad j = 1, 2. \quad (6.4.27)$$

Using (6.4.19), (6.4.24), we find that ν is replaced by $\nu(1+\tilde{K}_2)/(1+\tilde{K}_1)$, so that

$$\tilde{\kappa}(s) = \frac{3(1+\tilde{K}_1) - \nu(1+\tilde{K}_2)}{1+\tilde{K}_1 + \nu(1+\tilde{K}_2)}. \tag{6.4.28}$$

Now

$$\tilde{K}_j = k_j \int_0^\infty \exp(-st)\exp(-\delta t)dt = \frac{k_j}{s+\delta}$$

so that

$$\begin{aligned}\tilde{\kappa}(s) &= \left\{3 - \nu + \frac{3k_1 - \nu k_2}{s+\delta}\right\} / \left\{1 + \nu + \frac{k_1 + \nu k_2}{s+\delta}\right\} \\ &= \frac{3-\nu}{1+\nu}\left\{\frac{s+\lambda}{s+\lambda-\varepsilon}\right\} = \kappa \cdot \left(\frac{s+\lambda}{s+\lambda-\varepsilon}\right)\end{aligned} \tag{6.4.29}$$

where

$$\lambda = \delta + \frac{3k_1 - \nu k_2}{3-\nu}, \quad \lambda - \varepsilon = \delta + \frac{k_1 + \nu k_2}{1+\nu}. \tag{6.4.30}$$

The Laplace transform $P(x, s)$ is given by (6.4.17) as

$$\tilde{p}(x,s) = \frac{P}{4\pi\sqrt{a^2 - x^2}}\left\{(\tilde{\kappa}(s))^{\frac{1}{2}+i\phi} + (\tilde{\kappa}(s))^{\frac{1}{2}-i\phi} + (\tilde{\kappa}(s))^{-\frac{1}{2}+i\phi} + (\tilde{\kappa}(s))^{-\frac{1}{2}-i\phi}\right\}. \tag{6.4.31}$$

Now divide equation (6.4.31) by s and take the inverse Laplace transforms. We need to find the inverse Laplace transforms of a generic function

$$\begin{aligned}F(s,\gamma) &= \frac{1}{s}\left(\frac{s+\lambda-\varepsilon}{s+\lambda}\right)^{-\gamma} = \frac{1}{s}\left(1 - \frac{\varepsilon}{s+\lambda}\right)^{-\gamma} \\ &= \frac{1}{s}\sum_{k=0}^{\infty}\frac{\gamma(\gamma+1)\ldots(\gamma+k-1)}{k!}\left(\frac{\varepsilon}{s+\lambda}\right)^k.\end{aligned} \tag{6.4.32}$$

The identity

$$\frac{1}{1-x} = 1 + x + \cdots\ x^k + \frac{x^{k+1}}{1-x}$$

gives

$$G(s) \equiv \frac{1}{s(s+\lambda)^k} = \frac{1}{\lambda^k s} - \sum_{n=1}^{k}\frac{1}{\lambda^n (s+\lambda)^{k+1-n}}. \tag{6.4.33}$$

Consider the integral (Gradshteyn & Ryzhik (1965, 3.3814)

$$\int_0^\infty x^{\nu-1}\exp(-\mu x)dx = \frac{\Gamma(\nu)}{\mu^\nu}, \quad \mathrm{Re}(\mu) > 0, \quad \mathrm{Re}(\nu) > 0. \tag{6.4.34}$$

Put $\mu = s + \lambda,\ \nu = k + 1 - n$, then

$$\int_0^\infty x^{k-n}\exp(-\lambda x)\cdot\exp(-sx)dx = \frac{\Gamma(k+1-n)}{(s+\lambda)^{k+1-n}}. \tag{6.4.35}$$

Now use the identity

$$\frac{1}{\Gamma(k+1-n)} = \frac{k(k-1)\dots(k+1-n)}{k!}$$

to obtain the inverse transform of $G_k(s)$ as

$$g_k(t) = \frac{1}{\lambda^k} - \exp(-\lambda t)\sum_{n=1}^{k} \frac{k(k-1)\dots(k+1-n)}{k!}\frac{t^{k-n}}{\lambda^n}. \tag{6.4.36}$$

The inverse Laplace transform of $F(p, \gamma)$ is therefore

$$f(t, \gamma) = \sum_{k=0}^{\infty} \frac{\gamma(\gamma+1)\dots(\gamma+k-1)}{k!}\varepsilon^k g_k(t). \tag{6.4.37}$$

Now return to equation (6.4.31). After having divided the equation by s, we have on the left, the function $P(x, s)/s$. By equation (6.2.7), the inverse transform is $\int_0^t p(x, \tau)d\tau$. Thus

$$\int_0^t p(x,\tau)d\tau = \frac{P}{4\pi\sqrt{a^2-x^2}}\left\{\kappa^{\frac{1}{2}+i\phi} f\left(t, \frac{1}{2}+i\phi\right) + \kappa^{\frac{1}{2}-i\phi} f\left(t, \frac{1}{2}-i\phi\right)\right.$$
$$\left. +\kappa^{-\frac{1}{2}+i\phi} f\left(t, -\frac{1}{2}+i\phi\right) + \kappa^{-\frac{1}{2}-i\phi} f\left(t, -\frac{1}{2}-i\phi\right)\right\} \tag{6.4.38}$$

Mitkevich (1970) has $p(x, t)$ rather than the integral. That this must be incorrect, one can see by considering a small time interval; then the force must be considered as an impulse: $p(x, t)$ is unbounded as $t \to 0$ but the integral

$$\lim_{t\to 0}\int_0^t p(x,\tau)d\tau$$

is finite.

Since $f(t, \pm 1/2 - i\phi) = \overline{f(t, \pm\frac{1}{2} + i\phi)}$, we can write (6.4.38) as

$$\int_0^t p(x,\tau)d\tau = \frac{P}{2\pi\sqrt{a^2-x^2}}\,\text{Re}\left\{\kappa^{\frac{1}{2}+i\phi} f\left(t, \tfrac{1}{2}+i\phi\right) + \kappa^{-\frac{1}{2}+i\phi} f\left(t, -\tfrac{1}{2}+i\phi\right)\right\}.$$

We now examine this. For short time intervals, $t \to 0$. We note that $g_0(0) = 1$, $g_k(0) = 0$ for $k = 1, 2, \dots$ so that

$$f(0, \gamma) = 1$$

and thus

$$\lim_{t\to 0}\int_0^t p(x,\tau)d\tau$$

has precisely the value given in equation (6.4.15) or (6.4.17) for the elastic case.

For long time intervals, i.e., $t \to \infty$, equation (6.4.36) shows that $g_k(t) \to 1/\lambda^k$, and therefore

$$f(t,\lambda) \to \sum_{k=0}^{\infty} \frac{\gamma(\gamma+1)\dots(\gamma+k-1)}{k!}\left(\frac{\varepsilon}{\lambda}\right)^k = \left(1-\frac{\varepsilon}{\lambda}\right)^{-\gamma}. \tag{6.4.39}$$

This means that the effective value of κ is

$$\kappa^* = \frac{\kappa}{1-\frac{\varepsilon}{\lambda}} = \frac{(3-\nu)\delta + 3k_1 - \nu k_2}{(1+\nu)\delta + k_1 + \nu k_2} = \frac{3-\nu\eta}{1+\nu\eta}, \tag{6.4.40}$$

where

$$\eta = \frac{\delta + k_2}{\delta + k_1}. \tag{6.4.41}$$

Thus

$$\lim_{t\to 0}\int_0^t p(x,\tau)d\tau$$

has the value (6.4.15) or (6.4.17) where κ is replaced by κ^*. In particular, at the centre of the contact region, where $x = 0 = \phi$, we have

$$\int_0^\infty p(0,\tau)d\tau = \frac{P}{2\pi a}\frac{(\kappa^*+1)}{\sqrt{\kappa^*}}. \tag{6.4.42}$$

Thus, the integral will increase or decrease as t goes from 0 to ∞, depending on whether

$$L = \left(\frac{\kappa^*+1}{\sqrt{\kappa^*}}\right)^2 - \left(\frac{\kappa+1}{\sqrt{\kappa}}\right)^2$$

is positive or negative.

$$L = \frac{{\kappa^*}^2+1}{\kappa^*} - \frac{(\kappa^2+1)}{\kappa} = \frac{(\kappa-\kappa^*)(1-\kappa\kappa^*)}{\kappa^*\kappa}$$

and after some algebra we find

$$(\kappa-\kappa^*)(1-\kappa\kappa^*) = \frac{16\nu(\eta-1)(\nu+\nu\eta-2)}{[(1+\nu)(1+\nu\eta)]^2} \tag{6.4.43}$$

so that the integral increases or decreases depending on whether $(\eta-1)(\nu+\nu\eta-2)$ is positive or negative. If $\eta < 1$ then $\nu+\nu\eta-2 < 0$, and $L > 0$. If $\nu+\nu\eta-2 > 0$, i.e., $\eta > 2\nu^{-1}-1$, then $\eta > 1$ and again $L > 0$. If $1 < \eta < 2\nu^{-1}-1$ then $L < 0$.

6.5 Viscoelastic Rolling Contact

A body rolling on another experiences resistance. One of the principal causes of this resistance is the relative slip of the two surfaces in the contact region; see Reynolds (1876) and Pinegin and Orlow (1961). The other principal cause is the viscoelasticity of the contacting bodies. The viscoelasicity produces an asymmetric pressure distribution in the contact region, and this in turn produces resistance to rolling. The analysis follows that of Goryacheva (1973a).

In this section, we consider rolling contact of a viscoelastic cylinder on a viscoelastic base, under the assumption that there are stick and slip zones.

First we consider the purely elastic problem of an elastic cylinder rolling over an elastic base. Suppose, as in Section 4.3, that the cylinder rolls over the elastic base with constant speed c much smaller than the speed of sound in the body, so that we can neglect inertial effects. If the radius of the cylinder is large compared to the dimensions of the contact region, then we can replace the cylinder by the lower half plane. The problem is then that of the contact of two half planes, where the contact region shifts with speed c between them. As in Section 4.3, we introduce a moving coordiante system $x = x_1 - ct$, $y = y_1$, where $O(x_1, y_1)$ is the fixed system.

We assume that the contact region $(-a, b)$ is split into two parts, a slip zone $(-a, d)$ and a stick zone (d, b). We assume that the cylinder occupies the lower half plane and, before contact, has the equation $y = -x^2/(2R)$. During contact, the cylinder is shifted by an amount Δ into the upper half plane. Consider two points, one on the cylinder and the other on the boundary of the upper half plane, that occupy the same point after contact. The first is initially at $(x, -x^2/(2R))$, the other at $(x, 0^+)$. As the cylinder is shifted toward the half-plane, the first goes to $(x, -x^2/(2R) + \Delta + v^{(2)}(x, 0-))$ while the other goes to $(x, v^{(1)}(x, 0+))$. Therefore, in the contact region,

$$v^{(1)}(x, 0+) = -\frac{x^2}{2R} + \Delta + v^{(2)}(x, 0-), \quad x \in (-a, b). \tag{6.5.1}$$

By differentiating throughout by x, we find

$$v^{(1)\prime}(x, 0+) - v^{(2)\prime}(x, 0-) = -x/R, \quad x \in (-a, b). \tag{6.5.2}$$

Now consider the rolling. The horizontal component of the velocity of a point on the boundary of the upper half-plane is $\partial u^{(1)}(x_1, 0+, t)/\partial t$. That of the corresponding point on the boundary of the lower half-plane is $c - \omega r + \partial u^{(2)}(x, 0-, t)/\partial t$, where ω is the angular velocity of the cylinder. In the stick zone, these are equal:

$$\frac{\partial u^{(1)}}{\partial t}(x_1, 0+, t) = c - \omega r + \frac{\partial u^{(2)}}{\partial t}(x, 0+, t). \tag{6.5.3}$$

But $u^{(1)}$ and $u^{(2)}$ are both functions of $x_1 - ct$, so that

$$-\frac{c\partial u^{(1)}}{\partial x}(x, 0+) = c - \omega r - \frac{c\partial u^{(2)}}{\partial x}(x, 0-),$$

i.e.,

$$u^{(1)\prime}(x,0+) - u^{(2)\prime}(x,0-) = (\omega r - c)/c = \delta, \quad x \in (d,b). \tag{6.5.4}$$

In the slip zone, we assume that there is a state of limiting friction:

$$\rho\sigma_{yy} + \sigma_{xy} = 0, , \quad x \in (-a,d). \tag{6.5.5}$$

Now we use the complex variable formulation given in Section 2.2. In the upper half-plane, the derivatives of the displacements on the + side of the x-axis are given by equations (2.2.22), (2.2.23); thus

$$\frac{1}{\vartheta}\cdot u^{(1)\prime}(x,0+) = -\beta p(x) + \frac{1}{\pi}\oint_{-\infty}^{\infty}\frac{q(t)dt}{t-x}, \tag{6.5.6}$$

$$\frac{1}{\vartheta}\cdot v^{(1)\prime}(x,0+) = \frac{1}{\pi}\oint_{-\infty}^{\infty}\frac{p(t)dt}{t-x} + \beta q(x), \tag{6.5.7}$$

where we note that

$$\frac{4\mu}{\kappa+1} = \frac{\mu}{1-\nu} = \frac{1}{\vartheta}\cdot \tag{6.5.8}$$

We need to get the corresponding results for the displacements on the lower side of the x-axis. Equations (2.2.22), (2.2.23) were derived from equations (2.2.16), (2.2.17), which in turn were derived from equation (2.2.5). Recall that in that equation $f(z)$ is holomorphic in the upper half-plane, and the result is obtained by integrating around the contour in Figure 2.2.3, with a small circle around z. If, instead, $f(z)$ is holomorphic in the lower half-plane, then

$$\frac{1}{2\pi i}\int_{-\infty}^{\infty}\frac{f(t)dt}{t-z} = \begin{cases} 0 & \text{if } z \in s^+, \\ -f(z) & \text{if } z \in s^-. \end{cases} \tag{6.5.9}$$

Carrying this change through the analysis, as in Section 2.2, we find

$$\frac{1}{\vartheta}\cdot u^{(2)\prime}(x,0-) = -\beta p(x) - \frac{1}{\pi}\oint_{-\infty}^{\infty}\frac{q(t)dt}{t-x}, \tag{6.5.10}$$

$$\frac{1}{\vartheta}\cdot v^{(2)\prime}(x,0-) = -\frac{1}{\pi}\oint_{-\infty}^{\infty}\frac{p(t)dt}{t-x} + \beta q(x). \tag{6.5.11}$$

Now substitute these results in the boundary conditions (6.5.2), (6.5.4), (6.5.5). We obtain

$$-\frac{x}{\vartheta R} = \frac{2}{\pi}\oint_{-\infty}^{\infty}\frac{p(t)dt}{t-x}, \quad x \in (-a,b), \tag{6.5.12}$$

$$\frac{\delta}{\vartheta} = \frac{2}{\pi}\oint_{-\infty}^{\infty}\frac{q(t)dt}{t-x}, \quad x \in (d,b), \tag{6.5.13}$$

$$\rho p + q = 0 \quad x \in (-a,d) \tag{6.5.14}$$

$$p = 0 = q \qquad x \in (-\infty, -a) \cup (b, \infty). \tag{6.5.15}$$

Equations (6.5.12), (6.5.15) provide a Riemann–Hilbert problem for p:

$$\frac{2}{\pi}\oint_{-\infty}^{\infty} \frac{p(t)dt}{t-x} = -\frac{x}{\vartheta R} \qquad x \in (-a, b) \tag{6.5.16}$$

$$p = 0 \qquad x \in (-\infty, -a) \cup (b, \infty) \tag{6.5.17}$$

Equations (6.5.12)–(6.5.15) may be combined to form a second Riemann–Hilbert problem for $p + \rho q$:

$$\frac{2}{\pi}\oint_{-\infty}^{\infty} \frac{\rho p(t) + q(t)dt}{t-x} = -\frac{\rho x}{\vartheta R} + \frac{\delta}{\vartheta} \qquad x \in (d, b) \tag{6.5.18}$$

$$\rho p + q = 0, \qquad x \in (-\infty, d) \cup (b, \infty). \tag{6.5.19}$$

First, consider the problem for p. Equations (2.2.32) show that equations (6.5.16), (6.5.17) may be written

$$v_1^+(x) = \frac{x}{4\vartheta R}, \qquad x \in (-a, b), \tag{6.5.20}$$

$$u_1^+(x) = 0, \qquad x \in (-\infty, -a) \cup (b, \infty). \tag{6.5.21}$$

This is an example of equations (3.4.2), (3.4.3) with $f(x) = x/(4\vartheta R)$, and the solution is given by equation (3.5.4) with $w_*(z) = if(z)$:

$$w_1(z) = \frac{iz}{4\vartheta R} + \frac{i}{\sqrt{(z+a)(z-b)}}\frac{N(z)}{D(z)}, \tag{6.5.22}$$

where $N(z)$, $D(z)$ are chosen to make $w_1(z) = \frac{iP}{(2\pi z)}$ for large z. Clearly, $D(z) = 1$, and $N(z)$ is a quadratic in z. Thus

$$w_1(z) = \frac{i}{4\vartheta R}\left\{z + \frac{a_0 + a_1 z + a_2 z^2}{\sqrt{(z+a)(z-b)}}\right\}.$$

For large z, the expression in brackets has the form

$$z + \left(\frac{a_0}{z} + a_1 + a_2 z\right)\left(1 - \frac{1}{2}\frac{(a-b)}{z} + \left(\frac{3a^2}{8} + \frac{3b^2}{8} - \frac{ab}{4}\right)\frac{1}{z^2} + \cdots\right.$$

$$= (1 + a_2)z + \left\{a_1 - \frac{1}{2}(a-b)a_2\right\} + \left\{a_0 - \frac{1}{2}(a-b)a_1\right.$$

$$\left. + \left(\frac{3a^2}{9} + \frac{3b^2}{8} - \frac{ab}{4}\right)a_2\right\}\frac{1}{z} + \cdots$$

so that

$$a_2 = -1, \quad a_1 = \frac{b-a}{2}, \quad a_0 = \frac{(a+b)^2}{8} + \frac{2\vartheta R}{\pi} \cdot P. \tag{6.5.23}$$

We introduce the notation

$$\ell = \frac{a+b}{2}, \quad g = \frac{b-a}{2}, \tag{6.5.24}$$

so that

$$a_2 = -1, \quad a_1 = g, \quad a_0 = \ell^2/2 + 2\vartheta RP/\pi,$$

and

$$w_1(z) = \frac{i}{4\vartheta R}\left\{z + \frac{\ell^2 + 2gz - 2z^2}{2\sqrt{(z+a)(z-b)}}\right\} + \frac{iP}{2\pi\sqrt{(z+a)(z-b)}}. \tag{6.5.25}$$

According to equation (2.2.30), $p(x) = 2u_1^+(x)$. On the upper side of the x-axis, in $(-a, b)$,

$$\sqrt{(z+a)(z-b)} = i\sqrt{(x+a)(b-x)},$$

so that

$$p(x) = \frac{P}{\pi\sqrt{(x+a)(b-x)}} + \frac{1}{4\vartheta R} \cdot \frac{\ell^2 + 2gx - 2x^2}{\sqrt{(x+a)(b-x)}}. \tag{6.5.26}$$

Now consider the Riemann–Hilbert problem for $\rho p + q$. Put $w(z) = \rho w_1(z) + w_2(z)$, then equations (6.5.18), (6.5.19) can be written

$$v^+(x) = \frac{\rho x}{4\vartheta R} - \frac{\delta}{4\vartheta}, \qquad x \in (d, b) \tag{6.5.27}$$

$$u^+(x) = 0, \qquad x \in (-\infty, d) \cup (b, \infty). \tag{6.5.28}$$

This is a second example of equations (3.4.2), (3.4.3) with

$$f(x) = \frac{\rho x}{4\vartheta R} - \frac{\delta}{4\vartheta},$$

and the solution is given by equation (3.5.4):

$$w(z) = \frac{i\rho z}{4\vartheta R} - \frac{i\delta}{4\vartheta} + \frac{i}{\sqrt{(z-d)(z-b)}}\frac{N(z)}{D(z)}.$$

We must choose $N(z)$, $D(z)$ so that

$$w(z) = i(\rho P + Q)/(2\pi z)$$

for large z. Again $D(z) = 1$, and $N(z)$ is a quadratic, so that

$$w(z) = \frac{i\rho z}{4\vartheta R} - \frac{i\delta}{4\vartheta} + \frac{i(b_0 + b_1 z + b_2 z^2)}{\sqrt{(z-d)(z-b)}}.$$

Proceeding as before, we find

$$b_2 = -\frac{\rho}{4\vartheta R}, \quad b_1 = \frac{\delta}{4\vartheta} + \frac{\rho h}{4\vartheta R}, \quad b_0 = -\frac{\delta h}{4\vartheta} + \frac{\rho m^2}{8\vartheta R} + \frac{\rho P + Q}{2\pi},$$

where

$$m = (b-d)/2, \quad h = (b+d)/2$$

and

$$w(z) = \frac{i\rho}{8\vartheta R}\left\{2z + \frac{m^2 + 2hz - 2z^2}{\sqrt{(z-d)(z-b)}}\right\} + \frac{i\delta}{4\vartheta}\left\{-1 + \frac{z-h}{\sqrt{(z-d)(z-b)}}\right\} + \frac{i(\rho P + Q)}{2\pi\sqrt{(z-d)(z-b)}} \tag{6.5.29}$$

The quality $\rho p(x) + q(x)$ is twice the real part of $w(z)$ on the $+$ side of (d, b). Noting that, there, $\sqrt{(z-d)(z-b)} = i\sqrt{(x-d)(b-x)}$, we find

$$\rho p(x) + q(x) = \frac{\rho}{4\vartheta R}\cdot\frac{m^2 + 2hx - 2x^2}{\sqrt{(x-d)(b-x)}} + \frac{\delta}{2\vartheta}\cdot\frac{x-h}{\sqrt{(x-d)(b-x)}} + \frac{\rho P + Q}{\pi}\cdot\frac{1}{\sqrt{(x-d)(b-x)}}, \quad x \in (b, d). \tag{6.5.30}$$

Equations (6.5.26), (6.5.30) yield the contact stresses $p(x), q(x)$ in the general case. If the punch is rounded at the ends, so that $p(x)$ is zero at $x = -a$, $x = b$ then equation (6.5.26) yields

$$g = 0, \quad \frac{P}{\pi} = \frac{a^2}{4\vartheta R}, \tag{6.5.31}$$

so that $b = a$, and the contact width $2a$, is

$$2a = 4\sqrt{\frac{P\vartheta R}{\pi}}. \tag{6.5.32}$$

The condition $\rho p(x) + q(x) = 0$ and $p(-a) = 0$ yields $q(-a)$ automatically, so we must enforce

$$\rho p(d+) + q(d+) = 0 = \rho p(b) + q(b) = 0. \tag{6.5.33}$$

These conditions lead to

$$h = \frac{\delta R}{\rho}, \quad \frac{\rho P + Q}{\pi} = \frac{\rho m^2}{4\vartheta R}, \tag{6.5.34}$$

and the stresses are

$$p(x) = \frac{\sqrt{a^2 - x^2}}{2\vartheta R}, \quad x \in (-a, a). \tag{6.5.35}$$

$$q(x) = -\frac{\rho\sqrt{a^2 - x^2}}{2\vartheta R}, \quad x \in (-a, d). \tag{6.5.36}$$

$$q(x) = -\frac{\rho\sqrt{a^2 - x^2}}{2\vartheta R} + \frac{\rho}{2\vartheta R}\sqrt{(a - x)(x - d)}, \quad x \in (d, a). \tag{6.5.37}$$

We now consider the analogous viscoelastic problem. To obtain the stress strain relations for an isotropic viscoelastic body, we start from the elastic equations (2.1.7) and modify them:

$$2\mu\left(\varepsilon_{xx} + \frac{\alpha\partial\varepsilon_{xx}}{\partial t}\right) = (1 - \nu)\left(\sigma_{xx} + \frac{\gamma\partial\sigma_{xx}}{\partial t}\right) - \nu\left(\sigma_{yy} + \frac{\gamma\partial\sigma_{yy}}{\partial t}\right), \tag{6.5.38}$$

$$2\mu\left(\varepsilon_{yy} + \frac{\alpha\partial\varepsilon_{yy}}{\partial t}\right) = (1 - \nu)\left(\sigma_{yy} + \frac{\gamma\partial\sigma_{yy}}{\partial t}\right) - \nu\left(\sigma_{xx} + \frac{\gamma\partial\sigma_{xx}}{\partial t}\right), \tag{6.5.39}$$

$$2\mu\left(\varepsilon_{xy} + \frac{\alpha\partial\varepsilon_{xy}}{\partial t}\right) = \sigma_{xy} + \frac{\gamma\partial\sigma_{xy}}{\partial t}. \tag{6.5.40}$$

Here α and γ are quantities characterising the viscous properties of the medium.

Now proceed as in the elastic case: introduce moving coordinates so that $\partial/\partial t$ is replaced by $-c\partial/\partial x$. In the moving coordinates, the strains and stresses are

$$\varepsilon_{ij} - \frac{\alpha c\partial\varepsilon_{ij}}{\partial x} = \varepsilon^*_{ij}, \quad \sigma_{ij} - \frac{\gamma c\partial\sigma_{ij}}{\partial x} = \sigma^*_{ij} \tag{6.5.41}$$

and the strain-stress equations are

$$\begin{aligned} 2\mu\varepsilon^*_{xx} &= (1 - \nu)\sigma^*_{xx} - \nu\sigma^*_{yy} \\ 2\mu\varepsilon^*_{yy} &= (1 - \nu)\sigma^*_{yy} - \nu\sigma^*_{xx} \\ 2\mu\varepsilon^*_{xy} &= \sigma^*_{xy} \end{aligned} \tag{6.5.42}$$

and the corresponding displacements are

$$u - \frac{\alpha c\partial u}{\partial x} = u^*, \quad v - \frac{\alpha c\partial v}{\partial x} = v^* \tag{6.5.43}$$

so that equation (6.5.2) gives

$$v^{*(1)\prime}(x, 0+) - v^{*(2)\prime}(x, 0-) = -(x - \alpha c)/R, \quad x \in (-a, b) \tag{6.5.44}$$

and equations (6.5.12)–(6.5.15) change to

$$-\frac{(x - \alpha c)}{\vartheta R} = \frac{2}{\pi}\oint_{-\infty}^{\infty}\frac{p^*(t)dt}{t - x}, \quad x \in (-a, b), \tag{6.5.45}$$

$$\frac{\delta}{\vartheta} = \frac{2}{\pi}\oint_{-\infty}^{\infty}\frac{q^*(t)dt}{t - x}, \quad x \in (d, b), \tag{6.5.46}$$

$$\rho p^* + q^* = 0, \quad x \in (-a, d), \tag{6.5.47}$$

$$p^* = 0 = q^*, \quad x \in (-\infty, -a) \cup (b, \infty). \tag{6.5.48}$$

Proceeding as before, we find

$$w_1^*(z) = \frac{i(z - \alpha c)}{4\vartheta R} + \frac{i}{\sqrt{(z+a)(z-b)}} \cdot \frac{N(z)}{D(z)}$$

and, as before

$$w_1^*(z) = \frac{i}{8\vartheta R}\left\{2(z - \alpha c) + \frac{\ell^2 + 2(z - \alpha c)(g - z)}{\sqrt{(z+a)(z-b)}}\right\} + \frac{iP}{2\pi\sqrt{(z+a)(z-b)}} \tag{6.5.49}$$

and equation (6.5.26) is replaced by

$$p^*(x) = \frac{P}{\pi\sqrt{(x+a)(b-x)}} + \frac{1}{4\vartheta R}\frac{\ell^2 + 2(x - \alpha c)(g - x)}{\sqrt{(x+a)(b-x)}}, \quad x \in (-a, b). \tag{6.5.50}$$

We proceed to find $\rho p^* + q^*$. As before, we find

$$w^*(z) = \frac{i\rho}{8\vartheta R}\left\{2(z - \alpha c) + \frac{m^2 + 2(z - \alpha c)(h - z)}{\sqrt{(z-d)(z-b)}}\right\} +$$

$$-\frac{i\delta}{4\vartheta}\left\{1 + \frac{h - z}{\sqrt{(z-d)(z-b)}}\right\} + \frac{i(\rho P + Q)}{2\pi\sqrt{(z-d)(z-b)}}, \quad x \in (d, b). \tag{6.5.51}$$

The quality $\rho p^* + q^*$ is twice the real part of $w^*(z)$, so that

$$\rho p^*(x) + q^*(x) = \frac{\rho}{4\vartheta R}\frac{m^2 + 2(x - \alpha c)(h - x)}{\sqrt{(x-d)(b-x)}} + \frac{\delta}{2\vartheta}\frac{x - h}{\sqrt{(x-d)(b-x)}}$$

$$+\frac{\rho P + Q}{\pi\sqrt{(x-d)(b-x)}}, \quad x \in (d, b). \tag{6.5.52}$$

To find the corresponding stresses $p(x), q(x)$, we must solve the equations

$$p(x) - \gamma c\frac{dp(x)}{dx} = p^*(x), \quad q(x) - \gamma c\frac{dq(x)}{dx} = q^*(x). \tag{6.5.53}$$

We write the first as

$$\gamma c\frac{d}{dx}\left(p(x)\exp\left(-\frac{x}{\gamma c}\right)\right) = -p^*(x)\exp\left(-\frac{x}{\gamma c}\right)$$

so that, since $p(-a) = 0$,

$$p(x) = -\frac{1}{\gamma c}\exp\left(\frac{x}{\gamma c}\right)\int_{-a}^{x} p^*(t)\exp\left(-\frac{t}{\gamma c}\right)dt, \tag{6.5.54}$$

and the condition that $p(b) = 0$ reduces to

$$\int_{-a}^{b} p^*(t) \exp\left(-\frac{x}{\gamma c}\right) dx = 0. \tag{6.5.55}$$

Make successive changes of variables:

$$x = \ell\xi + g, \quad \xi = \cos\theta, \tag{6.5.56}$$

and use the modified Bessel function

$$I_n(z) = \frac{1}{\pi} \int_0^{\pi} \exp(z\cos\theta)\cos n\theta d\theta. \tag{6.5.57}$$

After considerable algebraic manipulation, we find

$$\left(\frac{P}{\pi} - \frac{\ell^2}{4\vartheta R}\right) I_0\left(\frac{\ell}{\gamma c}\right) + \frac{\ell}{2\vartheta R}\{g + c(\gamma - \alpha)\} I_1\left(\frac{\ell}{\gamma c}\right) = 0 \tag{6.5.58}$$

which agrees with equation (3.18) of Galin's text.

Now return to equation (6.5.45). Temporarily write

$$v^{(1)\prime}(x, 0+) - v^{(2)\prime}(x, 0-) = G(x)$$
$$v^{*(1)\prime}(x, 0-) - v^{*(2)\prime}(x, 0-) = G^*(x)$$

then

$$G(x) - \alpha c\frac{dG}{dx} = G^*(x). \tag{6.5.59}$$

For $x \in (-b, a)$

$$G^*(x) = -\frac{(x - \alpha c)}{R}$$

so that

$$G(x) - \alpha c\frac{dG}{dx} = -\frac{(x - \alpha c)}{R}$$

which may be written

$$G(x) + \frac{x}{R} - \alpha c\left(\frac{dG}{dx} + \frac{1}{R}\right) = 0$$

which implies

$$\left(\exp\left(-\frac{x}{\alpha c}\right)\left(G + \frac{x}{R}\right)\right)' = 0$$

and hence

$$\exp\left(-\frac{x}{\alpha c}\right)\left(G + \frac{x}{R}\right) = \text{const}, \quad x \in (-a, b). \tag{6.5.60}$$

If therefore we put $G(b) = -b/R$ then $G(x) = -x/R$ for $x \in (-a, b)$. Equation (6.5.59) states that

$$G(x) = \frac{1}{\alpha c} \exp\left(\frac{x}{\alpha c}\right) \int_x^\infty G^*(x) \exp\left(-\frac{t}{\alpha c}\right) dt,$$

so that we must enforce the condition

$$\int_b^\infty G^*(x) \exp\left(-\frac{x}{\alpha c}\right) dx = -\alpha c \frac{b}{R} \exp\left(-\frac{b}{\alpha c}\right). \tag{6.5.61}$$

Now equation (6.5.46) shows that

$$G^*(x) = \frac{2\vartheta}{\pi} \oint_{-\infty}^\infty \frac{p^*(t)dt}{t-x} = -4\vartheta v_1^*(x) \tag{6.5.62}$$

and equation (6.5.49) gives

$$G^*(x) = -\frac{(x-\alpha c)}{R} - \frac{1}{2R} \frac{\ell^2 + 2(x-\alpha c)(g-x)}{\sqrt{(x+a)(x-b)}} - \frac{2\vartheta P}{\pi\sqrt{(x+a)(x-b)}}, \quad x \in (-a, b) \tag{6.5.63}$$

and since

$$-\frac{1}{R}\int_b^\infty (x-\alpha c) \exp\left(-\frac{x}{\alpha c}\right) dx = -\frac{\alpha c b}{R} \exp\left(-\frac{b}{\alpha c}\right),$$

the condition (6.5.60) reduces to

$$\frac{1}{2R}\int_b^\infty \frac{\{\ell^2 + 2(x-\alpha c)(g-x)\}}{\sqrt{(x+a)(x-b)}} \exp\left(-\frac{x}{\alpha c}\right) dx + \frac{2\vartheta P}{\pi}\int_b^\infty \frac{1}{\sqrt{(x+a)(x-b)}} \cdot \exp\left(-\frac{x}{\alpha c}\right) dx = 0.$$

After routine manipulations, this reduces to

$$\left(\frac{P}{\pi} - \frac{\ell^2}{4\vartheta R}\right) K_0\left(\frac{\ell}{\alpha c}\right) - \frac{g\ell}{2\vartheta R} K_1\left(\frac{\ell}{\alpha c}\right) = 0 \tag{6.5.64}$$

where

$$K_n(x) = \int_0^\infty \exp(-x\cosh\theta)\cosh n\theta d\theta. \tag{6.5.65}$$

This agrees with equation (3.17) of Galin's text.

To this point, we have satisfied the conditions $p(b) = 0 = p(-a)$, and equation (6.5.45). We now go through a similar analysis to enforce the conditions $q(t) = 0 = \rho p(d+) + q(d+) = 0$ and equation (6.5.4). Since $p(-a) = 0$, and $\rho p + q = 0$ when $x \in (-a, d)$, we automatically have enforced $q(-a) = 0$; since $p(b) = 0$, we have to enforce only $\rho p(d+) + q(d+) = 0$, and $\rho p(b) + q(b) = 0$.

Since $\rho p^*(t) + q^*(t) = 0$ for $t \in (-a, d)$, we may write

$$\rho p(x) + q(x) = -\frac{1}{\gamma c}\exp\left(\frac{x}{\gamma c}\right)\int_d^x \{\rho p^*(t) + q^*(t)\}\exp\left(-\frac{t}{\gamma c}\right)dt$$

and must make

$$\int_d^b \{\rho p^*(x) + q^*(x)\}\exp\left(-\frac{x}{\gamma c}\right)dx = 0. \tag{6.5.66}$$

Routine calculations show that this equation reduces

$$\left(\frac{\rho P + Q}{\pi} - \frac{\rho m^2}{4\vartheta R}\right) I_0\left(\frac{m}{\gamma c}\right) + \left\{\frac{m}{2\vartheta}\left(\frac{\rho h}{R} - \delta\right) + \frac{m\rho}{2\vartheta R}c(\gamma - \alpha)\right\} I_1\left(\frac{m}{\gamma c}\right) = 0, \tag{6.5.67}$$

which agrees with Galin's equation (3.27).

The final equation to be satisfied is equation (6.5.4). We proceed as with equation (6.5.58). Let

$$u^{(1)\prime}(x, 0+) - u^{(2)\prime}(x, 0-) = F(x)$$
$$u^{*(1)\prime}(x, 0+) - u^{*(2)\prime}(x, 0-) = F^*(x)$$

then

$$F(x) - \alpha c\frac{dF}{dx} = F^*(x). \tag{6.5.68}$$

For $x \in (d, b)$, $F^*(x) = \delta$, so that

$$\left(\exp\left(-\frac{x}{\alpha c}\right)(F - \delta)\right)' = 0$$

which implies

$$\exp\left(-\frac{x}{\alpha c}\right)(F - \delta) = \text{ constant } x \in (d, b). \tag{6.5.69}$$

If therefore we put $F(b) = \delta$, then $F = \delta$ for $x \in (d, b)$. Equation (6.5.68) states that

$$F(x) = \frac{1}{\alpha c}\exp\left(\frac{x}{\alpha c}\right)\int_x^\infty F^*(t)\exp\left(-\frac{t}{\alpha c}\right)dt.$$

We must therefore enforce the condition

$$\int_b^\infty F^*(x)\exp\left(-\frac{x}{\alpha c}\right)dx = \alpha c\delta\exp\left(-\frac{b}{\alpha c}\right). \tag{6.5.70}$$

Equation (6.5.47) shows that

$$F^*(x) = -4\vartheta v_2^*(x)$$

and $v_2^* = v^* - \rho v_1^*$. Equations (6.5.49), (6.5.50) show that, for $x \in (b, \infty)$,

$$
\begin{aligned}
v_2^*(x) = {} & \frac{\rho}{8\vartheta R}\cdot\frac{m^2+2(x-\alpha c)(h-x)}{\sqrt{(x-d)(x-b)}} - \frac{\delta}{4\vartheta}\left\{1+\frac{h-x}{\sqrt{(x-d)(x-b)}}\right\} \\
& + \frac{\rho P + Q}{2\pi\sqrt{(x-d)(x-b)}} - \frac{\rho}{8\vartheta R}\cdot\frac{\ell^2+2(x-\alpha c)(g-x)}{\sqrt{(x+a)(x-b)}} \\
& - \frac{\rho P}{2\pi\sqrt{(x+a)(x-b)}}
\end{aligned} \tag{6.5.71}
$$

so that the condition (6.5.69) reduces to

$$
\left(\frac{\rho P+Q}{\pi} - \frac{\rho m^2}{4\vartheta R}\right) K_0\left(\frac{m}{\gamma c}\right) - \frac{m}{2\vartheta}\left(\frac{\rho h}{R} - \delta\right) K_1\left(\frac{m}{\gamma c}\right) = 0 \tag{6.5.72}
$$

where we have made use of equation (6.5.64).

We now have four equations: (6.5.58), (6.5.64), (6.5.67), (6.5.72). We may eliminate g from equations (6.5.58), (6.5.64) to obtain

$$
\left(\frac{P}{\pi} - \frac{\ell^2}{4\vartheta R}\right) D(\ell) + \frac{\ell c}{2\vartheta R}(\gamma-\alpha) K_1\left(\frac{\ell}{\alpha c}\right) I_1\left(\frac{\ell}{\gamma c}\right) = 0 \tag{6.5.73}
$$

where

$$
D(\ell) = K_1\left(\frac{\ell}{\alpha c}\right) I_0\left(\frac{\ell}{\gamma c}\right) + K_0\left(\frac{\ell}{\alpha c}\right) I_1\left(\frac{\ell}{\gamma c}\right).
$$

Given $P, \vartheta, R, c, \gamma, \alpha$, we can use this equation to compute the contact length 2ℓ. The solution of equation (6.5.72) was carried out numerically for values of the parameters α, γ of the order 10^{-1} sec to 10^6 sec; the speed c was taken to be 10, 10^2, 10^3 cm sec^{-1}. In all these cases, the contact length, 2ℓ, was close to the value

$$
2\ell = 4\sqrt{\frac{P\vartheta R\gamma}{\pi\alpha}}, \tag{6.5.74}
$$

which equation (6.5.32) shows is the elastic contact length for an elastic parameter $\vartheta^* = \vartheta\gamma/\alpha$.

Once ℓ is known, equation (6.5.64) gives the centre of the contact region as

$$
\frac{b-a}{2} = g = -c(\gamma-\alpha) K_0\left(\frac{\ell}{\alpha c}\right) I_1\left(\frac{\ell}{\gamma c}\right) / D(\ell). \tag{6.5.75}
$$

We may take various special cases. If $\gamma = \alpha$, then equation (6.5.74) shows that $g = 0$ and equation (6.5.72) shows that

$$
\frac{P}{\pi} = \frac{\ell^2}{4\vartheta R}. \tag{6.5.76}
$$

If $\alpha c \ll \ell$ and $\gamma c \ll \ell$, then we may take the asymptotic values of I_0, I_1, K_0, K_1:

$$I_0(x) \sim \frac{\exp(x)}{\sqrt{2\pi x}}, \quad I_1(x) \sim \frac{\exp(x)}{\sqrt{2\pi x}}$$

$$K_0(x) \sim \sqrt{\frac{\pi}{2x}}\exp(-x), \quad K_1(x) \sim \sqrt{\frac{\pi}{2x}}\exp(-x)$$

and find

$$g \sim -\frac{c(\gamma-\alpha)}{2}, \quad \frac{P}{\pi} \sim \frac{\ell^2}{4\vartheta R} - \frac{\ell c(\gamma-\alpha)}{4\vartheta R}. \tag{6.5.77}$$

We may treat equations (6.5.66), (6.5.71) similarly. On eliminating $(\rho h/R) - \delta$ we find

$$\left(\frac{\rho P + Q}{\pi} - \frac{\rho m^2}{4\vartheta R}\right) D(m) + \frac{m\rho c}{2\vartheta R}(\gamma-\alpha)K_1\left(\frac{m}{\alpha c}\right) I_1\left(\frac{m}{\gamma c}\right) = 0 \tag{6.5.78}$$

which we may solve to find m. Once m is known, equations (6.5.66), (6.5.71) give

$$\frac{\rho h}{R} - \delta = -\rho c(\gamma-\alpha)K_0\left(\frac{m}{\alpha c}\right) I_1\left(\frac{m}{\gamma c}\right)/D(m) \tag{6.5.79}$$

so that when $\gamma = \alpha$, we have

$$\frac{\rho P + Q}{\pi} = \frac{\rho m^2}{4\vartheta R}, \quad \delta = \frac{\rho h}{R} \tag{6.5.80}$$

and when $\alpha c \ll m$ and $\gamma c \ll m$, then

$$\frac{\rho h}{R} - \delta \sim \frac{\rho c(\gamma-\alpha)}{2}, \quad \frac{\rho P + Q}{\pi} \sim \frac{\rho m^2}{4\vartheta R} - \frac{\rho m c(\gamma-\alpha)}{4\vartheta R}. \tag{6.5.81}$$

The axis of the cylinder is subjected to the tangential reaction $-Q$. Also, since the resultant of the vertical stresses does not pass through the centre of the cylinder, there is a couple with moment

$$M_1 = \int_{-a}^{b} x p(x) dx. \tag{6.5.82}$$

Equation (6.5.52) gives

$$\int_{-a}^{b} p(x)dx - \gamma c \int_{-a}^{b} \frac{\partial p}{\partial x} dx = \int_{-a}^{b} p^*(x)dx$$

and

$$\int_{-a}^{b} \frac{\partial p}{\partial x} dx = [p(x)]_{-a}^{b} = 0$$

so that

$$P = \int_{-a}^{b} p(x)dx = \int_{-a}^{b} p^*(x)dx.$$

Similarly (6.5.52) gives

$$\int_{-a}^{b} xp(x)dx - \gamma c \int_{-a}^{b} x\frac{\partial p}{\partial x}dx = \int_{-a}^{b} xp^*(x)dx$$

and

$$\int_{-a}^{b} x\frac{\partial p}{\partial x}dx = [xp(x)]_{-a}^{b} - \int_{-a}^{b} p(x)dx = -P$$

and

$$\int_{-a}^{b} xp(x)dx = \int_{-a}^{b} xp^*(x)dx - \gamma cP.$$

Equation (6.5.50) gives

$$\int_{-a}^{b} xp^*(x)dx = Pg + \frac{\pi \ell^2}{4\vartheta R}(\alpha c - g),$$

and thus

$$M_1 = P(g - \gamma c) + \frac{\pi \ell^2}{4\vartheta R}(\alpha c - g), \tag{6.5.83}$$

which agrees with Galin's result. We note that $M_1 = 0$ when $\gamma = \alpha$.

The moment M_1 together with the moment $M_2 = QR$ about the centre of the cylinder, produces the rolling friction moment $M^* = M_1 + M_2$.

Plots of the contact pressure distribution, the contact width, 2ℓ; contact shift, g/ℓ; width of the stick zone, $2m$; the coefficient of rolling friction, $M^*/2P\ell$; may be found in Goryacheva (1998, section 3.5), where there are comments on the interpretation of the analysis.

6.6 The Limiting Case of Rolling of a Cylinder on a Viscoelastic Base

We consider the contact problem for a viscoelastic cylinder rolling on a viscoelastic base under the assumption that there is sliding in the whole of the contact region.

We assume that the materials of the cylinder and base are isotropic and linearly viscoelastic. The strain-stress relationships are taken as follows:

$$2\mu_i\left(\varepsilon_{xx} + \alpha\frac{\partial \varepsilon_{xx}}{\partial t}\right) = (1-\nu_i)\left(\sigma_{xx} + \gamma\frac{\partial \sigma_{xx}}{\partial t}\right) - \nu_i\left(\sigma_{yy} + \gamma\frac{\partial \sigma_{yy}}{\partial t}\right) \tag{6.6.1}$$

$$2\mu_i\left(\varepsilon_{yy} + \alpha\frac{\partial \varepsilon_{yy}}{\partial t}\right) = (1-\nu_i)\left(\sigma_{yy} + \gamma\frac{\partial \sigma_{yy}}{\partial t}\right) - \nu_i\left(\sigma_{xx} + \gamma\frac{\partial \sigma_{xx}}{\partial t}\right) \tag{6.6.2}$$

$$2\mu_i\left(\varepsilon_{xy} + \alpha\frac{\partial \varepsilon_{xy}}{\partial t}\right) = \sigma_{xy} + \gamma\frac{\partial \sigma_{xy}}{\partial t}. \tag{6.6.3}$$

Here α and γ are quantities characterising the viscous properties of the medium, and μ_1, ν_1 and μ_2, ν_2 are the elastic constants of the materials of the cylinder and base respectively.

The inertial terms are again neglected, so that the solution is valid for c much smaller than the speed of sound in the viscoelastic medium. As in Section 6.5, we suppose that the radius R of the cylinder is large compared to the contact length, so that the problem is the contact of two half-planes in which the contact region shifts with speed c.

As in Section 6.5, we assume that the displacements and stresses are functions of $x - ct$ and y, and we write

$$\varepsilon_{ij} + \alpha\frac{\partial\varepsilon_{ij}}{\partial t} = \varepsilon_{ij} - \alpha c\frac{\partial\varepsilon_{ij}}{\partial x} = \varepsilon_{ij}^*$$

$$\sigma_{ij} + \gamma\frac{\partial\sigma_{ij}}{\partial t} = \sigma_{ij} - \gamma c\frac{\partial\sigma_{ij}}{\partial x} = \sigma_{ij}^*$$

$$u - \alpha c\frac{\partial u}{\partial x} = u^*, \qquad v - \alpha c\frac{\partial v}{\partial x} = v^*$$

so that the starred quantities obey the usual elastic relations. Since the deformations are small, we satisfy them on the undeformed surface $y = 0$. We assume that the following relationship holds in the contact region $(-a, b)$:

$$\rho\sigma_{xx} + \sigma_{xy} = 0 \qquad x \in (-a, b).$$

The sign of the coefficient of friction depends on the direction of sliding, which in turn depends on the relative hardness of the contacting bodies (Reynolds, 1876).

Proceeding as in Section 6.5, we find the boundary conditions

$$\begin{aligned}
&v^{(1)*\prime}(x, 0+) - v^{(2)*\prime}(x, 0-) = -(x - \alpha c)/R && x \in (-a, b)\\
&\rho\sigma_{xx}^* + \sigma_{xy}^* = 0 && x \in (-a, b)\\
&\sigma_{xx}^* = 0 = \sigma_{xy}^* && x \in (-\infty, -a) \cup (b, \infty)
\end{aligned}$$

Since the cylinder and the base have different elastic constants, the equations (6.5.6), (6.5.7), (6.5.10), (6.5.11) must be modified:

$$\frac{u^{(1)*\prime}}{\vartheta_1}(x, 0+) = -\beta_1 p^*(x) + \frac{1}{\pi}\oint_{-\infty}^{\infty}\frac{q^*(t)}{t - x}dt, \tag{6.6.4}$$

$$\frac{v^{(1)*\prime}}{\vartheta_1}(x, 0+) = \frac{1}{\pi}\oint_{-\infty}^{\infty}\frac{p^*(t)}{t - x}dt + \beta_1 q^*(x), \tag{6.6.5}$$

$$\frac{u^{(2)*\prime}}{\vartheta_2}(x, 0-) = -\beta_2 p^*(x) - \frac{1}{\pi}\oint_{-\infty}^{\infty}\frac{q^*(t)}{t - x}dt, \tag{6.6.6}$$

$$\frac{v^{(2)*\prime}}{\vartheta_2}(x, 0-) = -\frac{1}{\pi}\oint_{-\infty}^{\infty}\frac{p^*(t)}{t - x}dt + \beta_2 q^*(x), \tag{6.6.7}$$

so that equations (6.5.2), (6.5.5) become

$$\frac{(\vartheta_1+\vartheta_2)}{\pi}\oint_{-\infty}^{\infty}\frac{p^*(t)}{t-x}dt+(\vartheta_1\beta_1-\vartheta_2\beta_2)q^*(x)=-\frac{(x-\alpha c)}{R},\qquad x\in(-a,b) \tag{6.6.8}$$

$$\rho p^*(x)+q^*(x)=0,\qquad x\in(-a,b) \tag{6.6.9}$$

$$p^*(x)=0=q^*(x),\qquad x\in(-\infty,-a)\cup(b,\infty). \tag{6.6.10}$$

On introducing the complex potential $w_1^*(z)$ from equation (2.2.28) we may write

$$-(\vartheta_1+\vartheta_2)v_1^{*+}(x)-\rho(\vartheta_1\beta_1-\vartheta_2\beta_2)u_1^{*+}(x)=-\frac{(x-\alpha c)}{2R},\qquad x\in(-a,b) \tag{6.6.11}$$

$$u_1^{*+}(x)=0,\qquad x\in(-\infty,-a)\cup(b,\infty). \tag{6.6.12}$$

We write equation (6.6.11) as

$$v^{*+}(x)+\rho\beta u^{*+}(x)=(x-\alpha c)/(2\vartheta R),\qquad x\in(-a,b) \tag{6.6.13}$$

where

$$v^{*+}(x)=v_1^{*+}(x),\quad u^{*+}(x)=u_1^{*+}(x), \tag{6.6.14}$$

$$\vartheta=(\vartheta_1+\vartheta_2),\quad \beta=(\vartheta_1\beta_1-\vartheta_2\beta_2)/\vartheta. \tag{6.6.15}$$

In the notation of equation (3.2.1), equation (6.6.12), (6.6.13) give a Riemann–Hilbert problem with

$$c(x)=\rho\beta,\quad d(x)=1,\quad f(x)=(x-\alpha c)/(2\vartheta R),\qquad x\in(-a,b)$$

$$c(x)=1,\quad d(x)=0,\quad f(x)=0,\qquad x\in(-\infty,-a)\cup(b,\infty).$$

Thus equation (3.2.4) gives

$$\omega(x)=\begin{cases}\pi\left(\frac{1}{2}-\theta\right) & x\in(-a,b)\\ 0 & x\in(-\infty,-a)\cup(b,\infty)\end{cases}$$

where $\pi\theta=\arctan(\rho\beta)$, and equation (3.2.7) gives

$$\begin{aligned}w_0^*(z)&=\exp\left[\left(\tfrac{1}{2}-\theta\right)\int_{-a}^{b}\tfrac{dt}{t-z}\right]\tfrac{iN(z)}{D(z)}\\&=\left(\tfrac{z-b}{z+a}\right)^{\frac{1}{2}-\theta}\cdot\tfrac{iN(z)}{D(z)}\end{aligned}$$

For a particular solution of equations (6.6.12), (6.6.13), we may take

$$w\,(z)=\frac{i(z-\alpha c)}{2\vartheta R}$$

so that

$$w^*(z) = \frac{i(z-\alpha c)}{2\vartheta R} + \left(\frac{z-b}{z+a}\right)^{\frac{1}{2}-\theta} \frac{iN(z)}{D(z)}$$

where we must choose $N(z)$ and $D(z)$ so that $w^*(z) \sim iP/(2\pi z)$ for large z. After some algebraic manipulation, we find

$$\begin{aligned} w^*(z) = {} & \frac{i}{4\vartheta R}\left\{2z + \frac{\ell^2(1-4\theta^2) + 2(g+2\ell\theta)z - 2z^2}{(z+a)^{\frac{1}{2}-\theta}(z-b)^{\frac{1}{2}+\theta}}\right\} \\ & - \frac{i\alpha c}{2\vartheta R}\left\{1 + \frac{g+2\ell\theta - z}{(z+a)^{\frac{1}{2}-\theta}(z-b)^{\frac{1}{2}+\theta}}\right\} + \frac{iP}{2\pi(z+a)^{\frac{1}{2}-\theta}(z-b)^{\frac{1}{2}+\theta}} \end{aligned} \tag{6.6.16}$$

The normal stress under the punch is twice the real part of $w(z+0)$, so that, on using (3.4.9) with θ replaced by $\frac{1}{2}-\theta$, we find

$$\begin{aligned} p^*(x) = {} & \frac{\cos\pi\theta}{2\vartheta R}\left\{\frac{\ell^2(1-4\theta^2) + 2(g+2\ell\theta)x - 2x^2}{(x+a)^{\frac{1}{2}-\theta}(b-x)^{\frac{1}{2}+\theta}}\right\} \\ & - \frac{\alpha c}{\vartheta R}\cos\pi\theta\left\{\frac{g+2\ell\theta - x}{(x+a)^{\frac{1}{2}-\theta}(b-x)^{\frac{1}{2}+\theta}}\right\} + \frac{P\cos\pi\theta}{\pi(x+a)^{\frac{1}{2}-\theta}(b-x)^{\frac{1}{2}+\theta}} \end{aligned} \tag{6.6.17}$$

To find the normal stress $p(x)$, we must solve

$$p(x) - \gamma c\frac{dp}{dx} = p^*(x)$$

so that, since $p(-a) = 0$, we have

$$p(x) = \frac{-1}{\gamma c}\cdot\exp\left(\frac{x}{\gamma c}\right)\int_{-a}^{x} p^*(t)\exp\left(\frac{-t}{\gamma c}\right)dt. \tag{6.6.18}$$

We use the function

$$\Phi(\sigma,\tau;x) = \frac{2^{1-\tau}\exp(x/2)}{B(\sigma,\tau-\sigma)}\int_{-1}^{1}(1-t)^{\tau-\sigma-1}(1+t)^{\sigma-1}\exp\left(\frac{xt}{2}\right)dt$$

so that

$$\begin{aligned} & \int_{-a}^{b}(x+a)^{\sigma}(b-x)^{\tau}\exp\left(\frac{-x}{\gamma c}\right)dx \\ & = \exp\left(\frac{-b}{\gamma c}\right)\ell^{\sigma+\tau+1}2^{\sigma+\tau+1}B(\sigma+1,\tau+1)\Phi\left(\sigma+1,\sigma+\tau+2;\frac{2\ell}{\gamma c}\right) \end{aligned}$$

where $B(x, y) = \Gamma(x)\Gamma(y)/\Gamma(x+y)$.

After algebraic simplification, we find the condition $p(b) = 0$ leads to the equation

$$\left\{\frac{P}{\pi} - \frac{2\ell^2}{\vartheta R}\left(\frac{1}{4} - \theta^2\right)\right\}\Phi\left(\frac{1}{2} - \theta, 1; \frac{2\ell}{\gamma c}\right)$$
$$+\frac{2\ell^2}{\vartheta R\alpha c}\left(\frac{1}{4} - \theta^2\right)\{g - 2\ell\theta - c(\alpha - \gamma)\}\Phi\left(\frac{3}{2} - \theta, 3; \frac{2\ell}{\gamma c}\right) = 0. \quad (6.6.19)$$

This agrees with equation (4.9) of Galin's text except that his equation has a factor of ℓ missing from the second term.

As a check, we note that when the roller and base have identical material, so that $\beta = 0$, $\vartheta = 2\vartheta_1$, and $\theta = 0$, we have

$$\Phi\left(\frac{1}{2}, 1; \frac{2\ell}{\gamma c}\right) = \exp\left(\frac{\ell}{\gamma c}\right) I_0\left(\frac{\ell}{\gamma c}\right)$$
$$\Phi\left(\frac{3}{2}, 1; \frac{2\ell}{\gamma c}\right) = \frac{2\gamma c}{\ell}\exp\left(\frac{\ell}{\gamma c}\right) I_1\left(\frac{\ell}{\gamma c}\right)$$

so that equation (6.6.19) reduces to equation (6.5.58).

The second equation is obtained as in (6.5.60) and is

$$\left\{\frac{P}{\pi} - \frac{2\ell^2}{\vartheta R}\left(\frac{1}{4} - \theta^2\right)\right\}\Psi\left(\frac{1}{2} - \theta, 1; \frac{2\ell}{\alpha c}\right)$$
$$-\frac{4\ell^2}{\vartheta R\alpha c}\left(\frac{1}{2} - \theta\right)(g - 2\ell\theta)\Psi\left(\frac{3}{2} - \theta, 1; \frac{2\ell}{\alpha c}\right) = 0 \quad (6.6.20)$$

where

$$\Psi(\sigma, \tau; x) = \frac{1}{\Gamma(\sigma)}\int_0^\infty \exp(-xt)t^{\sigma-1}(1+t)^{\tau-\sigma-1}dt.$$

When the base and roller have the same elastic properties, so that $\vartheta = 2\vartheta_1$ and $\theta = 0$, then

$$\Psi\left(\frac{1}{2}, 1; \frac{2\ell}{\alpha c}\right) = \pi^{-\frac{1}{2}}\exp\left(\frac{\ell}{\alpha c}\right) K_0\left(\frac{\ell}{\alpha c}\right) \quad (6.6.21)$$
$$\Psi\left(\frac{3}{2}, 1; \frac{2\ell}{\alpha c}\right) = \pi^{-\frac{1}{2}}\exp\left(\frac{\ell}{\alpha c}\right)\frac{\alpha c}{2\ell}\cdot K_1\left(\frac{\ell}{\alpha c}\right) \quad (6.6.22)$$

and equation (6.6.20) reduces to (6.5.64).

On eliminating $g - 2\ell\theta$ from equations (6.6.19), (6.6.20), we find that

$$\frac{P}{\pi} - \frac{2\ell^2}{\vartheta R}\left(\frac{1}{4} - \theta^2\right) = C(\theta)(\alpha - \gamma) \quad (6.6.23)$$

where $C(\theta) > 0$. This means that if $\alpha \geq \gamma$ and $|\theta| < 1/2$ then

$$\ell^2 \leq \frac{P}{\pi} \cdot \frac{\vartheta R}{2} / \left(\frac{1}{4} - \theta^2 \right) \tag{6.6.24}$$

with equality when $\alpha = \gamma$; in that case $g = 2\ell\theta$.

When the roller and the base have the same elastic parameters ($\vartheta = 2\vartheta_1, \theta = 0$) then equations (6.6.20)–(6.6.22) give

$$g = \frac{b-a}{2} = \frac{2}{\ell} \left(\frac{\vartheta_1 RP}{\pi} - \frac{\ell^2}{4} \right) K_0 \left(\frac{\ell}{\alpha c} \right) / K_1 \left(\frac{\ell}{\alpha c} \right). \tag{6.6.25}$$

When $\alpha = \gamma$, so that

$$\frac{P}{\pi} = \frac{2\ell^2}{\vartheta R} \left(\frac{1}{4} - \theta^2 \right), \quad g = 2\ell\theta \tag{6.6.26}$$

then (6.6.17) becomes

$$p^*(x) = \frac{\cos \pi\theta}{\vartheta R} \left\{ \frac{(x+a)(b-x) - \alpha c(2g - x)}{(x+a)^{\frac{1}{2}-\theta}(b-x)^{\frac{1}{2}+\theta}} \right\} \tag{6.6.27}$$

so that

$$p(x) = A(x+a)^{\frac{1}{2}+\theta}(b-x)^{\frac{1}{2}-\theta}.$$

The constant A may be found from the condition

$$\int_{-a}^{b} p(x)dx = \int_{-a}^{b} p^*(x)dx = P.$$

This gives

$$p(x) = \frac{1}{\vartheta R}(x+a)^{\frac{1}{2}+\theta}(b-x)^{\frac{1}{2}-\theta}. \tag{6.6.28}$$

As in equation (6.5.82), the moment applied by the loading is

$$M_1 = \int_{-a}^{b} xp(x)dx = \int_{-a}^{b} xp^*(x)dx - \gamma cP$$

which reduces to

$$M_1 = 2\pi \frac{\ell^2}{\vartheta R} \left(\frac{1}{4} - \theta^2 \right) \left(\alpha c + \frac{2\ell}{3}\theta - g \right) + P(g + 2\ell\theta - \gamma c) \tag{6.6.29}$$

which agrees with Galin's equation (4.18). The moment M_1 together with the moment $M_2 = QR$ of the force $Q = -\rho P$ with respect to the centre of the cylinder, produces the rolling friction moment $M^* = M_1 + M_2$.

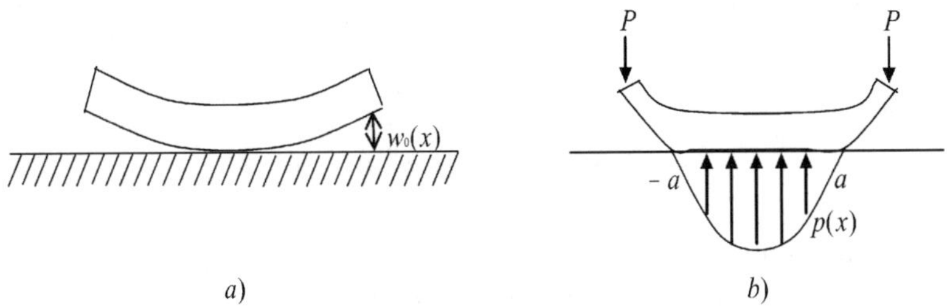

Fig. 6.7.1 The beam is in contact with a half-plane.

6.7 Contact Problems in the Presence of Wear

In this section we consider a problem in which an initially curved beam is dragged along a rigid half-space which is subject to wear.

We assume that the beam is slender, and its initial form is specified by a deflection $w_0(x)$ as in Figure 6.7.1a. After deformation caused by contact with the rigid half-plane, the beam assumes the form shown in Figure 6.7.1b. We assume that the beam is symmetric, and is in contact along the interval $(-a, a)$. We assume that the distance between a datum line in the beam, and the surface of the half-plane, remains constant. As a result of the wear, the pressure between the beam and the half-plane is gradually reduced. We assume that the wear is abrasive: the amount of material removed as a result of wear is proportional to the work done by the frictional forces. Experimental results relating to abrasive wear may be found in Khrushchov and Babichev (1960) and Hirst (1971). (If the wear is accompanied by a running-in process and, in particular, if the bodies in contact are constructed of the same material, the wear law may become nonlinear.)

Under these assumptions, the time rate of decrease, δ, of the deflection of the beam will be

$$\delta = K\sigma$$

where σ is the normal stress on the beam, and where K is the product of three factors: υ, the average speed of the beam (in the direction perpendicular to the plane of the diagram); K^*, a proportionality factor connecting the work done by the frictional forces and the amount of material removed; and ρ, the coefficient of friction: $K = \rho\upsilon K^*$.

We assume that the pressure, $\sigma \equiv p(x)$, is zero at the ends of the contact region. The wear can affect the positions of the end points of the contact region, but we neglect this since the amount of wear is small compared to the contact length.

The equation governing the deflection is

$$\frac{\partial w}{\partial t}(x, t) = -Kp(x, t) \tag{6.7.1}$$

where $p(x)$ is related to the deflection by

$$p(x,t) = EI\frac{\partial^4 w(x,t)}{\partial x^4},$$

so that

$$\frac{\partial w}{\partial t} = -KEI\frac{\partial^4 w}{\partial x^4}. \tag{6.7.2}$$

The initial condition is

$$w(x,0) = w_0(x). \tag{6.7.3}$$

We first seek separated solutions of equation (6.7.2), (6.7.3):

$$w(x,t) = \exp(-\alpha t)w_\alpha(x), \tag{6.7.4}$$

and find that $w_\alpha(x)$ satisfies

$$\alpha w_\alpha(x) = KEI\frac{d^4 w_\alpha}{dx^4} \tag{6.7.5}$$

which has the general solution

$$w_\alpha(x) = A_1 \sin\lambda x + B_2 \cos\lambda x + A_3 \sinh\lambda x + B_4 \cosh\lambda x \tag{6.7.6}$$

where

$$\lambda^4 = \frac{\alpha}{KEI}. \tag{6.7.7}$$

We now seek solutions $w_\alpha(x)$ that satisfy the end conditions; the beam may be either supported at $x = \pm a$, or clamped at $\pm a$.

$$w_\alpha(\pm a) = 0 = w_\alpha''(\pm a) \text{ (supported ends)}$$
$$w_\alpha(\pm a) = 0 = w_\alpha'(\pm a) \text{ (clamped ends)}$$

In either case, these conditions yield four homogeneous equations

$$\sum_{i=1}^{4} c_{ik}(\lambda)A_i = 0 \qquad k = 1, 2, 3, 4 \tag{6.7.8}$$

for the four unknowns $(A_i)_1^4$. These equations will have a non-trivial solution if and only if the determinant of coefficients is zero, and this happens if and only if λ takes one of an infinite sequence of eigenvalues $(\lambda_i)_1^\infty$. Each eigenvalue will give rise to a solution $w_{\alpha_i}(x) \equiv \phi_i(x)$; it may be shown that they yield a complete system of orthogonal functions on $(-a, a)$. (The $\phi_i(x)$ may be identified as normal vibrational modes of the beam.) To solve equation (6.7.1), we express $w_0(x)$ in the form

$$w_0(x) = \sum_{i=1}^{\infty} c_i\phi_i(x), \tag{6.7.9}$$

and find

$$w(x,t) = \sum_{i=1}^{\infty} c_i \exp(-\alpha_i t)\phi_i(x), \tag{6.7.10}$$

where $\alpha_i = KEI\lambda_i^4$.

Since the eigenvalues λ_i form an increasing sequence, we deduce that in studying the asymptotic solution for large time, it is sufficient to restrict the solution to only the first one or two terms.

We now consider the problem in which a stamp is dragged over an elastic layer in the direction perpendicular to the plane of the diagram in Figure 6.7.1. We investigate the contact problem under the following assumptions:

1. The layer rests on a rigid support, and there is no friction in the x-direction between the layer and the support.
2. The layer is rigidly bound to the support.
3. The modulus of elasticity of the layer material is a function of y, and is not zero when $y = 0$. The lower boundary of the layer may lie on the support without friction, or be rigidly bound to it.
4. The distance between a datum line in the stamp and the surface of the layer remains constant. This means that, as a result of wear, the pressure between the stamp and the layer gradually decreases.

Under these conditions, the displacement of the surface under the stamp is related to the pressure by an integral equation of the form

$$w(x) = \int_{-a}^{a} K(|x-\xi|)p(\xi)d\xi. \tag{6.7.11}$$

We note that the kernel is a function of only the distance between the field point x and source point ξ. The simplest example of this type is that for the half-plane (see equation (2.2.25)):

$$w(x) = -\frac{1}{\vartheta\pi}\int_{-a}^{a} \ell n\left(\frac{|x-\xi|}{a}\right) p(\xi)d\xi. \tag{6.7.12}$$

For a layer, under the stated conditions, the kernel will have at most a logarithmic singularity, which means that it will be square integrable. Its form will depend on the conditions existing between the layer and the support, and on the end conditions at $\pm a$.

The governing equations are

$$\frac{\partial w}{\partial t}(x,t) = -kp(x,t), \tag{6.7.13}$$

$$w(x,t) = \int_{-a}^{a} K(|x-\xi|)p(\xi,t)d\xi, \tag{6.7.14}$$

with the initial condition is

$$p(x, 0) = p_0(x), \qquad x \in (-a, a). \tag{6.7.15}$$

Again, we seek separated solutions of (6.7.13) - (6.7.15):

$$w(x, t) = w(x) \exp(-\alpha t) \qquad p(x, t) = p(x) \exp(-\alpha t) \tag{6.7.16}$$

and find

$$\alpha w(x) = k p(x),$$
$$w(x) = \int_{-a}^{a} K(|x - \xi|) p(\xi) d\xi,$$

which yield the eigenvalue problem

$$\int_{-a}^{a} K(|x - \xi|) p(\xi) d\xi = \lambda p(x), \qquad \lambda = \frac{k}{\alpha}. \tag{6.7.17}$$

Since the work done by the pressure $p(x)$ over the displacement must be positive,

$$W = \int_{-a}^{a} w(x) p(x) dx = \int_{-a}^{a} \int_{-a}^{a} K(|x - \xi|) p(x) p(\xi) dx d\xi > 0,$$

the kernel $K(|x - \xi|)$ will be positive definite, and the eigenvalues λ_n will be positive. The general theory of integral equations states that the sequence $\{\lambda_n\}_1^\infty$ will tend to zero, and the eigenfunctions $\{\phi_n(x)\}_1^\infty$ will form a complete orthogonal set in $(-a, a)$. We therefore write

$$p_0(x) = \sum_{n=1}^{\infty} A_n \phi_n(x) \tag{6.7.18}$$

and find

$$p(x, t) = \sum_{n=1}^{\infty} A_n \exp\left(-\frac{kt}{\lambda_n}\right) \phi_n(x). \tag{6.7.19}$$

Note that Galin's analysis in Galin (1976) is incorrect; he fails to realise that the eigenvalues of (6.7.17) will all be positive; he corrects this error in III.§6 of the book.

6.8 Axisymmetric Contact Problems in the Presence of Wear

The analysis of this section is based on Galin and Goryacheva (1977).

A cylindrical stamp, with end surface $z = f(r)$ in cylindrical polar coordinates, is situated on an elastic half-space and is acted on by a normal force P and a moment M rotating it about the z-axis with constant angular velocity ω. The force P and moment M can both vary with time. Contact occurs over a circle S of radius a, where $a \leq b$ and b is the radius of the cylinder. As the cylinder rotates, frictional forces $\sigma_{z\theta}$ act over the contact region, and $\sigma_{z\theta} = \alpha\sigma_{zz}$, where α is the coefficient of friction.

Since the problem is axisymmetric, all the displacements, strains and stresses are functions of r, z, t only: there is no dependence on θ. The displacement $w(r, 0, t)$ of the surface of the half-space is the difference between the initial displacement $f(r)$ and the displacement due to wear. This yields the boundary conditions

$$w = w(r, 0, t), \quad \sigma_{z\theta} = \alpha\sigma_{zz}, \quad \sigma_{zr} = 0 \text{ on } S, \tag{6.8.1}$$

$$\sigma_{zr} = 0 = \sigma_{z\theta} = \alpha\sigma_{zz} \text{ outside } S. \tag{6.8.2}$$

We recall from Section 5.8 that, for axisymmetry, the equilibrium equations are equation (5.8.16)–(5.8.18), of which the first and second involve only u and w, while the third involves only v; the equations are

$$\frac{\partial\sigma_{rr}}{\partial r} + \frac{\partial\sigma_{zr}}{\partial z} + \frac{\sigma_{rr} - \sigma_{\theta\theta}}{r} = 0, \tag{6.8.3}$$

$$\frac{\partial\sigma_{zr}}{\partial r} + \frac{\partial\sigma_{zz}}{\partial z} + \frac{\sigma_{zr}}{r} = 0; \tag{6.8.4}$$

and

$$\frac{\partial\sigma_{r\theta}}{\partial r} + \frac{\partial\sigma_{\theta z}}{\partial z} + \frac{2\sigma_{r\theta}}{r} = 0. \tag{6.8.5}$$

This means that the stress state for the rotating punch can be separated into two independent stress states * and **. The state * is determined from eqation (6.8.3), (6.8.4) with the boundary conditions

$$w^*(r, 0, t) = w(r, 0, t), \quad \sigma^*_{z\theta} = 0 = \sigma^*_{zr} \text{ on } S,$$

$$\sigma^*_{zr} = 0 = \sigma^*_{z\theta} = \sigma^*_{zz} \text{ outside } S,$$

while the state ** is determined from equation (6.8.5) with the boundary conditions

$$w^{**} = 0, \quad \sigma^{**}_{z\theta} = \alpha\sigma^{**}_{zz}, \quad \sigma^{**}_{zr} = 0 \text{ on } S,$$

$$\sigma^{**}_{zr} = 0 = \sigma^{**}_{z\theta} = \sigma^{**}_{zz} \text{ outside } S.$$

The sum of these two states, i.e., $\sigma_{z\theta} = \sigma^*_{z\theta} + \sigma^{**}_{z\theta}$, etc. will satisfy equations (6.8.1), (6.8.2).

The stress state * is thus obtained by solving the frictionless punch problem. This problem was solved in Section 5.1: the solution is given in equation (5.1.32). Written in polar coordinates this is

$$w^*(r, 0, t) = \frac{\vartheta}{2\pi}\int_0^a\int_0^{2\pi}\frac{p(\rho, t)\rho d\rho d\phi}{\sqrt{r^2 + \rho^2 - 2r\rho\cos\phi}}, \tag{6.8.6}$$

where

$$\sigma^*_{zz}(r, 0, t) = -p(r, t). \tag{6.8.7}$$

We now find the displacement of the boundary as the result of wear. When the wear is abrasive, the amount of material removed can be assumed to be proportional to the work done by the frictional forces, see Khrushchov and Babichev (1960) and Hirst (1971). Thus, taking the condition

$$\sigma_{z\theta} = \alpha\sigma_{zz} = -\alpha p(r, t) \tag{6.8.8}$$

into account, we find the rate of decay of the displacement of the boundary as follows

$$-\frac{\partial w}{\partial t}(r, t) = krw|\sigma_{z\theta}| \equiv k\alpha r\omega p(r, t)$$

where k is the wear coefficient. Thus

$$w(r, 0, t) = w(r, 0, 0) - k\alpha r\omega \int_0^t p(r, t)d\tau. \tag{6.8.9}$$

Combining this with equation (6.8.6), and noting that $w(r, t) = w^*(r, t)$ in S, we find

$$\frac{\vartheta}{2\pi}\int_0^a\int_0^{2\pi}\frac{p(\rho, t)\rho d\rho d\phi}{\sqrt{r^2 + \rho^2 - 2r\rho\cos\phi}} = w(r, 0, 0) - k\alpha r\omega\int_0^t p(r, \tau)d\tau. \tag{6.8.10}$$

The integral over ϕ is

$$\int_0^{2\pi}\frac{d\phi}{\sqrt{r^2 + \rho^2 - 2r\rho\cos\phi}} = \frac{4}{(r + \rho)}\cdot K\left(\frac{2\sqrt{r\rho}}{r + \rho}\right) \tag{6.8.11}$$

where $K(k)$ is the complete elliptic integral of the first kind. Equation (6.8.10) may be written

$$\frac{2\vartheta}{\pi}\int_0^a\frac{\rho}{(r + \rho)}K\left(\frac{2\sqrt{r\rho}}{r + \rho}\right)p(\rho, t)d\rho = w(r, 0) - k\alpha\omega r\int_0^t p(r, \tau)d\tau. \tag{6.8.12}$$

Write

$$\frac{2\vartheta}{\pi k\alpha\omega}\cdot\frac{1}{(r + \rho)}K\left(\frac{2\sqrt{r\rho}}{r + \rho}\right) = H(r, \rho), \tag{6.8.13}$$

so that equation (6.8.12) becomes

$$\int_0^a H(r, \rho)\rho p(\rho, t)dt = \frac{w(r, 0)}{k\alpha\omega} - \int_0^t rp(r, \tau)d\tau. \tag{6.8.14}$$

We seek the solution in the form

$$p(\rho, t) = \sum_{n=1}^{\infty} A_n p_n(\rho) \exp(-\beta_n t). \tag{6.8.15}$$

This will satisfy (6.8.14) if

$$\beta_n \int_0^a H(r, \rho)\rho p_n(\rho)d\rho - r p_n(r) = 0 \tag{6.8.16}$$

and

$$w(r, 0) = k\alpha\omega \sum_{n=1}^{\infty} \beta_n^{-1} A_n r p_n(r). \tag{6.8.17}$$

Equation (6.8.16) is a Fredholm integral equation of the first kind, and will have positive eigenvalues. To prove this, we need to show that the operator $\rho H(r, \rho)$ is positive definite, i.e., the integral

$$J(p) = \int_0^a r \int_0^a \rho H(r, \rho) p(r) p(\rho) d\rho dr > 0 \tag{6.8.18}$$

for arbitrary $p(r)$. Equation (5.1.32) states that the displacement $w(r)$ associated with the pressure $p(r)$ is

$$w(r) = k\alpha\omega \int_0^a H(r, \rho)\rho p(\rho) d\rho$$

so that

$$J(p) = \frac{1}{k\alpha\omega} \int_0^a r p(r) w(r) dr. \tag{6.8.19}$$

This displays $J(p)$ as a positive multiple of the work done by the pressure $p(r)$ on the displacement $w(r)$, a positive quantity.

Since $H(r, \rho)$ is symmetric in r and ρ, the eigenfunctions $\rho p_n(\rho)$ will be orthogonal, so that the coefficients A_n may be found from equation (6.8.17) in the form

$$\int_0^a r p_n(r) w(r, 0) dr = k\alpha\omega\beta_n^{-1} A_n \int_0^a r^2 p_n^2(r) dr. \tag{6.8.20}$$

Note that Galin, following Galin and Goryacheva (1977), finds the A_n by first reconstructing the initial pressure distribution $p(r, 0)$ associated with the initial displacement $w(r, 0)$ by using the analysis in Section 5.5, but this is unnecessary.

Equation (6.8.15) shows that the pressure $p(r, t)$ has a form similar to that in the two-dimensional contact problem (see equation (6.7.10), and Galin, 1976).

6.9 Plane Contact Problems for Rough Elastic Bodies

Contact problems in the linear theory of elasticity for bodies with rough surfaces have been considered by Shtaerman (1949), Popov and Savchuk (1971) and Mitrofanov (1970). Shtaerman (1949) was the first to suggest including an additional term corresponding to the displacement of the surface due to deformation of asperities, proportional to the normal pressure. However, experimental results by Demkin (1970), Kragelskii, Dobychin and Kombalav (1977) show that the displacement of rough contacting surfaces due to deformation of asperities is a power function of pressure with an exponent $\alpha \leq 1$. Rabinowich (1974a), Rabinowich (1974b) and Mkhitaryan and Schekyan (1977) considered such non-linear contact problems. A solution of a rough axially symmetric problem is given by Rabinowich (1975).

Following Goryacheva (1978) we consider the plane contact problem for a rough elastic layer of thickness h lying on a rigid foundation. The layer occupies the region $|x| < \infty$, $0 < y < h$. A rigid die with shape $y = g(x)$, $(g(0) = 0)$ is pressed against the upper surface $y = h$. The tangential stress is taken to be zero on the upper surfaces, as is the normal stress outside the contact region. The conditions at the lower surface of the layer may be one of two types:

1. The layer lies on the rigid foundation without friction. The governing equations are then

$$\sigma_{xy}(x,0) = 0, \quad v(x,0) = 0, \quad |x| < \infty, \tag{6.9.1}$$

$$\sigma_{xy}(x,h) = 0 = \sigma_{yy}(x,h), \quad a < |x| < \infty, \tag{6.9.2}$$

$$\sigma_{xy}(x,h) = 0 \quad v(x,h) = g(x) + \delta, \quad |x| < a. \tag{6.9.3}$$

2. The layer is bonded to the foundation. Now equation (6.9.1) is replaced by

$$u(x,0) = 0 = v(x,0), \quad |x| < \infty. \tag{6.9.4}$$

We assume that the displacement $v(x,h)$ of the upper surface of the layer is the sum of two terms: the first is due to the roughness

$$v_1 = A[p(x)]^{\alpha}. \tag{6.9.5}$$

Here $p(x)$ is the contact pressure, A is the coefficient characterising the deformation properties of the rough layer, and α is an exponent, $\alpha \leq 1$. The second is

$$v_2 = \frac{\vartheta}{\pi} \int_{-a}^{a} k\left(\frac{\xi - x}{h}\right) p(\xi) d\xi, \tag{6.9.6}$$

where the kernel k may be written

$$k(x) = \int_0^{\infty} \frac{L(\omega)}{\omega} \cos \omega x d\omega. \tag{6.9.7}$$

The kernel $L(\omega)$ depends on the conditions that hold at the layer/foundation interface:

1. $$L(\omega) = \frac{2\sinh^2\omega}{2\omega + \sinh 2\omega} \tag{6.9.8}$$

2. $$L(\omega) = \frac{2\kappa \sinh 2\omega - 4\omega}{2\kappa \cosh 2\omega + 4\omega^2 + 1 + \kappa^2}, \qquad \kappa = 3 - 4\nu. \tag{6.9.9}$$

(See the monograph *Developments in the Theory of Contact Problems in the USSR* (1976).) Inserting (6.9.5), (6.9.6) into the boundary condition (6.9.3), we find

$$A[p(x)]^\alpha + \frac{\vartheta}{\pi}\int_{-a}^{a} k\left(\frac{\xi - x}{h}\right) p(\xi)d\xi = g(x) + \delta, \qquad |x| < a. \tag{6.9.10}$$

This is an integral equation of Hammerstein type. We reduce it to standard dimensionless form.

Write

$$\frac{x+a}{2a} = s, \qquad \frac{\xi + a}{2a} = t, \qquad \frac{\delta}{2a} = \eta, \qquad \frac{2a}{h} = \lambda \tag{6.9.11}$$

$$p(x) = \frac{\pi}{\vartheta} q(s), \quad g(x) = 2af(s), \quad B = \frac{A}{2a}\left(\frac{\pi}{\vartheta}\right)^\alpha, \quad \gamma = \frac{1}{\alpha} \tag{6.9.12}$$

then

$$\int_0^1 k[\lambda(s-t)]q(t)dt + B[q(s)]^\alpha = f(s) + \eta. \tag{6.9.13}$$

Put

$$\psi(s) = B[q(s)]^\alpha - f(s) - \eta$$

so that

$$q(s) = B^{-\gamma}\{\psi(s) + f(s) + \eta\}^\gamma, \tag{6.9.14}$$

and equation (6.9.10) becomes

$$B^{-\gamma}\int_0^1 k[\lambda(t-s)]\{\psi(t) + f(t) + \eta\}^\gamma ds + \psi(s) = 0. \tag{6.9.15}$$

In this notation, the total dimensionless force applied to the punch is

$$Q = \int_0^1 q(s)ds = B^{-\gamma}\int_0^1 \{\psi(s) + f(s) + \eta\}^\gamma ds. \tag{6.9.16}$$

We solve equation (6.9.15) by iteration starting with $\psi_0(s) = 0$ and

$$\psi_{n+1}(s) = -B^{-\gamma}\int_0^1 k[\lambda(t-s)]\{\psi_n(t) + f(t) + \eta\}^\gamma dt. \tag{6.9.17}$$

We now prove the convergence of the iterative sequence (6.9.17).

1. The kernel $k(t)$ may be expressed in the form

$$k(t) = -\ell n|t| + F(t)$$

where $F(t)$ is continuous; this follows from equations (6.9.7)–(6.9.9), and shows that the kernel is square integrable.

2. The function

$$q(s,u) = B^{-\gamma}\gamma\{u + f(s) + \eta\}^{\gamma}$$

satisfies a uniform Lipschitz condition

$$|q(s,u_1) - q(s,u_2)| < c(s)|u_1 - u_2| \tag{6.9.18}$$

in the interval of variation of $\psi(s)$. To prove this, we note that the normal pressure $q(s)$ is non-negative, while u is non-positive:

$$-f(s) - \eta \le u \le 0.$$

We use the identity

$$1 - x^{\gamma} < \gamma(1-x), \quad x > 0, \quad \gamma > 1$$

to give (6.9.18) with

$$c(s) = B^{-\gamma}\gamma|f(s) + \eta|^{\gamma-1} \tag{6.9.19}$$

3. The function

$$q(s,0) = B^{-\gamma}\{f(s) + \eta\}^{\gamma}$$

is square integrable.

We now examine the sequence (6.9.17):

$$\begin{aligned}&\psi_{n+1}(s) - \psi_n(s)\\&\quad = -B^{-\gamma}\int_0^1 k[\lambda(t-s)]\{(\psi_n(t) + f(t) + \eta)^{\gamma} - (\psi_{n-1}(t) + f(t) + \eta)^{\gamma}\}dt\end{aligned}$$

so that, on using (6.9.19), we have

$$|\psi_{n+1}(s) - \psi_n(s)| \le \int_0^1 k[\lambda(t-s)]c(t)|\psi_n(t) - \psi_{n-1}(t)|dt$$

and thus, in terms of the Euclidian norm

$$||\psi||^2 = \int_0^1 \psi^2(s)ds$$

we find

$$||\psi_{n+1} - \psi_n||^2 = \int_0^1 \left\{\int_0^1 k[\lambda(t-s)]c(t)|\psi_n(t) - \psi_{n-1}(t)|dt\right\}^2 ds.$$

Now use the Cauchy–Schwarz inequality

$$\left\{\int_0^1 f(t)g(t)dt\right\}^2 \le \int_0^1 f^2(t)dt \cdot \int_0^1 g^2(t)dt$$

to obtain

$$||\psi_{n+1} - \psi_n||^2 \le C||\psi_n - \psi_{n-1}||^2$$

where

$$C = \int_0^1 \int_0^1 k^2[\lambda(s-t)]c^2(t)dsdt. \tag{6.9.20}$$

Thus the sequence will converge if $C < 1$.

We now show that equation (6.9.15) has a unique solution, so that the sequence $\{\psi_n\}$ will converge to this solution. Suppose, if possible, that equation (6.9.15) had two solutions $\psi_1(s)$ and $\psi_2(s)$ and write $\Psi(s) = \psi_1(s) - \psi_2(s)$, so that on subtracting the equations for ψ_1 and ψ_2 we find

$$\Psi(s) + B^{-\gamma} \int_0^1 k[\lambda(t-s)]\varphi[t, \Psi(t)]dt = 0 \tag{6.9.21}$$

where

$$\varphi[t, \Psi(t)] = \{\psi_2(t) + \Psi(t) + f(t) + \eta\}^\gamma - \{\psi_2(t) + f(t) + \eta\}^\gamma.$$

Multiply equation (6.9.21) by $\varphi[s, \Psi(s)]$ and integrate over (0,1) to obtain

$$\int_0^1 \Psi(s)\varphi[s, \Psi(s)]ds + B^{-\gamma} \int_0^1 \int_0^1 k[\lambda(t-s)]\varphi[t, \Psi(t)]\varphi[s, \Psi(s)]dsdt = 0. \tag{6.9.22}$$

The function $\varphi[s, \Psi(s)]$ is positive for $\Psi(s) > 0$ and negative if $\Psi(s) < 0$, so that the first integral in equation (6.9.22) is positive. On the other hand, the kernel $k[\lambda(t-s)]$ is positive semi-definite, i.e.,

$$J(\omega) \equiv \int_0^1 \int_0^1 k[\lambda(t-s)]\omega(s)\omega(t)dsdt \ge 0$$

because (see (6.9.6)), apart from a positive factor, it is the work done by a pressure on a displacement; it is zero if and only if $\omega(t) \equiv 0$. Thus, both terms on the right of equation (6.9.22) are positive, implying that $\Psi(s) = 0$, i.e., $\psi_1(s) = \psi_2(s)$.

Thus, we can obtain the solution $\psi(s)$ by finding the limit of the sequence $\{\psi_n\}$. The pressure $p(x) = \pi q(s)/\vartheta$ is then obtained from equation (6.9.14).

Note that the contact pressure does not tend to infinity at the ends of the contact region. To show this, we assume that $q(t)$ in equation (6.9.13) had an integrable singularity $t^{-\theta}$ $(0 < \theta < 1)$ at zero. The first term in (6.9.13) will not be singular at $s = 0$ while the second will have the form $s^{-\alpha\theta}$, and there will be no singularity on the right hand side.

If a die with smooth macroshape penetrates the rough layer, the additional conditions $p(\pm a) = 0$, which reflect the continuity of pressure at the ends of the punch, can be used to find the extent of the contact region.

The method described above can be used to solve different plane contact problems for a rough layer, and to analyse the dependence of the contact pressure distribution on the layer thickness, roughness parameters, elastic characteristics of the layer, etc.

As an example, we consider the problem of frictionless contact between a punch with a flat base ($f(x) = 0$) and a rough layer. To determine the contact pressure, we use the integral equation (6.9.1) with $g(s) \equiv 0$ and

$$k(t) = -\ell n|t| + a_0$$

where $a_0 = -0.352$ for case 1, $a_0 = -0.527$ for case 2 ($\nu = 0.3$). This asymptotic representation of the kernel holds for comparitively thick layers, i.e., $\lambda^2 = (2a/h)^2 = o(1)$ (see the monograph *Development of the Theory of Contact Problems in the USSR*).

The problem is attacked by solving equation (6.9.15) by iteration. Then we obtain the contact pressure as

$$q(s) = B^{-\gamma}\{\psi(s) + \eta\}^{\gamma}$$

where $\psi(s)$ is the limit of the sequence $\{\psi_n(s)\}$ determined by

$$\psi_{n+1}(s) = B^{-\gamma}\int_0^1 (\ell n|t-s| + c_0)\{\psi_n(s) + \eta\}^{\gamma}\,dt$$

$$c_0 = \ell n\left(\frac{2a}{h}\right) - a_0.$$

This limit exists if the condition (6.9.20) holds, which has the following form in this case

$$\gamma^2 B^{-2\gamma}\eta^{2\gamma-2}\left(c_0^2 - 3c_0 + 3.5\right) < 1.$$

Based on experimental results due to Demkin (1970) and Kragelskii et al. (1982), we used the following values for the numerical calculations: $\alpha = 0.4$, $B = 1$, $c_0 = -3$.

Figure 6.9.1 illustrates the pressure distribution for two values of the dimensionless load Q:

$$Q^{(1)} = 0.006 \text{ (curve 1) and } Q^{(2)} = 0.0075 \text{ (curve 2).}$$

The values Q are related to the actual loads P by

$$P = \frac{2\pi a}{\vartheta}\cdot Q.$$

Penetrations for the cases $Q^{(1)}$ and $Q^{(2)}$ are $\eta^{(1)} = 0.15$ and $\eta^{(2)} = 0.17$, where $\eta = \delta/(2a)$. The results indicate that for the same roughness parameter, the pressure increases especially at the periphery of the contact region, as the load increases. For

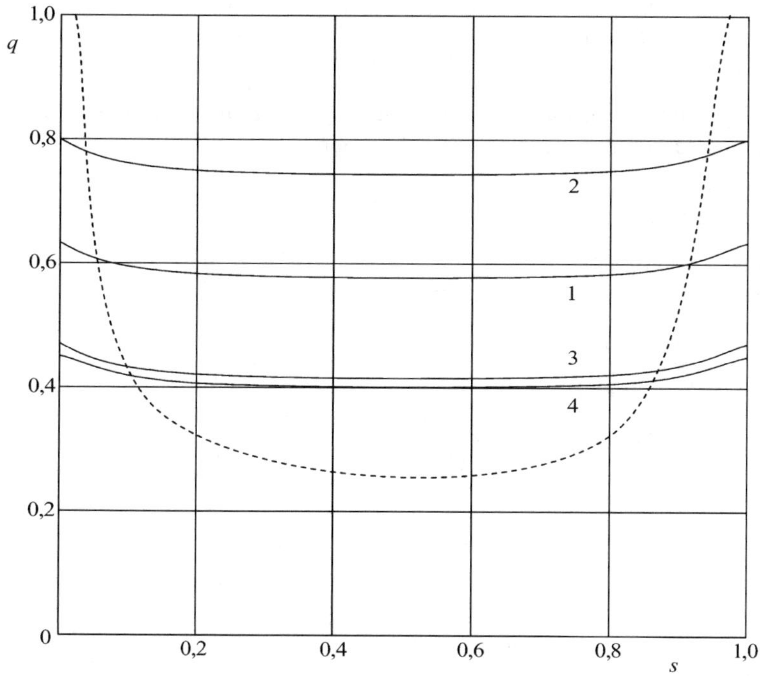

Fig. 6.9.1 The pressure distribution under the punch.

a fixed load $Q = 0.0041$, the penetration and the pressure distribution depend on the roughness parameters B and α. For $B = 0.75$, $\alpha = 0.4$ the penetration is $\eta = 0.1$; for $B = 0.35$, $\alpha = 0.4$ (the smoother surface) the penetration is smaller, $\eta = 0.06$. The graphs of pressure distributions for the two cases are shown by curves 3 and 4 in Figure 6.9.1 respectively; the pressure distribution for the smooth punch is shown by a broken line; it is singular at the ends of the contact region. The calculation showed the fast convergence of the iterative method: for an accuracy of 10^{-5} it is sufficient to take $15 \sim 20$ iterations for $\psi(x)$. The process converges for a wide range of roughness parameters and elastic constants of the layer.

6.10 Contact Problems for Rough Axisymmetric Elastic Bodies

Following Goryacheva (1978), consider the contact of an axisymmetric punch and a rough elastic half-space $z > 0$. The macroshape of the contacting surface of the punch is given by the function $z = g(r)$, where $g(0) = 0$. The punch is pressed into the half-space by a force P. The contact region is a disc of radius a. The elastic displacements of the surface of the half-space due to the pressure from the punch with given macroshape and given penetration δ are the sum of two parts: the displacement

$$w_1(r) = \frac{\vartheta}{2\pi} \int_0^a \int_0^{2\pi} \frac{\rho p(\rho) d\rho d\theta}{\sqrt{r^2 + \rho^2 - 2r\rho\cos\theta}}, \tag{6.10.1}$$

given in equation (5.1.32) as the displacement by the nominal pressure $p(\rho)$ distributed over a disc of radius a, and the displacement

$$w_2(r) = A[p(r)]^\alpha, \quad A \geq 0 \tag{6.10.2}$$

given in equation (6.9.5).

We introduce dimensionless qualitites:

$$s = \frac{r}{a}, \quad \sigma = \frac{\rho}{a}, \quad p(r) = \frac{2\pi}{\vartheta} q(s), \quad g(r) = af(s)$$

$$\frac{\delta}{a} = \eta, \quad B = \frac{A}{a}\left(\frac{2\pi}{\vartheta}\right)^\alpha, \quad Q = \frac{\vartheta}{2\pi a^2} \cdot P, \quad \gamma = \frac{1}{\alpha}$$

so that the equation

$$w_1(r) + w_2(r) = g(r) + \delta$$

becomes

$$\int_0^1 \int_0^{2\pi} \frac{\sigma q(\sigma) d\sigma d\theta}{\sqrt{s^2 + \sigma^2 - 2s\sigma\cos\theta}} + B[q(s)]^\alpha = f(s) + \eta. \tag{6.10.3}$$

Put

$$\psi(s) = B[q(s)]^\alpha - f(s) - \eta$$

then

$$q(s) = B^{-\gamma}\{\psi(s) + f(s) + \eta\}^\gamma$$

and equation (6.10.3) becomes

$$B^{-\gamma} \int_0^1 k(s, \sigma)\{\psi(\sigma) + f(\sigma) + \eta\}^\gamma \sigma d\rho + \psi(s) = 0 \tag{6.10.4}$$

where

$$k(s, \sigma) = \int_0^{2\pi} \frac{d\theta}{\sqrt{s^2 + \sigma^2 - 2s\sigma\cos\theta}}.$$

The Hammerstein integral equation (6.10.4) is solved by iteration as in Section 6.9. The kernel $k(s, \rho)$ is square integrable, and the function

$$h(s, u) = B^{-\gamma} s\{u + f(s) + \eta\}^\gamma$$

satisfies a uniform Lipschitz condition

$$|h(s, u_1) - h(s, u_2)| < c(s)|u_1 - u_2|$$

where

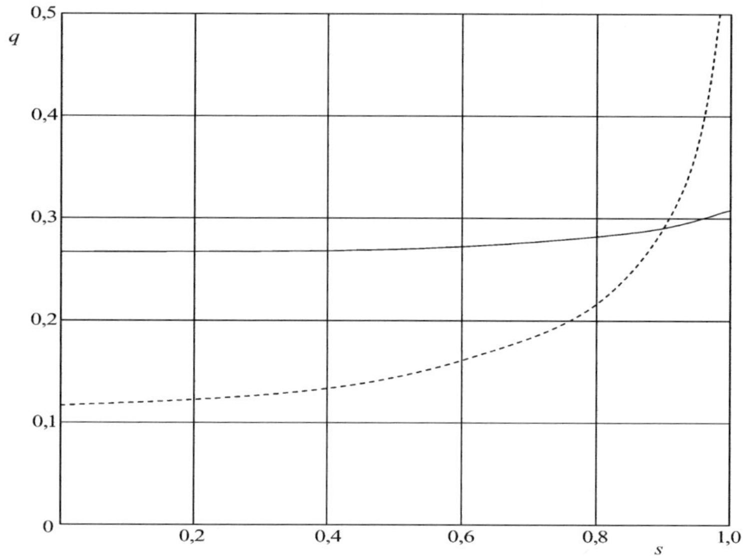

Fig. 6.10.1 The pressure distribution for a rough (full line) and smooth (broken line) flat circular punch.

$$c(s) = \alpha B^{-\gamma} s\{f(s) + \eta\}^{\gamma-1}.$$

The function $h(s, 0) = B^{-\gamma} s(f(s) + \eta)^{\gamma}$ is square integrable, so that the sequence $\psi_n(s)$ converges almost everywhere to the solution of equation (6.10.4). See Tricomi (1957) if the parameters of the problem satisfy

$$\alpha^2 B^{-2\gamma} \int_0^1 \int_0^1 s^2 (f(s) + \eta)^{2\gamma-2} k^2(s, \sigma) ds d\sigma < 1.$$

The uniqueness of the solution may be proved as in Section 6.9.

As an example, we consider the problem of contact between a rough elastic half-space and a cylindrical punch with a flat base ($g(r) \equiv 0$). The following values of the parameters were used in the numerical calculation: $\alpha = 0.4$, $\beta = 0.9$, $\eta = 0.1$. The calculation showed the fast convergence of the iterative method. For an accuracy of 10^{-5} it was sufficient to take 12 iterations for $\psi(x)$. Figure 6.10.1 shows the dimensionless pressure distribution $q(s)$. The pressure distribution for a smooth punch is shown by the broken line. For both cases, the punch is acted on by a dimensionless load $Q = 0.008625$.

References

There are three journals containing many of the references:

[**1**] *Doklady Akad. Nauk SSSR = Reports of the USSR Academy of Sciences.*
[**2**] *Prikl. Matematika i Mekhanika = Journal of Applied Mathematics and Mechanics.*
[**3**] *Journal of Applied Mathematics and Mechanics*, English translation of [**2**].

This bibliography complements and is complemented by that appearing in Gladwell (1980).

Abramov, V.M. (1937) The problem of the contact of an elastic half-plane with an absolutely rigid foundation in the presence of friction. [**1**], **17**, no. 4 [in Russian].

Abramov, V.M. (1939) Investigation of the case of non-symmetric pressure of a stamp of circular cross-section on an elastic half space. [**1**], **23**, no. 3 [in Russian].

Abramowitz, M. and Stegun, I. (1972) *Handbook of Mathematical Functions*, 2nd Ed. New York: Dover Publications.

Afonkin, M.N. (1941) Calculation of foundations of the cut type. *Gidrotekhnicheskoye stroitelstvo* no. 6 [in Russian].

Arscott, F.M. (1962) *Periodic Differential Equations: an Introduction to Mathieu, Lamé and Allied Functions*. New York: MacMillan.

Begiashvili, A.I. (1940) Solution of the problem of the pressure of a system of rigid profiles on the straight boundary of an elastic half-plane. [**1**], **27**, 914-916 [in Russian].

Belyaev, N.M. (1917) Application of Hertz's theory to the calculation of local stresses at the point of contact of a wheel and a rail. *Vestnik izhenerov i tekhnikov*, no. 2 [in Russian].

Belyaev, N.M. (1924) Local stresses as a result of the pressure between bodies in contact. *Inzhenernye sooruzheniya i stroitelnaya mekhanika*. Leningrad [in Russian].

Belyaev, N.M. (1929a) Calculation of the maximum design stresses resulting from the pressure between bodies in contact. *Sbornik Leningradskogo instituta inzhenerov putei soobscheniya* no. 102 [in Russian].

Belyaev, N.M. (1929b) On the question of local stresses in connection with the resistance of rails to rushing. *Sbornik Leningradskogo instituta inzhenerov putei soobscheniya*, no. 90 [in Russian].

Borowicka, H. (1943) On eccentrically loaded rigid plates on an elastic isotropic foundation. *Ingenieur Archiv*. **14**, 1–8 [in German].

Boussinesq, J. (1885) *Application des Potentials à l'Étude de l'Equilibre et du Mouvement des Solides Élastiques*. Gauthier-Villars: Paris.

Byrd, P.F. and Friedman (1971) *Handbook of Elliptic Integrals*. Berlin: Springer Verlag.

Chaplygin, S.A. (1950) Pressure of a rigid stamp on an elastic foundation. *Collected Works*, **3**, 317–323 [in Russian].

Courant, R. and Hilbert, D. (1953) *Methods of Mathematical Physics*. 2 Vols. New York: Interscience Publishers.

Demkin (1970) *Contact of Rough Surfaces*, Moscow: Nauka.

Dinnik, A.N. (1906) Hertz's formula and its experimental verification. *Zhurnal russk. fiz.-khim. ob-va, fiz. otd*, **38**, part 1, no. 4, 242–249 [in Russian].

Dundurs, J. and Stippes, M. (1970) Role of elastic constants in certain contact problems. *Journal of Applied Mechanics*, **37**, 965–970.

Erdelyi, A. (Ed) (1955) *Higher Transcendental Functions.* (3 Vols.) New York: McGraw Hill.

Falkovich, S.V. (1946) On the pressure of a rigid stamp on an elastic half-space in the presence of local adhesion and slip. [**2**], **9**, no. 5, 425–432 [in Russian].

Filippov, A.P. (1942) Infinitely long beam on an elastic half-space. [**2**], **6**, no. 2–3 [in Russian].

Florin, V.A. (1936a) Determination of the state of stress in an elastic foundation. *Sbornik Gidrostroyproekta*, no. 1 [in Russian].

Florin, V.A. (1936b) Determination of the state of stress in an elastic foundation. *Sbornik Gidrostroyproekta*, no. 1 [in Russian].

Florin, V.A. (1948) *Design of the Bases of Hydrotechnical Buildings*. Stroyizdat [in Russian].

Fromm, H. (1927) Calculation of the slipping in the case of rolling deformable bars. *ZAAM*, **7**, no. 1 [in German].

Galin, L.A. (1943a) The mixed problem of the theory of elasticity with frictional forces for the half-plane. [**1**], **39**, no. 3, 88–93 [in Russian].

Galin, L.A. (1943b) On the Winkler-Zimmerman hypothesis for beams. [**2**], **7**, 293–300 [in Russian].

Galin, L.A. (1945) Pressure of a stamp in the presence of friction and adhesion. [**2**], **9**, no. 5, 413–424 [in Russian].

Galin, L.A. (1946) Spatial contact problems of the theory of elasticity for a stamp with circular planform. [**2**], **10**, 425–448 [in Russian].

Galin, L.A. (1947a) On the pressure of a stamp of elliptical planform on an elastic half-space. [**2**], **11**, 281–284 [in Russian].

Galin, L.A. (1947b) Pressure of a stamp in the form of an infinite wedge on an elastic half-space. [**1**], **58**, no. 2 [in Russian].

Galin, L.A. (1948a) Estimate of the displacements in spatial contact problems of the theory of elasticity. [**2**], **12**, 241–250 [in Russian].

Galin, L.A. (1948b) On the pressure of a rigid body on a plate. [**1**], **12**, no. 3 [in Russian].

Galin, L.A. (1967) Deformation of an orthotropic viscoelastic body in the 2-D formulation. [**1**], **77**, no. 4 [in Russian].

Galin, L.A. (1976) Contact problems of the theory of elasticity in the presence of wear. [**3**], **40**, 981–986.

Galin, L.A. and Goryacheva, I.G. (1977) Axisymmetric contact problems of the theory of elasticity in the presence of wear. [*3*], *41*, 826–831.

Gastev, V.A. (1937) On the stresses in an elastic medium bounded by a plane as a result of the loading of an infinite rigid plate. *Sbornik Leningrad. in-ta zh. d. transp.* no. 127 [in Russian].

Gladwell, G.M.L. (1980) *Contact Problems in the Classical Theory of Elasticity*. Alphen aan den Rijn: Sijthoff and Noordhoff.

Glagolev, N.I. (1942) Elastic stresses along the bases of dams. [**1**], **34**, no. 7, 204–208 [in Russian].

Glagolev, N.I. (1943) Determination of the stresses arising from a system of rigid stamps. [**2**], **7**, no. 5, 383–388 [in Russian].

Glagolev, N.I. (1945) Resistance to rolling of cylindrical bodies. [**2**], **9**, no. 4 [in Russian].

Gorbunov-Posadov, M.I. (1937) Design of a beam on an elastic foundation under the conditions of the plane problem of the theory of elasticity. *Sbornik NIS Fundamentstroya*, no. 8 [in Russian].

Gorbunov-Posadov, M.I. (1939a) Beams and rectangular plates lying on an elastic half space foundtion. [**1**], **24**, no. 5, 421–425 [in Russian].

Gorbunov-Posadov, M.I. (1939b) *Tables for the Design of Beams on an Elastic Foundation.* Moscow: Gosstroyizdat [in Russian].

Gorbunov-Posadov, M.I. (1940) Design of beams and plates on an elastic half-space. [**2**], **4**, 61–80 [in Russian].

Gorbunov-Posadov, M.I. (1941) *Slabs on Elastic Foundations*. Moscow: Gosstroyizdat [in Russian].

Gorbunov-Posadov, M.I. (1946a) Design of centrally-loaded foundation plates. *Inzhenernyi Sbornik*, **3**, no. 1 [in Russian].
Gorbunov-Posadov, M.I. (1946b) Semi-infinite beam on an elastic foundation. *Inzhenernyi Sbornik*, **4**, no. 1 [in Russian].
Gorbunov-Posadov, M.I. (1946c) *Settling of a Foundation on a Layer of Soil Supported by a Rock Base*. Moscow: Stroiizdat [in Russian].
Gorbunov-Posadov, M.I. (1948) An infinite beam on an elastic half-space. *Inzhenernyi Sbornik*, **3**, no. 2 [in Russian].
Gorbunov-Posadov, M.I. (1949) *Beams and Plates on an Elastic Foundation*. Izd. Ministerstva stroitelstva predpriyatiy mashinustroeniya [in Russian].
Goryacheva, I.G. (1970) A contact problem for a viscoelastic half-plane. **[3]**, **34**, 765-767.
Goryacheva, I.G. (1973a) Contact problem of rolling of a viscoelastic cylinder on a base of the same material. **[2]**, **37**, 925-933; **[3]**, **37**, 877-885.
Goryacheva, I.G. (1973b) The limiting case of rolling of a cylinder on a viscoelastic base. *Proceedings of the Postgraduate Students of Moscow State University*, Moscow: MSU [in Russian].
Goryacheva, I.G. (1978) Plane and axisymmetric contact problems for rough elastic bodies. [2], **43**, 99-105; **[3]**, **43**, 104-111.
Goryacheva, I.G. (1998a) *Contact Mechanics in Tribology*. Dordrecht: Kluwer.
Gradshteyn, I.S. and Ryzhik, I.M. (1965) *Tables of Integrals, Series and Products*. 4th Ed. Moscow 1963. English translation, Jeffrey, A. (Ed.) New York: Academic Press.
Harding, J.W. and Sneddon, I.N. (1945) The elastic stresses produced by the indentation of the plane surface of a semi-infinite elastic solid by a rigid punch. *Proc. Camb. Phil. Soc.* **41**, 16–26.
Hertz, H. (1882a) Über die Berührung fester elastischer Körper. *Journal für die reine und angewandte Mathematik*. **92**, 156-171. In Hertz, H., *Gesammelte Werke*, Band 1. Leipzig: Johann Ambrosius Barth (1895), 155–173.
Hertz, H. (1882b) Über die Berührung fester elastischer Körper und über die Harte. In Hertz, H., *Gesammelte Werke*, Band 1. Leipzig: Johann Ambrosius Barth (1895), 174–196.
Hertz, H. (1895) *Collected Works*, pp. 179–185 [in German].
Hirst, V. (1971) Wear of brittle materials. In collection. *Contact Interaction of Solids, Computation of Frictional Forces and Wear*. Moscow: Nauka [in Russian].
Hobson E.W. (1900) On Green's function for a circular disc, with applications to electrostatic problems. *Proc. Camb. Phil. Soc.* **18**, 277-291.
Hobson, E.W. (1931) *The Theory of Spherical and Ellipsoidal Harmonics*. London: Cambridge University Press.
Ishkova, A.G. (1947) An exact solution of the problem of bending of a round plate supported by an elastic half-space and subjected to an axially symmetrical uniformly distributed load. *Inzhenernyi Sbornik*, **56**, 129–132 [in Russian].
Kartsivadze, I.N. (1943) The fundametal problems of the theory of elasticity for the elastic disc. *Trudy Tbilissk. Mat. Inst.* **12**, 95–104 [in Georgian with Russian summary].
Kartsivadze, I.N. (1946) Effective solution of the fundamental problems of the theory of elasticity for certain regions. *Soobshcheniya A.N. Gruz SSR*. **7**, no. 8, 507–513 [in Russian].
Keer, L.M. and Parihar, K.S. (1978) A not on the singularity at the corner of a wedge-shaped punch or crack. *SIAM J. Appl. Math.* **34**, 297–302.
Klubin, P.I. (1938) State of stress in an elastic medium loaded by a perfectly rugid strip of constant width. *Trudy Leningrad. in-ta inzh. prom. stroit.* no. 6 [in Russian].
Kochin, N.E. (1940) Theory of a wing of finite span with circular form in a plane. [2], **4**, 3–32 [in Russian].
Kragelskii, I.V., Dobychin, M.N., and Kombalav, V.S. (1977) *Foundations of Calculus on Friction and Wear*. Moscow: Mashinostroenie [in Russian].
Kragelskii, I.V., Dobychin, M.N., and Kombalav, V.S. (1982) *Friction and Wear: Calculation Methods*. Oxford: Pergamon Press.
Krushchov, M.M. and Babichev, M.A. (1960) Wear of metals. *Izd. Akad. Nauk. SSSR. Mekhan.* [in Russian].

Kuznetsov, V.I. (1938) *Beams on a Continuous Elastic Foundation*. Transzheldorizdat [in Russian].

Kuznetsov, V.I. (1939) Design of beams lying on a continuous elastic foundation, treated as an isotropic elastic half-space. *Trudy NIIT*, no. 55 [in Russian].

Kuznetsov, V.I. (1940) *Questions on the Design of the Upper Structure of a Railroad*. Transzheldorizdat [in Russian].

Lebedev, N.N. (1937) The functions associated with a ring of oval cross-section. *Technical Physics of the USSR*. **4**, 1–24.

Leonov, M.Ia. (1939) On the theory of design of elastic foundations. [**2**], **3**, no. 2, 52–78 [in Russian].

Leonov, M.Ia. (1940) On the analysis of foundation plates. [**2**], **4**, no. 3, 80–98 [in Russian].

Lomidze, B.M. (1947) Design of rigid ribbon foundations. *Gidrotekhnicheskoye stroitelstvo* no. 10 [in Russian].

Love, A.E.H. (1927) A Treatise on the Mathematical Theory of Elasticity, 4th Ed. London: Cambridge University Press.

Love, A.E.H. (1939) Boussinesq's problem for a rigid cone. *Quart. Math. Oxford*. **10**, 161–175.

Lur'e, A.I. (1939) Investigation of the case of the non-symmetrical pressure of a rigid flat stamp of elliptical cross-section on an elastic half-space. [**1**], **23**, 759–763 [in Russian].

Lur'e, A.I. (1941) Some contact problems of the theory of elasticity. [**2**], **5**, 383–408 [in Russian].

Lvin, Ya.B. (1950) Stability of rigid walls and columns on an elastic and elasto-plastic foundation. *Inzheneryi Sbornik*. **7** [in Russian].

Mikhlin, S.G. (1945) Problems on the contact of two elastic half-spaces. [**2**], **9**, no. 2 [in Russian].

Mindlin, R.D. (1949) Compliance of elastic bodies in contact. *J. Appl. Mech*. **16**, no. 3.

Mintsberg, B.L. (1948) Mixed boundary value problems of the theory of elasticity for a plate with a circular hole. [**2**], **12**, no. 4, 416–422 [in Russian].

Mitkevich, I.G. (1970) On a contact problem for a viscoelastic half-plane. [3], **34**, 765–767; [3], **34**, 732–734.

Mitrofanov, B.P. (1970) Plane contact problem for an elastic body with the influence of the surface layer taken into account. *Izv. Tomsk. Polytech. Inst.* [in Russian] **157**.

Mkhitarian, S.M. and Schekyan, L.A. (1977) Plane contact problem for two rough elastic solids fabricated from gradually hardening materials. *Izv. Akad. Nauk. ArmSSR, Mekhanica* **30**, No. 3. [in Russian].

Muskhelishvili, N.I. (1935) The solution of the fundamental mixed problem of the theory of elasticity for the half-plane. [**1**], **8**, no. 2, 51–54 [in Russian].

Muskhelishvili, N.I. (1941) The fundamental boundary value problems of the theory of elasticity for a half plane. *Soobshcheniya A.N. Gruz SSR*. **2**, no. 10, 873–880 [in Russian].

Muskhelishvili, N.I. (1942) On the problem of the equilibrium of a rigid stamp on the boundary of an elastic half-plane in the presence of friction. [**1**], **3**, no. 5, 413–418.

Muskhelishvili, N.I. (1946) *Singular Integral Equations*. Gostekhizdat [in Russian]. English translation, JRM Radok, Ed. Groningen: Noordhoff.

Muskhelishvili, N.I. (1953a) *Some Basic Problems of the Mathematical Theory of Elasticity*. Noordhoff: Gröningen.

Muskhelishvili, N.I. (1953b) *Singular Integral Equations*. English Translation, Radok, J.R.M. (Ed.) Groningen: Noordhoff.

Narodetskii, M.Z. (1943) On a certain contact problem. [**1**], **12**, no. 6 [in Russian].

Narodetskii, M.Z. (1947) On the Hertz problem on the contact of two cylinders. [**1**], **53**, 463–466.

Nehari, Z. (1952) *Conformal Mapping*. New York: McGraw Hill.

Neuber, H. (1934) Ein neuer Ansatz zur Lösung räumlicher Probleme der Elastizitätstheorie. Der Hohlkegel unter Einzellast als Beispiel. *ZAMM*, **14**, 203–212.

Okubo, H. (1940) The stress distribution in a sem-infinite domain having a plane boundary and compressed by a rigid body. *ZAAM*, **20**, 271–276; correction **21**, 383–384.

Orlov, A.B. and Pinegrin, S.V. (1971) *Residual Deformations in Contact Loading*. Moscow: Nauka [in Russian].

Papkovich, P.F. (1932a) Solution générale des équations différentielles fondamentales d'elasticité exprimée par trois fonctions harmoniques. *Comptes Rendues de l'Academie des Sciences Paris*, **195**, 513–515.

Papkovich, P.F. (1932b) Expression for the general integral of the principal equations of the theory of elasticity by harmonic functions. *A.N. SSR. Izvestiya. Otdelenie matematicheskikh i estestvennykh nauk.* no. 10, 1425–1435 [in Russian].

Petrusevich, A.I. (1951) Basic deductions from the contact-hydrodynamical theory of lubriction. *Izvest. A.N. SSSR. Otd. Tekhn. Nauk.*, no 2 [in Russian].

Pólya and Szegö (1945) Inequalities for the capacitance of a condenser. *American Journal of Mathematics*, *67*, 1–32.

Pólya and Szegö (1951) *Isosperimetric Inequalities in Mathematical Physics*. Princeton University Press.

Popov, G.Ya. and Savchuk, V.V. (1971) Contact problem of elasticity theory in the presence of a circular contact area taking into account of the surface configuration of the bodies making contact. *Izv. Akad. Nauk. SSSR, Mekhan, Tverd. Tela.* **3**, 80–87.

Poritsky, H. (1950) Stresses and deflections of cylindrical bodies in contact, with application to contact of gears and locomotive wheels. *J. Appl. Mech.* **17**, 191–201.

Proktor, G.E. (1922) Dissertation: On bending of beams lying on an elastic foundation without the Winkler–Zimmerman hypothesis. (Short statement in the book. Kuznetzov (1940).

Rabinovich, A.S. (1974a) Plane contact problem for rough elastic bodies. *Izv. Akad. Nauk. SSSR. Mekhan. Tverd. Tela.* no. 3. [in Russian].

Rabinovich, A.S. (1974b) Plane contact problem on the pressure of a stamp with a rectangular base on a rough elastic half-space. *Izv. Akad. Nauk. ArmSSR, Mekhanika.* **27**, no. 4 [in Russian[.

Rabinovich, A.S. (1975) Axisymmetric contact problems for rough elastic bodies. *Izv. Akad. Nauk. SSSR, Mekhan. Tverd. Tela* **4**.

Reissner, E. and Sagoci, H.F. (1944) Forced torsional oscillation of an elastic half-space. *Journal of Applied Physics*. **15**, 652–654.

Reynolds, O. (1876) On rolling friction. *Phil. Trans. Roy. Soc. London A*, **166**, part 1.

Sadovski, M.A. (1928) Zweidimensionale Probleme der Elastizitätstheorie. *ZAMM* **8**, 107–121.

Saverin, M.M. (1946) Contact strength of materials under the conditions of the simultaneous action of normal and shear loading. *Mashgiz* [in Russian].

Savin, G.N. (1939) The pressure of an absolutely rigid stam on an elastic anisotropic medium. *Dopovidi, A.N. URSR viddil tekhn. nauk.* no. 6, 27–34 [in Russian].

Savin, G.N. (1940a) On the additional pressure transferred by a perfectly rigid stamp on an elastic anisotropic base caused by a neighbouring load. [**1**], no. 7 [in Russian].

Savin, G.N. (1940b) Pressure of a system of perfectly rigid stamps on an elastic anisotropic half plane. Soobshcheniya A.N. Gruz SSR, **1**, no. 10 [in Russian].

Sen, B. (1946) Note on the stresses in a semi-infinite plate produced by a rigid punch on the straight boundary. *Bull. Calcutta Math. Soc.*, **38**, no. 3.

Shail R. (1978) Lamé polynomial solutions to some elliptic crack and punch problems. *Int. J. Engng. Sci.* **16**, 551-563.

Shekhter, O.Ya (1940a) On the influence of the thickness of soil on the distribution of stresses in a foundation beam. Piles and natural foundations. *Sbornik NIS Tresta glubin. rabot.* no. 10 [in Russian].

Shekhter, O.Ya (1940b) Design of infinite foundation plates lying on an elastic foundation of finite or infinite thickness and loaded by a concentrated force: piles and natural foundations. *Sbornik NIS Tresta glubin. rabot.* [in Russian].

Sherman, D.I. (1938) The plane problem of the theory of elasticity with mixed boundary conditions. *Trudy Seismologichesk. Inst. A.N. SSSR*, no. 88 [in Russian].

Shtaerman, I.Ya. (1939) On the Hertz theory of local deformations resulting from the pressure of elastic bodies. [**1**], **25**, no. 5, 360–361 [in Russian].

Shtaerman, I. Ya. (1941a) Local deformations resulting from the pressure of elastic circular cylinders, the radii of which are almost equal. [**1**], **29**, no. 3, 182–184 [in Russian].

Shtaerman, I.Ya. (1941b) On the question of local deformations resulting from the pressure of elastic bodies. [**1**], [**2**], **31**, 738–741 [in Russian].

Shtaerman, I.Ya. (1941c) On a generalization of the Hertz problem. [**2**], **5**, no. 3 [in Russian].

Shtaerman, I.Ya. (1949) *The Contact Problem of the Theory of Elasticity*. Moscow-Leningrad [in Russian].

Sneddon, I.N. (1944) The stress distribution due to a force in the interior of a semi-infinite elastic medium. *Proc. Camb. Phil. Soc.* **40**, 229–238.

Sneddon, I.N. (1946) Boussinesq's problem for a flat-ended cylinder. *Proc. Camb. Phil. Soc.* **42**, 29–39.

Sneddon, I.N. (1947) Note on a boundary value problem of Reissner and Sagoci. *J. Appl. Phys.* **18**, 130–132.

Sneddon, I.N. (1948) Boussinesq's problem for a rigid cone. *Proc. Camb. Phil. Soc.* **44**, 492–507.

Sneddon, I.N. (1951) *Fourier Transforms*. New York: McGraw Hill.

Sokolovsky, V.V. (1950) *Theory of Plasticity*. Moscow: Gostekhizdat [in Russian].

Sommerfeld A. (1897) Über verzweigte Potentiale im Raum. *Proc. Lond. Math. Soc.* **28**, 395–429.

Tricomi, F. (1957) *Integral Equations*, New York: Interscience.

Whittaker, E.T. and Watson, G.N. (1952) *A Course of Modern Analysis*. 4th ed. London: Cambridge University Press.

Zhemochkin, B.N. (1937) Plane problems of the design of infinitely long beams on an elastic half-space and half-plane. *Izdat. voen.-inzh. akad.* Moscow [in Russian].

Zhemochkin, B.N. (1938) *Design of Circular Plates on an Elastic Foundation under Symmetrical Loading*. Moscow [in Russian].

Zhemochkin, B.N. and Sinitsin, A.P. (1948) *Practical Design of Foundation Beams and Plates on an Elastic Foundation without the Winkler Hypothesis*. Moscow: Stroyizdat [in Russian].

Index

Developments of Galin's Research in Contact Mechanics

I.G. Goryacheva

1 Two-Dimensional Sliding Contact of Elastic Bodies

We consider problems similar to those described in Sections 3.7 and 3.8 of the text,[1] for the more general Coulomb law of friction , and various shapes of punch. The results presented in this part were obtained by Goryacheva (1998a). The first section repeats, in different notation, some of the results presented in Sections 3.7 and 3.8 of the text.

1.1 Problem Formulation

We consider sliding contact of a rigid cylinder and an elastic half-space (Figure 1). The shape of the rigid body is described by the function $y = f(x)$. External forces also are independent of the z-coordinate. This problem is considered as a two-dimensional (plane) problem for a punch and an elastic half-plane. The two-term friction law established by Coulomb (1785) is assumed to hold within the contact zone $(-a, b)$:

$$q(x) = (\tau_0 + \rho p(x)) \operatorname{sgn} V, \tag{1}$$

where $p(x) = -\sigma_{yy}(x, 0)$ and $q(x) = -\sigma_{xy}(x, 0)$ are the normal pressure and shear stress at the surface of the elastic half-plane ($y = 0$) within the contact region, and V is the velocity of the cylinder, τ_0 and ρ are the parameters.

Applied shear Q and normal P forces cause the body to be in the limiting equilibrium state, or to move with a constant velocity. This motion occurs so slowly that dynamic effects may be neglected.

In the moving coordinate system connected with the rigid cylinder, the following boundary conditions hold ($y = 0$)

[1] In this article, 'the text' refers to the translation of Galin's books given in Chapters 1–6.

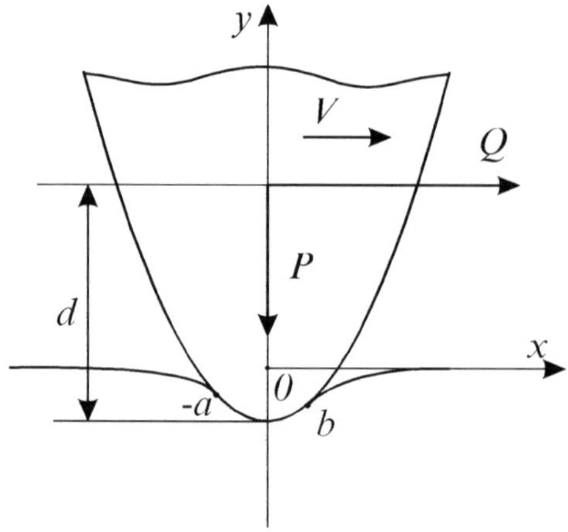

Fig. 1 Sliding contact of a cylindrical punch and an elastic half-space.

$$\begin{aligned} &\sigma_{yy} = 0, \quad \sigma_{xy} = 0, \quad (-\infty < x < -a, \quad b < x < +\infty), \\ &v = f(x) - D, \quad q = (\tau_0 + \rho p) \operatorname{sgn} V, \quad (-a \leq x \leq b), \end{aligned} \tag{2}$$

where v is the normal displacement of the half-plane surface, D is the approach of the contacting bodies.

The relationship between stresses and the normal displacement gradient at the boundary $y = 0$ of the lower half-plane has the form:

$$\frac{\pi E}{2\left(1 - \nu^2\right)} \cdot \frac{\partial v}{\partial x} = \int_{-\infty}^{+\infty} \sigma_{yy} \frac{dt}{t - x} - \frac{1 - 2\nu}{2 - 2\nu} \pi \tau_{xy}. \tag{3}$$

Using Galin's method (see Section 2.2 of the text), we introduce a function $w_1(z)$ of a complex variable in the lower half-plane $y \leq 0$

$$w_1(z) = u_1 - i v_1 = \int_{-\infty}^{+\infty} \sigma_{yy} \frac{dt}{t - z}. \tag{4}$$

Using (2), (3) and the limiting values of the Cauchy integral (4) as $z \to x - i0$, we can derive the following boundary conditions for the function $w_1(z)$

$$\begin{aligned} &v_1 = 0, \quad (-\infty < x < -a, \quad b < x < +\infty), \\ &u_1 + \rho\beta v_1 \operatorname{sgn} V = \pi F(x), \quad (-a \leq x \leq b), \end{aligned} \tag{5}$$

where

$$\begin{aligned} F(x) &= \beta \tau_0 \operatorname{sgn} V + \frac{f'(x)}{\pi K}, \\ K &= \frac{2\left(1 - \nu^2\right)}{\pi E}, \quad \beta = \frac{1 - 2\nu}{2(1 - \nu)}. \end{aligned} \tag{6}$$

So the problem is reduced to the determination of the analytic function $w_1(z)$ (4) based on the relationships (5) between its real and imaginary parts u_1, v_1 at the boundary of the region of its definition. This is a particular case of the Riemann–Hilbert problem (see Section 3.2 of the text).

The solution of this problem that satisfies the condition $w_1(z) \sim P/z$ as $z \to \infty$ and has the integrable singularities at the boundary is the following function

$$w_1(z) = \frac{\varrho}{X(z)} \int_{-a}^{b} F(t) X^+(t) \frac{dt}{t-z} + \frac{P}{X(z)}, \tag{7}$$

where

$$X(z) = (z+a)^{1/2+\eta}(z-b)^{1/2-\eta},$$
$$X^+(t) = (t+a)^{1/2+\eta}(b-t)^{1/2-\eta},$$

$$\varrho = \frac{1}{\sqrt{1+\rho^2\beta^2}}, \qquad \eta = \frac{1}{\pi}\arctan(\rho\beta)\operatorname{sgn} V, \qquad |\eta| < \frac{1}{2}. \tag{8}$$

Using the function (7), we can determine the stress-strain state of the elastic half-plane. For example, Eq. (4) implies that the normal stress at the x-axis $\sigma_{yy}(x, o)$ is the imaginary part of the function (7) as $z \to x - i0$. The limiting value of the Cauchy integral

$$\Phi(z) = \frac{1}{2\pi i} \int_{-\infty}^{+\infty} \varphi(t) \frac{dt}{t-z}$$

as $z \to x - i0$ can be determined by the Plemelj (1908) formula (see (2.2.11), (2.2.12) of the text)

$$\Phi^-(x) = \frac{1}{2\pi i} \int_{-\infty}^{+\infty} \varphi(t) \frac{dt}{t-x} - \frac{1}{2}\varphi(x).$$

The limiting value of the function $1/X(z)$ as $z \to x - i0$ is determined by the formula

$$\left[\frac{1}{(z+a)^{1/2+\eta}(z-b)^{1/2-\eta}}\right]_{z=x-i0} =$$

$$= \begin{cases} -\dfrac{1}{(-a-x)^{1/2+\eta}(b-x)^{1/2-\eta}}, & (-\infty < x < -a), \\ \dfrac{\sin\pi\eta + i\cos\pi\eta}{(x+a)^{1/2+\eta}(b-x)^{1/2-\eta}}, & (-a \le x \le b), \\ \dfrac{1}{(x+a)^{1/2+\eta}(x-b)^{1/2-\eta}}, & (b < x < +\infty). \end{cases}$$

So the contact pressure $p(x) = -\sigma_{yy}(x, 0) = -\frac{1}{l}\pi v_1(x, 0)$, where $v_1(x, 0)$ is the imaginary part of the function $w_1(z)$ as $z \to x - i0$, is given by

$$p(x) = -F(x)\varrho \sin \pi\eta +$$

$$\frac{\varrho}{\pi} \cdot \frac{\cos \pi\eta}{X^+(x)} \int_{-a}^{b} F(t) X^+(t) \frac{dt}{t-x} + \frac{P \cos \pi\eta}{\pi X^+(x)}, \qquad x \in (-a, b). \tag{9}$$

1.2 Contact Problem for a Cylinder

We consider the particular case of a sliding contact of a rigid cylinder and an elastic half-space. For this case $f(x) = \dfrac{x^2}{2R}$ and the function $F(x)$ (6) becomes

$$F(x) = \tau_0 \beta \operatorname{sgn} V + \frac{x}{\pi K R}. \tag{10}$$

Substituting (10) in (9) and using the following relationships (Gradshteyn and Ryzhik, 1963)

$$\int_{-a}^{b} (a+t)^{-1/2+\eta} (b-t)^{-1/2-\eta} \frac{dt}{t-x} =$$

$$= \pi \tan \pi\eta (a+x)^{-1/2+\eta} (b-x)^{-1/2-\eta}, \qquad (-a < x < b),$$

$$\int_{-a}^{b} (a+x)^{\mu-1} (b-x)^{\nu-1}\, dx = (a+b)^{\mu+\nu-1} B(\mu, \nu), \qquad (\mu > 0,\ \nu > 0),$$

we obtain the expression for the contact pressure

$$p(x) = \frac{\varrho L(x)}{(x+a)^{-1/2+\eta} (b-x)^{-1/2-\eta}},$$

$$L(x) = \frac{P}{\pi} + \frac{(a+b)^2 \left(\frac{1}{4} - \eta^2\right)}{2\pi K R} + \tau_0 \beta (a+b) \left(\frac{1}{2} - \eta\right) \operatorname{sgn} V \tag{11}$$

$$+ \frac{(a+b)x}{\pi K R} \left(\frac{1}{2} - \eta\right) - \tau_0 \beta (x+a) \operatorname{sgn} V + \frac{(a+x)x}{\pi K R}.$$

The contact pressure (11) has to be bounded at the ends of the contact zone. Equation (11) shows that if it is bounded there, it must in fact be zero there, i.e. $p(-a) = p(b) = 0$ and

$$l^2 = (a+b)^2 = \frac{2RPK}{\frac{1}{4} - \eta^2}, \tag{12}$$

$$\frac{a-b}{2} = l\eta + \tau_0 \beta K R \pi \operatorname{sgn} V. \tag{13}$$

So that

$$p(x) = \frac{Q}{\pi K R}(x+a)^{1/2-\eta}(b-x)^{1/2+\eta}. \tag{14}$$

The relationships (12), (13) and (14) determine the contact width, the shift of the contact zone and the contact pressure, respectively. Equations (12) and (14) coincide with those obtained by Galin (1953), where the contact problem in the analogous formulation with Amontons'(1699) law of friction $q = \rho p$ was considered (see Section 3.7 of the text).

The results indicate that the magnitude τ_0 in the law (1) influences only the contact displacement (13).

It follows from Eq. (14) that the contact pressure is an unsymmetrical function. It provides the moment M

$$M = \int_{-a}^{b} p(x)x\,dx = -P\left[\frac{4}{3}l\eta + \pi\beta K R\tau_0 \operatorname{sgn} V\right], \tag{15}$$

where

$$P = \int_{-a}^{b} p(x)\,dx.$$

If there is no active moment applied to the cylinder, the moment M is equal to the moment of the tangential force Q

$$T = \int_{-a}^{b} q(x)dx = (\tau_0 l + \rho P) \operatorname{sgn} V. \tag{16}$$

In this case, it follows from the equilibrium conditions that the force Q must be applied at the point $(0, d)$ (Figure 1): $d = |M/T|$.

Note that in most cases $\rho\beta \ll 1$, so that we may approximate Eq. (8) by

$$|\eta| \approx \frac{\rho}{\pi}\beta \ll 1.$$

Based on this estimation, it follows from Eqs. (12), (13) and (14) that the friction coefficient ρ has no essential influence on the contact pressure, the shift or the width of contact zone.

The analysis of subsurface stresses revealed that the effect of the parameter τ_0 on the stress-strain state in an elastic body is similar to a friction coefficient ρ: it moves the point where the maximum principal shear stress $(\tau_1)_{max}$ takes place closer to the surface, and it increases the magnitude of $(\tau_1)_{max}$ (Figure 2).

Eqs. (12)–(15) can be used to determine contact characteristics (contact width and displacement, contact pressure etc.) for sliding contact of two elastic bodies with radii of curvature R_1 and R_2. We replace the parameters K, β, R and η (see Eqs. (6) and (8)) by the parameters K^*, β^*, R^*, η^*. For plane stress

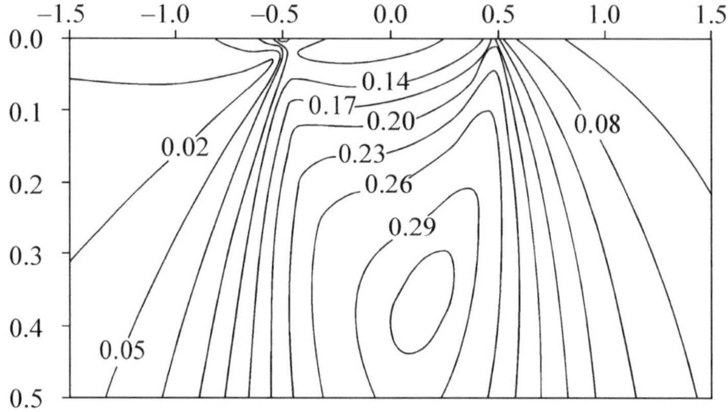

Fig. 2 Contours of the principal shear stress beneath a sliding contact ($\rho = 0$, $\tau_0/p_0 = 0.1$).

$$K^* = \frac{2}{\pi}\left(\frac{1}{E_1} + \frac{1}{E_2}\right), \quad \beta^* = \frac{1}{K\pi}\left(\frac{1-\nu_2}{E_2} - \frac{1-\nu_1}{E_1}\right), \tag{17}$$

and for plane strain

$$\begin{gathered} K^* = \frac{2}{\pi}\left(\frac{1-\nu_1^2}{E_1} + \frac{1-\nu_2^2}{E_2}\right), \\ \beta^* = \frac{1}{K\pi}\left[\frac{(1+\nu_2)(1-2\nu_2)}{E_2} - \frac{(1+\nu_1)(1-2\nu_1)}{E_1}\right], \end{gathered} \tag{18}$$

and

$$\frac{1}{R^*} = \frac{1}{R_1} + \frac{1}{R_2}, \quad \eta^* = \frac{1}{\pi}\arctan(\rho\beta^*)\,\mathrm{sgn}\,V.$$

Provided that $l \ll R_i$, $(i = 1, 2)$ we can consider the cylinders as half-planes. So we use Eq. (3) to determine the gradient of normal displacement for both cylinders, taking into account the relationship: $\sigma_{xy}^{(1)} = -\sigma_{xy}^{(2)}$.

1.3 Contact Problem for a Flat Punch

We consider sliding contact of a punch with a flat base (Figure 3). Under the applied forces, the punch has the inclination α. So the equation for the punch shape is $f(x) = -\alpha x - D$.

The function $F(x)$ (6) has the following form

$$F(x) = \tau_0\beta\,\mathrm{sgn}\,V - \frac{\alpha}{\pi K} \quad (-a \le x \le b). \tag{19}$$

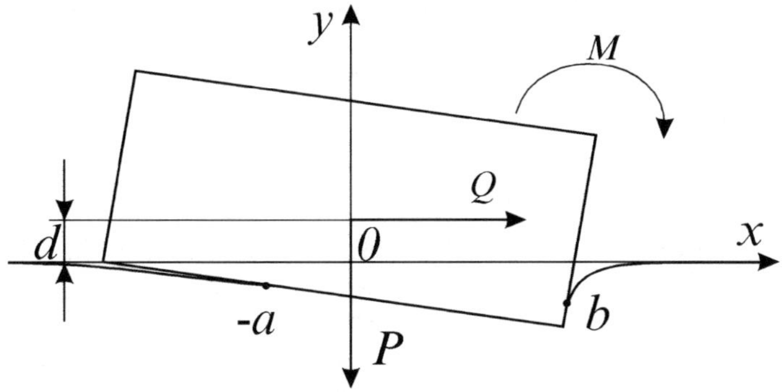

Fig. 3 Sliding contact of a flat punch and an elastic half-plane.

We introduce the dimensionless parameter

$$\kappa = \frac{b}{P}\left(\tau_0 \beta \operatorname{sgn} V - \frac{\alpha}{\pi K}\right). \tag{20}$$

Substituting Eq. (19) in Eq. (9) and transforming this equation, we have

$$p(x) = \frac{P\varrho}{\pi b} \cdot \frac{b + \pi\kappa\left[b - x - \left(\frac{1}{2} + \eta\right)(a+b)\right]}{(x+a)^{1/2+\eta}(b-x)^{1/2-\eta}}. \tag{21}$$

Eq. (21) shows that the contact pressure near the ends of contact zone ($x \to +0$) can be represented as

$$\begin{aligned} p(-a+x) &= \frac{A_1}{x^{1/2+\eta}} + O\left(x^{1/2-\eta}\right), \\ A_1 &= \frac{P\varrho}{\pi b} \cdot \frac{\left[b + \pi\kappa(a+b)\left(\frac{1}{2} - \eta\right)\right]}{(a+b)^{1/2-\eta}}, \end{aligned} \tag{22}$$

$$\begin{aligned} p(b-x) &= \frac{A_2}{x^{1/2-\eta}} + O\left(x^{1/2+\eta}\right), \\ A_2 &= \frac{P\varrho}{\pi b} \cdot \frac{\left[b - \pi\kappa(a+b)\left(\frac{1}{2} + \eta\right)\right]}{(a+b)^{1/2+\eta}}. \end{aligned} \tag{23}$$

We consider the case of complete contact of a flat punch and an elastic half-plane. Setting $a = b$ in Eq. (21) we have

$$p(x) = \frac{P\varrho}{\pi b} \cdot \frac{[b - \pi\kappa(x + 2b\eta)]}{(x+b)^{1/2+\eta}(b-x)^{1/2-\eta}}. \tag{24}$$

The contact pressure is a nonnegative function, $p(x) \geq 0\,(-b \leq x \leq b)$, and hence

$$\kappa_1 \leq \kappa \leq \kappa_2, \tag{25}$$

where

$$\kappa_1 = -\frac{1}{\pi(1-2\eta)}, \qquad \kappa_2 = \frac{1}{\pi(1+2\eta)}. \tag{26}$$

The contact pressure $p(x)$ given by Eq. (24) tends to infinity at the edges of the punch ($x = \pm b$), if $\kappa \in (\kappa_1,\ \kappa_2)$. If $\kappa = \kappa_1$ or $\kappa = \kappa_2$, the contact pressure is zero at the left end or at the right end of the contact zone, respectively.

If the parameter $\kappa \notin [\kappa_1,\ \kappa_2]$, there is only partial contact. If $\kappa \leq \kappa_1 < 0$ the separation of the punch base from the half-plane appears at the left-hand end of the contact zone at the point $x = -a$. The contact width is found according to Eq. (22)

$$a + b = -\frac{b}{\pi\kappa\left(\dfrac{1}{2} - \eta\right)}. \tag{27}$$

Using Eqs. (21) and (27), we obtain the contact pressure

$$p(x) = \frac{P\varrho}{\pi b} \cdot \frac{\left[b + \pi\kappa(b-x)\left(\dfrac{1}{2} - \eta\right)\right]}{\left(\dfrac{1}{2} - \eta\right)(x+a)^{1/2+\eta}(b-x)^{1/2-\eta}}. \tag{28}$$

If $\kappa > \kappa_2 > 0$ the contact pressure is zero at the right-hand end of the contact zone at the point $x = b$, where $|b| < a$ (a is the half-width of the punch in this case). Using (21) and (23), we find the equation for the contact pressure

$$p(x) = \frac{P\kappa\varrho}{b}\left(\frac{b-x}{x+a}\right)^{1/2+\eta}.$$

It follows from Eqs. (20) and (23) that the coordinate $x = b$ is determined by the formula

$$b = -a + \frac{P}{\pi\left(\tau_0\beta\,\mathrm{sgn}\,V - \dfrac{\alpha}{\pi K}\right)\left(\dfrac{1}{2} + \eta\right)}. \tag{29}$$

The contact pressure distributions for different values of the parameter κ are shown in Figure 4. The curves 1–4 correspond to the cases of complete contact and pressure approaching to infinity at the ends of contact zone ($\kappa = 0$), complete contact when $p(-b) = 0$ ($\kappa = \kappa_1$, see Eqs. (24) and (26)), and partial contact ($\kappa = -0.5$ and $\kappa = -0.75$), respectively. For the calculations we used $|\rho\beta| = 0.057$ ($\rho = 0.2,\ \nu = 0.3$). Note that for frictionless contact ($\rho = 0,\ \tau_0 = 0$) the results obtained in this part coincide with those obtained by Galin (1953).[2]

[2] See Section 3.5 of the text.

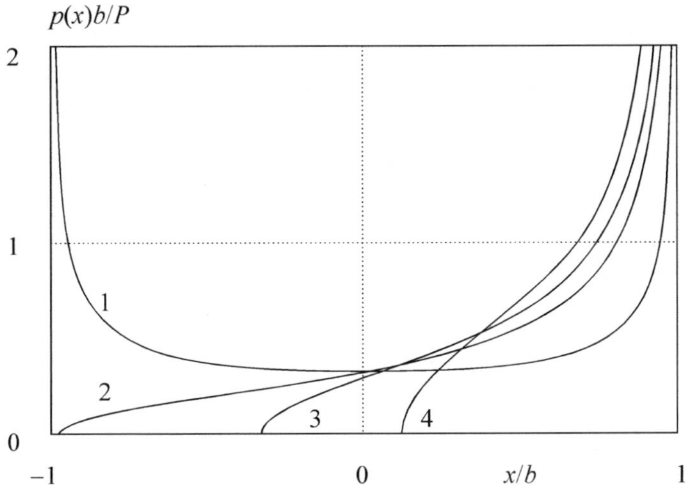

Fig. 4 Contact pressure under a flat inclined punch sliding on an elastic half-plane ($\rho\beta = 0.057$); $\kappa = 0$ (curve 1); $\kappa = \kappa_1 = -0.33$ (curve 2); $\kappa = -0.5$ (curve 3); $\kappa = -0.75$ (curve 4).

The parameter κ depends on the inclination α (see Eq. (20)). For definiteness, let us consider the punch moving in the x-axis direction ($V > 0$). The parameter α can be found using the equilibrium conditions for the punch. The normal load P, the tangential force Q, and the active moment M are applied to the punch (see Figure 3). The contact pressure $p(x)$ and the shear stress $q(x)$ form the resistance forces which satisfy the following equilibrium conditions:

$$P = \int_{-a}^{b} p(x)dx,$$
$$T = \int_{-a}^{b} q(x)dx = \tau_0(a+b) + \rho P, \tag{30}$$

$$\int_{-a}^{b} (b-x)p(x)dx - Pb + Td - M = 0, \tag{31}$$

where $(0, d)$ are the coordinates of the point where the force Q is applied, and M is the active moment relative to the point $x = b$.

Using Eqs. (21) and (30), we can transform Eq. (31) to the following relation

$$-P(a+b)\left(\frac{1}{2} - \eta\right) + Pa + \frac{\pi P\kappa(a+b)^2}{2b}\left(\frac{1}{4} - \eta^2\right)$$
$$+\tau_0(a+b)d + \rho P d - M = 0. \tag{32}$$

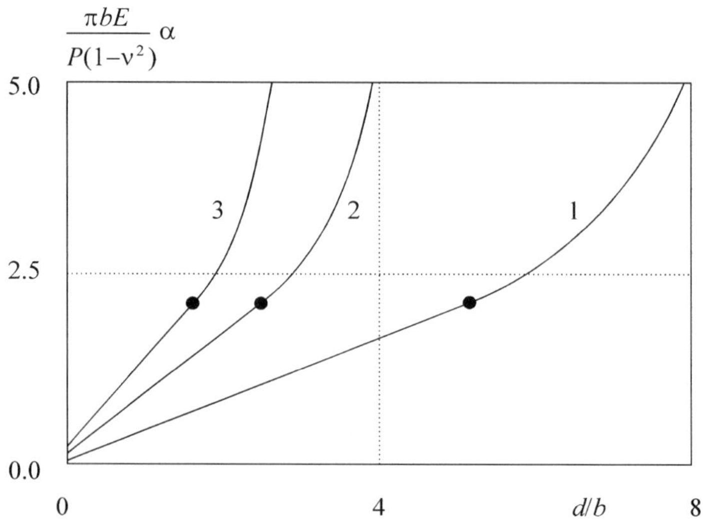

Fig. 5 The effect of the position of the point of application of the tangential force Q on the inclination of a punch ($\nu = 0.3$, $\tau_0 = 0$); $\rho = 0.1$ (curve 1), $\rho = 0.2$ (curve 2), $\rho = 0.3$ (curve 3); $d_1^{(i)}$, ($i = 1, 2, 3$) indicates the transition point from complete to partial contact.

Eqs. (19) and (32) are used to determine the inclination α, which depends on both quantities d and M.

Let us consider the particular case $M = 0$ and analyze the dependence of the inclination α on the distance d. Using Eqs. (20), (25) and (32) we conclude that the complete contact occurs for $d \in (0, d_1)$, where

$$d_1 = \frac{Pb\left(\dfrac{1}{2} - \eta\right)}{2b\tau_0 + \rho P}. \tag{33}$$

The inclination α for this case is

$$\alpha = \frac{2bP\eta + (2b\tau_0 + \rho P)d}{2b^2\left(\dfrac{1}{4} - \eta^2\right)}K + \tau_0\pi\beta K. \tag{34}$$

If $d \in (d_1, d_2)$, the partial contact occurs with the separation point $x = -a$, where $|a| < b$; d_2 is determined by the condition $-a = b$, i.e. there is point contact. It follows from Eq. (32), that $d_2 = b/\rho$. The inclination α of the punch for the case $d_1 \le d < d_2$ is determined from Eqs. (20), (27) and (32)

$$\alpha = \frac{P\left(\dfrac{1}{2} + \eta\right) + 2\tau_0 d}{2\left(\dfrac{1}{2} - \eta\right)(b - \rho d)}K + \tau_0\pi\beta K. \tag{35}$$

It follows from Eq. (35) that $\alpha \to +\infty$ (the punch is overturned) as $d \to d_2 - 0$. Figure 5 illustrates the dependence of the inclination α on the distance $d \in [0, d_2)$ for different magnitudes of the coefficient ρ and $\tau_0 = 0$. The Eqs. (34) and (35) have been used to plot the curves.

The results of this analysis can be used in the design of devices for tribological tests. If two specimens with flat surfaces come into contact, the hinge is used to provide their complete contact. The results show that the hinge must be fixed at a distance $d \in (0, d_1)$ from the specimen base. The limiting distance d_1 essentially depends on the friction coefficient ρ. If $\tau_0 = 0$, we obtain from Eq. (33)

$$\frac{d_1}{b} \approx \frac{1}{2\rho} - \frac{1-2\nu}{2\pi(1-\nu)}.$$

2 Contact Problem with Partial Slip for the Inclined Punch with Rounded Edges

Contact with partial slip arises if the tangential load applied to the bodies is less than that necessary to cause complete sliding between the contacting bodies. The analysis of the stress distribution in partial slip contact is very important in investigating and preventing fretting fatigue.

The problem with partial slip can be classified as one of the most difficult problems in contact mechanics because the additional unknown values such as the bounds between the slip and stick zones must be found precisely. The most famous solution of the partial slip problem for the general Hertzian contact was found by Cattaneo (1938) and Mindlin (1949). A partial solution of the 2-D contact problem for a punch with flat base indenting an elastic half-space when both stick and slip zones exist within the contact region was obtained by Galin and is presented in Section 4.7 of the text. See also the article by O. Zhupanska in this volume.

Review of the known solutions of the contact problem with partial slip for various types of the contact geometry and elastically similar and dissimilar properties of contacting bodies has been made by Hills and Sosa (1999).

One contact geometry typical for gas turbine engine hardware configurations, in particular, in interaction between the dovetail segment of the blade and attachment slot of disk, can be modelled by the contact of an indenter with a flat base and rounded edges, and a half-space (2-D contact problem). This problem has been investigated by Ciavarella et al. (1998) analytically and McVeigh et al. (1999) using quasi-analytical and numerical techniques. They analyzed the symmetrical contact stress distribution and also the variation of pressure due to surface profile deviations.

An asymmetric contact pressure distribution arises if the punch is acted upon by a normal force and a moment. The 2-D contact problem for the inclined punch with flat base and the elastic half-space was solved in closed form by Galin (see Chapter 3 of the text) for the frictionless contact and by Goryacheva (1998a) (see also Section 1) for the case of sliding contact with friction described by the Coulomb law. The results show that the contact can be complete or incomplete, depending on the moment applied to the punch. For incomplete contact the contact pressure singularity occurs only at one end of the contact. This model is used to evaluate the influence of the asymmetrical loading on the location of the contact zone and the contact and internal stress distribution.

In this part the 2-D contact of the inclined indenter having a flat base and rounded edges and a half-space is considered under the assumption that the applied forces provide for conditions of partial slip within the contact region. The model is used to evaluate the stresses within and near the contact region depending on the normal and tangential forces and the moment applied to the indenter and the particular indenter geometry. The location of the contact region and the stick and slip zones are analyzed for various external conditions. This part reproduces the paper by I.G. Goryacheva, H. Murthy and T. Farris (2002).

2.1 Problem Formulation

Figure 6 illustrates the contact of an elastic punch having a flat base with rounded edges with an elastic half-plane. The straight part of the punch base is of length $2c$, and the corners are of radius R. The punch is acted upon by normal force P, tangential force Q and moment M. If the tangential force Q is applied at some distance r from the base of the punch it also gives the moment $M' = Qr$. The last moment exists even if $M = 0$. The shape of the inclined punch is described by the function:

$$f(x) = \begin{cases} \dfrac{(x+c)^2}{2R} - \alpha x, & \text{if} \quad -a \le x \le -c, \\ -\alpha x, & \text{if} \quad -c < x < c, \\ \dfrac{(x-c)^2}{2R} - \alpha x, & \text{if} \quad c \le x \le b, \end{cases} \tag{36}$$

The contact condition within the contact zone $[-a, b]$ is written in the form:

$$h'(x) \equiv u'_{y1} + u'_{y2} = -f'(x).$$

where u_{yi} is a displacement of the boundary of the half-plane ($i = 1$) or the punch ($i = 2$) which is parallel to the y-axis and measured positive into each body, ($'$) means the derivative with respect to x.

We assume that the tangential force Q satisfies the condition:

$$|Q| \le \rho P,$$

where ρ is the friction coefficient so that partial slip occurs within the contact. The boundary conditions at the slip and stick regions are as follows:

- in the slip region L_{slip}:

$$|q(x)| = \rho p(x), \tag{37}$$

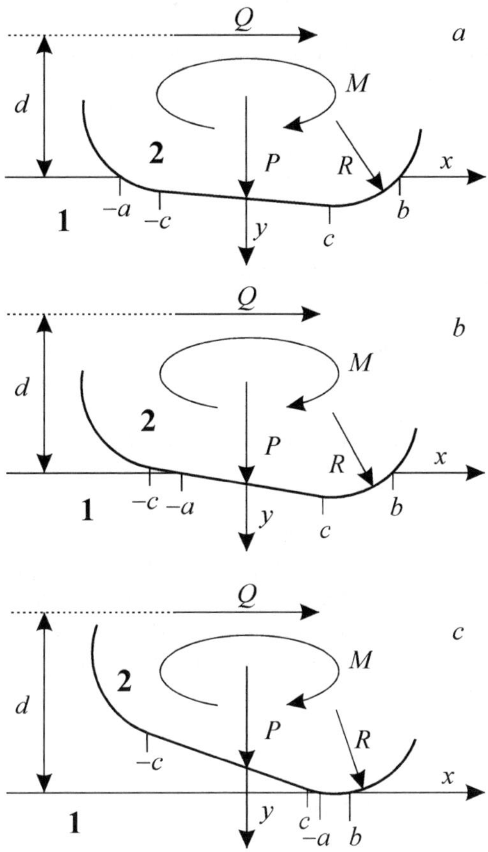

Fig. 6 Schematic diagrams of contact for different cases; (a) corresponds to $\alpha < \alpha_1$, (b) corresponds to $\alpha_1 < \alpha < \alpha_2$, (c) corresponds to $\alpha > \alpha_2$.

and the direction of the frictional traction q must oppose the direction of slip, i.e.

$$\frac{q(x)}{|q(x)|} = -\frac{s(x)}{|s(x)|}, \tag{38}$$

where

$$s(x) = (u_{x1}(x) - u_{x2}(x)) - \delta_x, \quad (\delta_x = \text{const}), \tag{39}$$

($u_{xi}(x)$ is the boundary displacement of each body ($i = 1, 2$) along the x-axis, δ_x is the relative displacement in the x-axis direction of the points of bodies which are located at some distance from the interface);

- in the stick region L_{stick}:

$$u_{x1}(x) - u_{x2}(x) = \delta_x, \tag{40}$$

$$|q(x)| \leq \rho p(x). \tag{41}$$

The contact pressure $p(x)$ and the contact shear stress $q(x)$ satisfy the following equilibrium conditions:

$$\int_{-a}^{b} p(x)dx = P, \tag{42}$$

$$\int_{-a}^{b} q(x)dx = Q, \tag{43}$$

If the inclination α is unknown, we also use the moment equation

$$-\int_{-a}^{b} xp(x)dx + Qr + M = 0, \tag{44}$$

In what follows we assume that the contacting bodies have similar elastic properties so that the shear contact stress does not influence the contact pressure. The contact pressure analysis below can also be used for frictionless contact.

The second assumption is that both contacting bodies are modeled by elastic half-planes. The justification of the last assumption for different values of parameter c/R is discussed in detail by Ciavarella et al. (1998).

2.2 Contact Pressure Analysis

Under these assumptions, the contact pressure $p(x)$ within the contact zone $[-a, b]$ is obtained from the equation:

$$\int_{-a}^{b} \frac{p(x')dx'}{x - x'} = -\frac{\pi E^*}{2} h'(x) \tag{45}$$

where

$$h'(x) = \begin{cases} -\dfrac{x+c}{R} + \alpha, & \text{if} \quad -a \leq x \leq -c, \\ \alpha, & \text{if} \quad -c < x < c, \\ -\dfrac{x-c}{R} + \alpha, & \text{if} \quad c \leq x \leq b, \end{cases} \tag{46}$$

and

$$E^* = \frac{E}{2(1-\nu^2)} = \frac{1}{\pi K}.$$

The solution of Eq. (45) which satisfies the condition $p(-a) = p(b) = 0$ is given by Muskhelishvili (1949) in the following form:

$$p(x) = -\frac{E^*}{2\pi}\sqrt{(x+a)(b-x)}\int_{-a}^{b}\frac{h'(t)\,dt}{(t-x)\sqrt{(t+a)(b-t)}}, \qquad x \in [-a,b], \quad (47)$$

where the ends of the contact zone can be obtained from the following equations:

$$\int_{-a}^{b}\frac{t\,h'(t)\,dt}{\sqrt{(t+a)(b-t)}} = -\frac{2P}{E^*}. \qquad (48)$$

$$\int_{-a}^{b}\frac{h'(t)\,dt}{\sqrt{(a+t)(b-t)}} = 0. \qquad (49)$$

The contact zone can have a different location in respect to the flat base $|x| \leq c$ of the punch. The location of the contact zone depends on the inclination angle of the punch. The different cases of the contact zone location are presented in Figure 1. In what follows we consider each of these cases.

2.2.1 The Case $-a \leq -c$, $c \leq b$ (Figure 6a)

First we assume that the contact region includes the flat base of the punch (see Figure 6a). Substituting Eq. (46) into (47), we obtain

$$p(x) = \frac{E^*}{2\pi}\sqrt{(x+a)(b-x)}\left[-\alpha\int_{-a}^{b}\frac{dt}{(t-x)\sqrt{(t+a)(b-t)}}+\right.$$
$$+\frac{x+c}{R}\int_{-a}^{-c}\frac{dt}{(t-x)\sqrt{(t+a)(b-t)}}+\frac{x-c}{R}\int_{c}^{b}\frac{dt}{(t-x)\sqrt{(t+a)(b-t)}}+ \qquad (50)$$
$$\left.+\frac{1}{R}\int_{-a}^{-c}\frac{dt}{\sqrt{(t+a)(b-t)}}+\frac{1}{R}\int_{c}^{b}\frac{dt}{\sqrt{(t+a)(b-t)}}\right].$$

Using the values of the following integrals

$$\int\frac{dt}{\sqrt{(t+a)(b-t)}} = \arcsin\frac{2t+a-b}{a+b},$$

$$\int_{x_1}^{x_2}\frac{dt}{(t-x)\sqrt{(t+a)(b-t)}} = \frac{2(1+y^2)}{(a+b)}\int_{y_1}^{y_2}\frac{d\tau}{(\tau-y)(1-\tau y)} =$$

$$= \frac{2(1+y^2)}{(a+b)(1-y^2)}\ln\left|\frac{(y_2-y)(1-yy_1)}{(y_1-y)(1-yy_2)}\right|.$$

where the last integral has been calculated using the substitutions:

$$x = \frac{b-a}{2} + \frac{y}{1+y^2}(a+b),$$
$$t = \frac{b-a}{2} + \frac{\tau}{1+\tau^2}(a+b).$$

we reduce Eq. (50) to the following expression for the contact pressure

$$p(y) = \frac{E^*(a+b)}{2\pi R(1+y^2)} \left\{ \frac{(1-y^2)}{2} \left[\pi - 2\arctan y_1 + 2\arctan y_2\right] + \right.$$
$$\left. + \frac{(y-y_2)(1-yy_2)}{(1+y_2^2)} \ln\left|\frac{y_2-y}{yy_2-1}\right| - \frac{(y-y_1)(1-yy_1)}{(1+y_1^2)} \ln\left|\frac{y_1-y}{yy_1-1}\right| \right\}, \tag{51}$$
$$|y| \le 1.$$

where

$$y = \tan\left(\frac{1}{2}\arcsin\frac{2x-b+a}{a+b}\right),$$
$$y_1 = \tan\left(\frac{1}{2}\arcsin\frac{2c-b+a}{a+b}\right), \tag{52}$$
$$y_2 = \tan\left(\frac{1}{2}\arcsin\frac{-2c-b+a}{a+b}\right).$$

Substitution of Eq. (46) into equations (48) and (49) gives the system of the equations to determine the ends of the contact $-a$ and b:

$$-\frac{2PR}{E^*} = -\alpha\pi \cdot \frac{(a+b)R}{2} + \sqrt{(a+c)(b-c)}\frac{2c+3a+b}{4} -$$
$$- \sqrt{(a-c)(b+c)}\frac{3a+b-2c}{4} + \frac{(a+b)}{8}(b-3a-4c)\times$$
$$\times\left(\frac{\pi}{2} - \arcsin\frac{a+2c-b}{a+b}\right) + \tag{53}$$
$$+ \frac{a+b}{8}(b-3a+4c)\left(\frac{\pi}{2} - \arcsin\frac{b+2c-a}{a+b}\right),$$

$$(b+2c-a)\arcsin\frac{b+2c-a}{a+b} - (a+2c-b)\arcsin\frac{a+2c-b}{a+b} =$$
$$= \pi(b-a-2\alpha R) - 2\sqrt{(a-c)(b+c)} + 2\sqrt{(a+c)(b-c)}. \tag{54}$$

Here the values of load P and inclination α are assumed to be given.

The system of equations (53) and (54) can be used to calculate the inclination α_1 corresponding to the case $a = c$.

If the inclination α is unknown we use Eq. (44) to determine its value. Substitution of Eq. (47) into Eq. (44) gives

$$
\begin{aligned}
M + Qr &= \int_{-a}^{b} x p(x)\, dx \\
&= \frac{E^*}{2\pi} \int_{-a}^{b} x\sqrt{(x+a)(b-x)}\, dx \int_{-a}^{b} \frac{h'(t)\, dt}{(t-x)\sqrt{(t+a)(b-t)}} = \\
&= \frac{E^*}{2\pi} \int_{-a}^{b} \frac{h'(t)\, dt}{\sqrt{(t+a)(b-t)}} \int_{-a}^{b} \frac{x\sqrt{(x+a)(b-x)}}{t-x} dx = \\
&= -\frac{E^*}{2\pi} \int_{-a}^{b} \frac{h'(t)\, dt}{\sqrt{(t+a)(b-t)}} \times \\
&\times \left[\int_{-a}^{b} \sqrt{(x+a)(b-x)}\, dx + t \int_{-a}^{b} \frac{\sqrt{(x+a)(b-x)}}{x-t} dx \right].
\end{aligned} \tag{55}
$$

Using Eq. (48) and (49) we reduce Eq. (55) to the following form:

$$
M + Qr = \frac{E^*}{2} \int_{-a}^{b} \frac{t^2 h'(t)\, dt}{\sqrt{(t+a)(b-t)}} + \frac{P(b-a)}{2}. \tag{56}
$$

Substituting Eq. (46) into (56) and calculating the integrals, we reduce the relationship to obtain the inclination α:

$$
\begin{aligned}
&-\frac{2R}{E^*}\left(M + Qr - \frac{P(b-a)}{2}\right) = \\
&= R\alpha\pi \frac{3a^2 - 2ab + 3b^2}{8} - \frac{\pi(b-a)}{16}\left(5a^2 + 2ab + 5b^2\right) + \\
&+ \frac{1}{24}\left(4c^2 + 8c(b-a) - 15(b-a)^2 - 16ab\right)\sqrt{(a+c)(b-c)} + \\
&+ \frac{1}{24}\left(-4c^2 + 8c(b-a) + 15(b-a)^2 + 16ab\right)\sqrt{(a-c)(b+c)} + \\
&+ \frac{(b-a)(5a^2 + 2ab + 5b^2) + 2c(3a^2 - 2ab + 3b^2)}{16} \arcsin \frac{2c + b - a}{a+b} + \\
&+ \frac{(b-a)(5a^2 + 2ab + 5b^2) - 2c(3a^2 - 2ab + 3b^2)}{16} \arcsin \frac{2c - b + a}{a+b}.
\end{aligned} \tag{57}
$$

2.2.2 The Case $-c \le -a \le c < b$ (Figure 6b)

A similar technique can be used to reduce the relationship for pressure and the system of equations for calculating the ends of contact zone in the case $-c \le -a < c < b$, i.e. when the inclination α satisfies the condition $\alpha > \alpha_1$.

As the result we obtain the following relationships:

$$p(y) = \frac{E^*(a+b)}{2\pi R(1+y^2)}\left\{\frac{(1-y^2)}{2}\left[\frac{\pi}{2} - 2\arctan y_1\right] - \right.$$
$$\left. - \frac{(y-y_1)(1-yy_1)}{(1+y_1^2)}\ln\left|\frac{y_1-y}{yy_1-1}\right|\right\}, \quad |y| \le 1, \tag{58}$$

where y and y_1 are determined by Eq. (52) and

$$-\frac{2P}{E^*} = -\alpha\pi\frac{a+b}{2} + \frac{1}{R}\left[\sqrt{(b-c)(a+c)}\frac{3a+b+2c}{4} + \right.$$
$$\left. + \frac{a+b}{8}(b-3a-4c)\left(\frac{\pi}{2} + \arcsin\frac{b-2c-a}{a+b}\right)\right]. \tag{59}$$

$$\alpha\pi R = \sqrt{(b-c)(a+c)} - \frac{(2c+a-b)}{2}\left(\frac{\pi}{2} + \arcsin\frac{b-2c-a}{a+b}\right). \tag{60}$$

The system of equations (59) and (60) coincides with Eqs. (53) and (54) if $a = c$.

Substitution of $-a = c$ in Eqs. (59) and (60) allows us to determine the value of α_2. Eqs. (59) and (60) are valid if $\alpha_1 < \alpha < \alpha_2$.

If the inclination α is unknown, we add to the system of Eqs. (59) and (60) the following equation which is reduced from Eq. (44):

$$-\frac{2R}{E^*}\left(M + Qr - \frac{P(b-a)}{2}\right) = R\alpha\pi\frac{3a^2-2ab+3b^2}{8} +$$
$$+ \frac{1}{24}\left(4c^2 + 8c(b-a) - 15(b-a)^2 - 16ab\right)\sqrt{(a+c)(b-c)} -$$
$$- \frac{(b-a)(5a^2+2ab+5b^2) - 2c(3a^2-2ab+3b^2)}{16} \times$$
$$\times \left(\frac{\pi}{2} - \arcsin\frac{2c-b+a}{a+b}\right). \tag{61}$$

2.2.3 The Case $c \le -a < b$ (Figure 6c)

In the case we obtain the contact pressure from Eq. (47) in the form which coincides with the solution for the parabolic punch:

$$p(x) = \frac{E^*}{2R}\sqrt{(x+a)(b-x)}. \tag{62}$$

The relationships for calculation of the ends of contact follow from Eqs. (48) and (49):

$$\alpha R = \frac{b-a}{2} - c, \tag{63}$$

$$\frac{4PR}{\pi E^*(a+b)} = c + \alpha R - \frac{1}{4}(b-3a) \tag{64}$$

in particular, it follows from these equations that

$$P = \frac{\pi E^*(a+b)^2}{16R}.$$

The last formula coincides with the solution obtained by Muskhelishvili (1949).

Eqs. (63) and (64) can also be used to calculate the inclination α_2 corresponding to the case $-a = c$. For this case Eqs. (63) and (64) coincide with Eqs. (59) and (60).

Note that the case $[-a, b] \in (-c, c]$ is impossible. This conclusion follows from Eqs. (46) and (47) which give zero pressure for the case, i.e.

$$p(x) = \frac{\alpha E^*}{2\pi}\sqrt{(b-x)(x+a)} \int_{-a}^{b} \frac{dt}{\sqrt{(b-t)(t+a)(t-x)}} = 0.$$

2.3 Shear Stress Analysis

We present the function $q(x)$ within the stick region L_{stick} as

$$q(x) = \rho p(x) - q^*(x). \tag{65}$$

For similar materials of contacting bodies the following integral equation is used to determine the function $q(x)$:

$$s'(x) = -\frac{2}{\pi E^*} \int_{-a}^{b} \frac{q(t)dt}{x-t}. \tag{66}$$

where the function $s(x)$ is determined by Eq. (39).

If the function $q(x)$ satisfies the condition $q(x) = \rho p(x)$ within the slip region, we obtain from Eqs. (65) and Eq. (66)

$$s'(x) = -\frac{2}{\pi E^*} \int_{-a}^{b} \frac{\rho p(t)dt}{x-t} + \frac{2}{\pi E^*} \int_{L_{\text{stick}}} \frac{q^*(t)dt}{x-t} \tag{67}$$

or using Eq. (45)

$$s'(x) = \rho h'(x) + \frac{2}{\pi E^*} \int\limits_{L_{\text{stick}}} \frac{q^*(t)dt}{x-t}. \tag{68}$$

Taking into account the condition $s'(x) = 0$, valid within a stick zone, and Eq. (68), we reduce the following equation to determine the function $q^*(x)$:

$$\int\limits_{L_{\text{stick}}} \frac{q^*(t)dt}{x-t} = -\frac{\pi \rho E^*}{2} h'(x), \quad x \in L_{\text{stick}} \tag{69}$$

where the function $h'(x)$ is presented by Eq. (46).

Note that only one stick zone exists within the contact i.e. $L_{\text{stick}} = [d_1, d_2]$. To prove it, let us assume that

$$L_{\text{stick}} = \bigcup_k [d_1^k, d_2^k].$$

Consider the functions $s'(x)$ and $s''(x)$ within a slip zone located between two stick zones, i.e. $x \in (d_2^k, d_1^{k+1})$. Since the function $q^*(x)$ satisfies the condition $q^*(d_1^k) = q^*(d_2^k) = 0$, the function $s'(x)$ is continuous. Differentiation of Eq. (68) for $d_2^k < x < d_1^{k+1}$ gives

$$s''(x) = \rho h''(x) - \frac{2}{\pi E^*} \int\limits_{L_{\text{stick}}} \frac{q^*(t)dt}{(x-t)^2}. \tag{70}$$

Since $h''(x) \leq 0$ (see Eq. (46)) and $q^*(x) \geq 0$ we conclude that $s''(x) < 0$. So we obtain the following inequality

$$s'(d_1^{k+1}) - s'(d_2^k) = \int\limits_{d_2^k}^{d_1^{k+1}} s''(x)dx < 0 \tag{71}$$

On the other hand, $s'(d_1^{k+1}) = s'(d_2^k) = 0$. This contradiction proves that there is no slip zone between two stick zones and so, there is only one stick zone within the contact region.

The solution of Eq. (69) within the stick region $L_{\text{stick}} = [d_1, d_2]$ satisfying the condition

$$q^*(d_1) = q^*(d_2) = 0$$

has the following form:

$$q^*(x) = -\frac{\rho E^*}{2\pi} \sqrt{(d_2 - x)(x - d_1)} \int\limits_{d_1}^{d_2} \frac{h'(t)dt}{\sqrt{(d_2 - t)(t - d_1)}(t - x)}. \tag{72}$$

The conditions for determination of the constants d_1 and d_2 are:

$$\int_{d_1}^{d_2} \frac{h'(t)dt}{\sqrt{(d_2-t)(t-d_1)}} = 0, \tag{73}$$

and

$$\rho P - Q = \frac{\rho E^*}{2} \int_{d_1}^{d_2} \sqrt{\frac{d_2-t}{t-d_1}} h'(t)dt. \tag{74}$$

The last relationship follows from the equilibrium condition (43).

We analyse the different dispositions of the stick zone within the contact region $[-a, b]$.

2.3.1 The Case $-a \le d_1 < -c, c < d_2 \le b$

Substituting Eq. (46) into (72), and using the technique described in the previous section, we obtain

$$\begin{aligned} q^*(\eta) = {} & \frac{\rho E^*(d_2-d_1)}{2\pi R(1+\eta^2)} \left\{ \frac{(1-\eta^2)}{2} [\pi - 2\arctan\eta_1 + \right. \\ & + 2\arctan\eta_2] + \frac{(\eta-\eta_2)(1-\eta\eta_2)}{(1+\eta_2^2)} \ln\left|\frac{\eta_2-\eta}{\eta\eta_2-1}\right| - \\ & \left. - \frac{(\eta-\eta_1)(1-\eta\eta_1)}{(1+\eta_1^2)} \ln\left|\frac{\eta_1-\eta}{\eta\eta_1-1}\right| \right\}, \quad |\eta| \le 1, \end{aligned} \tag{75}$$

where

$$\begin{aligned} \eta &= \tan\left(\frac{1}{2}\arcsin\frac{2x-d_2-d_1}{d_2-d_1}\right), \\ \eta_1 &= \tan\left(\frac{1}{2}\arcsin\frac{2c-d_2-d_1}{d_2-d_1}\right), \\ \eta_2 &= \tan\left(\frac{1}{2}\arcsin\frac{-2c-d_2-d_1}{d_2-d_1}\right). \end{aligned} \tag{76}$$

From Eqs. (46), (73) and (74) we obtain the following system of equations to determine the ends of the stick zone d_1 and d_2 for the given values of α, P and Q:

$$
\begin{aligned}
-\frac{2R(\rho P - Q))}{\rho E^*} &= -\alpha\pi \cdot \frac{(d_2 - d_1)R}{2} + \\
&+ \sqrt{(c-d_1)(d_2-c)}\frac{2c - 3d_1 + d_2}{4} + \\
&+ \sqrt{(-d_1-c)(d_2+c)}\frac{3d_1 - d_2 + 2c}{4} + \frac{(d_2-d_1)}{8}(d_2 + 3d_1 - 4c)\times \\
&\times\left(\frac{\pi}{2} - \arcsin\frac{2c - d_1 - d_2}{d_2 - d_1}\right) + \frac{d_2 - d_1}{8}(d_2 + 3d_1 + 4c)\times \\
&\times\left(\frac{\pi}{2} - \arcsin\frac{d_2 + 2c + d_1}{d_2 - d_1}\right)
\end{aligned}
\tag{77}
$$

and

$$
\begin{aligned}
\alpha\pi R = {}& \frac{\pi(d_1 + d_2)}{2} - \frac{d_1 + d_2 - 2c}{2}\arcsin\frac{2c - d_1 - d_2}{d_2 - d_1} - \\
&- \frac{d_1 + d_2 + 2c}{2}\arcsin\frac{2c + d_1 + d_2}{d_2 - d_1} + \\
&+ \sqrt{(d_2-c)(c-d_1)} - \sqrt{(-c-d_1)(d_2+c)}.
\end{aligned}
\tag{78}
$$

2.3.2 The Case $-c \le d_1 < c < d_2 < b$

From Eqs. (46) and (72) we obtain the function $q^*(x)$ in the form:

$$
\begin{aligned}
\rho q^*(\eta) = {}& \frac{\rho E^*(d_2 - d_1)}{2\pi R(1+\eta^2)}\left\{\frac{(1-\eta^2)}{2}\left[\frac{\pi}{2} - 2\arctan\eta_1\right] - \right. \\
&\left. - \frac{(\eta - \eta_1)(1 - \eta\eta_1)}{(1+\eta_1^2)}\ln\left|\frac{\eta_1 - \eta}{\eta\eta_1 - 1}\right|\right\}, \quad |\eta| \le 1,
\end{aligned}
\tag{79}
$$

where η and η_1 are determined by Eq. (76), and the following equations are reduced to calculate the ends of the stick zone:

$$
\begin{aligned}
-\frac{2R(\rho P - Q))}{\rho E^*} &= -\alpha\pi\frac{d_2 - d_1}{2} + \\
&+ \frac{1}{R}\left[\sqrt{(d_2-c)(c-d_1)}\frac{-3d_1 + d_2 + 2c}{4} + \right. \\
&\left. + \frac{d_2 - d_1}{8}(d_2 + 3d_1 - 4c)\left(\frac{\pi}{2} + \arcsin\frac{d_2 - 2c + d_1}{d_2 - d_1}\right)\right].
\end{aligned}
\tag{80}
$$

$$
\alpha\pi R = \sqrt{(d_2-c)(c-d_1)} - \frac{(2c - d_1 - d_2)}{2}\left(\frac{\pi}{2} + \arcsin\frac{d_2 - 2c + d_1}{d_2 - d_1}\right) \tag{81}
$$

2.3.3 The Case $c \le d_1 < d_2 < b$

From Eqs. (46) and (72) we obtain:

$$q^*(x) = \frac{\rho E^*}{2R}\sqrt{(d_2 - x)(x - d_1)}. \tag{82}$$

The equations to determine d_1 and d_2 are reduced from Eqs. (46), (73) and (74). They are:

$$\alpha R = \frac{d_2 + d_1}{2} - c, \tag{83}$$

$$\frac{4(\rho P - Q)R}{\rho \pi E^*(d_2 - d_1)} = c + \alpha R - \frac{1}{4}(d_2 + 3d_1). \tag{84}$$

The case $[d_1, d_2] \in (-c, c]$ is impossible since it follows from Eqs. (46) and (72):

$$q^*(x) = \frac{\rho\alpha E^*}{2\pi}\sqrt{(d_2 - x)(x - d_1)}\int_{d_1}^{d_2}\frac{dt}{\sqrt{(d_2 - t)(t - d_1)}(t - x)} = 0.$$

Thus, the shear stresses within the contact zone $(-a, b)$ are determined by

$$q(x) = \begin{cases} \rho p(x) - q^*(x), & \text{if} \quad d_1 < x < d_2, \\ \rho p(x), & \text{if} \quad -a \le x \le d_1, \quad d_2 \le x \le b, \end{cases} \tag{85}$$

where the functions $q^*(x)$ are determined by Eqs. (75), (79) and (82) for different dispositions of the stick zone.

The shear stress $q(x)$ satisfies Eq. (41) within a stick zone as far as the function $q^*(x)$ satisfies the condition:

$$0 < q^*(x) < \rho p(x), \quad x \in (d_1, d_2).$$

The solution obtained above also satisfies Eq. (38). To prove it let us consider the function $s'(x)$ within the slip zones $(-a, d_1)$ and (d_2, b). It follows from Eqs. (46) and (68) that

- in the slip zone $(-a, d_1)$, i.e. if $x = d_1 - \varepsilon$ $(\varepsilon > 0)$

$$s'(x) = s'(d_1 - \varepsilon) - s'(d_1) =$$

$$= h'(d_1 - \varepsilon) - h'(d_1) + \frac{2\varepsilon}{\pi E^*}\int_{d_1}^{d_2}\frac{q^*(t)dt}{(d_1 - t - \varepsilon)(d_1 - t)} > 0$$

- in the slip zone (d_2, b), i.e. if $x = d_2 + \varepsilon$ $(\varepsilon > 0)$

$$s'(x) = s'(d_2 + \varepsilon) - s'(d_2) =$$

$$= -\frac{\rho\varepsilon}{R} - \frac{2\varepsilon}{\pi E^*}\int_{d_1}^{d_2}\frac{q^*(t)dt}{(d_2 - t + \varepsilon)(d_2 - t)} < 0.$$

Taking also into account the condition $s(x) = s'(x) = 0$ if $x \in [d_1, d_2]$ we conclude that $s(x) < 0$ within the slip zones, i.e. slip has the opposite sign to the shear stress $q(x) > 0$.

The function $s(x)$ within the slip zones can be determined by the relationships which follows from Eqs. (39), (40) and (66):

- in the slip region $(-a, d_1)$:

$$\begin{aligned} s(x) &= -\frac{2}{\pi E^*}\int_{-a}^{b} q(x')\ln\left|\frac{x-x'}{d_1-x'}\right|dx' = \\ &= -\frac{2\rho}{\pi E^*}\int_{-a}^{b} p(x')\ln\left|\frac{x-x'}{d_1-x'}\right|dx' + \frac{2}{\pi E^*}\int_{d_1}^{d_2} q^*(x')\ln\left|\frac{x-x'}{d_1-x'}\right|dx', \end{aligned} \tag{86}$$

- in the slip region (d_2, b):

$$\begin{aligned} s(x) &= -\frac{2}{\pi E^*}\int_{-a}^{b} q(x')\ln\left|\frac{x-x'}{d_2-x'}\right|dx' = \\ &= -\frac{2\rho}{\pi E^*}\int_{-a}^{b} p(x')\ln\left|\frac{x-x'}{d_2-x'}\right|dx' + \frac{2}{\pi E^*}\int_{d_1}^{d_2} q^*(x')\ln\left|\frac{x-x'}{d_2-x'}\right|dx'. \end{aligned} \tag{87}$$

2.4 Tensile Stress Analysis

We use the Muskhelishvili method (see Muskhelishvili, 1949 or Chapter 3 of the text) to calculate the normal stress in the x-axis direction σ_{xx} at $y = 0$. As far as we calculated the contact pressure $p(x)$ and the shear stress $q(x)$ within the contact region $(-a, b)$, we can determine the Muskhelishvili's function $\Phi(z)$:

$$\Phi(z) = \frac{1}{2\pi i}\int_{-a}^{b}\frac{p(x') + iq(x')}{x' - z}\,dx'. \tag{88}$$

which is a Cauchy integral.

The stress components within the half-plane and at the boundary can be calculated based on Eq. (88). In particular, we have the following relationship:

$$\sigma_{xx} + \sigma_{yy} = 2\left[\Phi(z) + \bar{\Phi}(\bar{z})\right]. \tag{89}$$

At the boundary $y = 0$ Eq. (89) takes the form:

$$\left(\sigma_{xx} + \sigma_{yy}\right)_{y=0} = 2\left[\Phi^-(x) + \bar{\Phi}^+(x)\right], \tag{90}$$

where $\Phi^{-}(x)$ is the boundary value of the function $\Phi(z)$ if z tends to x from the lower half-plane, and $\bar{\Phi}^{+}(x)$ is the boundary value of the function $\bar{\Phi}(z)$ if z tends to x from the upper half-plane.

Using Eq. (88) and the Plemelj formulae, we obtain

$$\Phi^{-}(x) = -\frac{1}{2}\left[p(x) + iq(x)\right] + \frac{1}{2\pi i}\int_{-a}^{b} \frac{p(x') + iq(x')}{x' - x}\,dx', \tag{91}$$

$$\bar{\Phi}^{+}(x) = \frac{1}{2}\left[-p(x) + iq(x)\right] + \frac{1}{2\pi i}\int_{-a}^{b} \frac{-p(x') + iq(x')}{x' - x}\,dx'. \tag{92}$$

Then from Eqs. (90), (91) and (92), and taking into account the relationship $\sigma_{yy}(x,0) = -p(x)$, we finally obtain:

$$\sigma_{xx} = -p(x) + \frac{2}{\pi}\int_{-a}^{b} \frac{q(x')dx'}{x' - x}, \qquad -\infty < x < +\infty. \tag{93}$$

We consider Eq. (93) for different values of x. If $x \in [d_1, d_2]$, then using Eqs. (40) and (66), we conclude that

$$\sigma_{xx} = -p(x), \quad d_1 \le x \le d_2. \tag{94}$$

If $-a \le x < d_1$ and $d_2 < x \le b$, using Eqs. (65) we can present Eq. (93) in the following form:

$$\sigma_{xx} = -p(x) + \frac{2\rho}{\pi}\int_{-a}^{b} \frac{p(x')dx'}{x' - x} - \frac{2}{\pi}\int_{d_1}^{d_2} \frac{q^*(x')dx'}{x' - x}. \tag{95}$$

Then using Eq. (45) we obtain

$$\sigma_{xx} = -p(x) + \rho E^* h'(x) - \frac{2}{\pi}\int_{d_1}^{d_2} \frac{q^*(x')dx'}{x' - x}. \tag{96}$$

The integral in the left-hand side of Eq. (96) is non-singular and can be calculated for different functions $q^*(x)$ given by Eqs. (75), (79) and (82).

If $x \notin (-a, b)$ we use Eq. (95) for calculation σ_{xx} at $y = 0$ out of contact zone which takes the form

$$\sigma_{xx} = \frac{2\rho}{\pi}\int_{-a}^{b} \frac{p(x')dx'}{x' - x} - \frac{2}{\pi}\int_{d_1}^{d_2} \frac{q^*(x')dx'}{x' - x}. \tag{97}$$

The functions $p(x)$ are determined by Eqs. (51), (58) or (62). Both integrals are nonsingular.

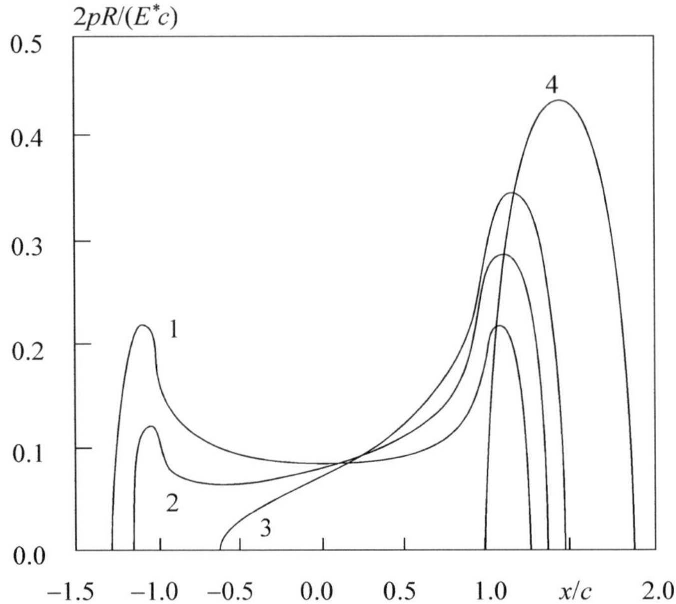

Fig. 7 Pressure distribution for $\bar{\alpha} = 0$ (1), $\bar{\alpha} = 0.05$ (2), $\bar{\alpha} = 0.1$ (3), $\bar{\alpha} = 0.44$ (4).

2.5 Results and Discussions

The expressions obtained above give the solution of the problem in closed form. In what follows we present some results which illustrate the stress distribution within and near the contact.

It follows from Eqs. (51), (58) and (62) that the dimensionless contact pressure $\bar{p} = \frac{2pR}{E^*c}$ depends on the dimensionless normal load $\bar{P} = \frac{2PR}{E^*c^2}$, and the dimensionless inclination $\bar{\alpha} = \frac{\alpha R}{c}$. The contact pressure distributions for various values of $\bar{\alpha}$ are presented in Figure 7. The main feature of the pressure distribution for inclined punch is an asymmetry of the pressure in respect to the axis of symmetry of the punch. If the contact zone includes the flat base $(-c, c)$, i.e. $\alpha < \alpha_1$ (see Figure 6a), the contact pressure has the local maximum in two points. For the larger absolute value of α the contact pressure has only one maximum near the right-hand end of the contact. The value of the maximum pressure increases with increasing of α. The left-hand end of the contact is located at the flat base for $\alpha_1 < \alpha < \alpha_2$, and at the rounded edge if $\alpha > \alpha_2$. The dependencies of the contact width $\frac{a+b}{c}$ and the contact shift $\frac{b-a}{b+a}$ on the indenter inclination $\bar{\alpha}$ for two values of the dimensionless load $\bar{P}$ are presented in Figure 8. The contact shift increases and the contact width decreases as the inclination increases. Here and in what follows the points corresponding to the transition from the contact condition presented at the Figure 6a to

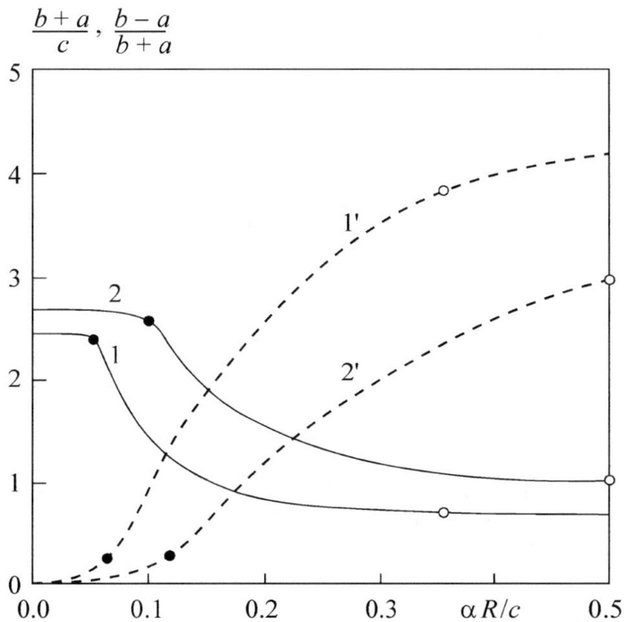

Fig. 8 Variation of contact width (solid lines) and contact shift (broken lines) with $\bar{\alpha}$ at $\bar{P} = 0.2$ (1, 1′) and $\bar{P} = 0.4$ (2, 2′).

Figure 6b ($\alpha = \alpha_1$) and from Figure 6b to Figure 6c ($\alpha = \alpha_2$) are indicated at the plots.

Figure 9 illustrates the contact shear stress distribution at different values of the dimensionless tangential force

$$\bar{Q} = \frac{2(\rho P - Q)R}{\rho E^* c^2}$$

for the case when the contact region includes the flat base (Figure 9a, $\bar{\alpha} = 0.12$) and for the case when the left-hand end of the contact region is located at the flat base (Figure 9b, $\bar{\alpha} = 0.2$). For both cases, the width of the stick zone depends on the tangential force. The shear stress has two or three local maxima depending on the contact region and stick zone location.

A slip function for different values of the inclination $\bar{\alpha}$ is presented in Figure 10. The value of slip at the ends of the contact zone increases as the inclination increases.

Figure 11 illustrates the distribution of σ_{xx} at $y = 0$ within a lower half-space (Figure 11a) and within an upper half-space (indenter, Figure 11b) for various values of $\bar{\alpha}$. The increase of the inclination decreases the tensile stress at the left-hand side of the contact region within the lower half-space. The maximum value of the tensile

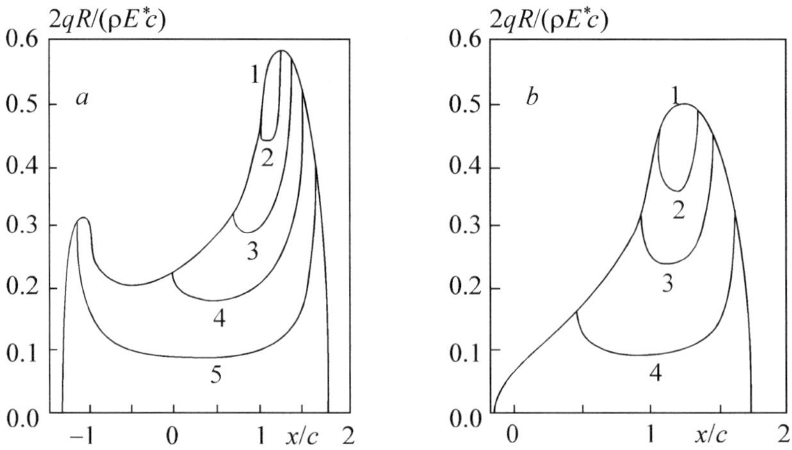

Fig. 9 Shear stress distribution for $\bar{\alpha} = 0.12$, $\bar{P} = 1$ and $\bar{Q} = 0$ (1), $\bar{Q} = 0.42$ (2), $\bar{Q} = 0.75$ (3), $\bar{Q} = 0.9$ (4), $\bar{Q} = 0.98$ (5) (a) and for $\bar{\alpha} = 0.2$, $\bar{P} = 0.5$ and $\bar{Q} = 0$ (1), $\bar{Q} = 0.22$ (2), $\bar{Q} = 0.40$ (3), $\bar{Q} = 0.47$ (4) (b).

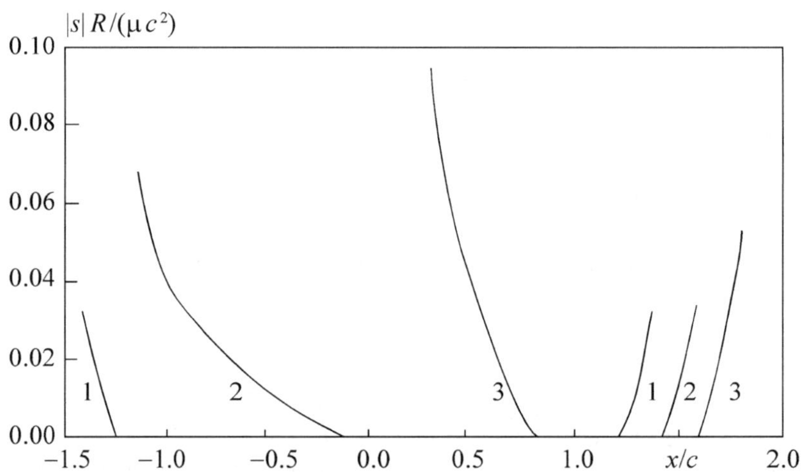

Fig. 10 Slip function for $\bar{\alpha} = 0$ (1), $\bar{\alpha} = 0.1$ (2), $\bar{\alpha} = 0.25$ (3).

stress within the indenter occurs at the right-hand side of the contact and it increases as the inclination increases.

The dependence of the inclination $\bar{\alpha}$ on the dimensionless moment

$$\bar{M} = \frac{2(M + Qr)R}{E^* c^3}$$

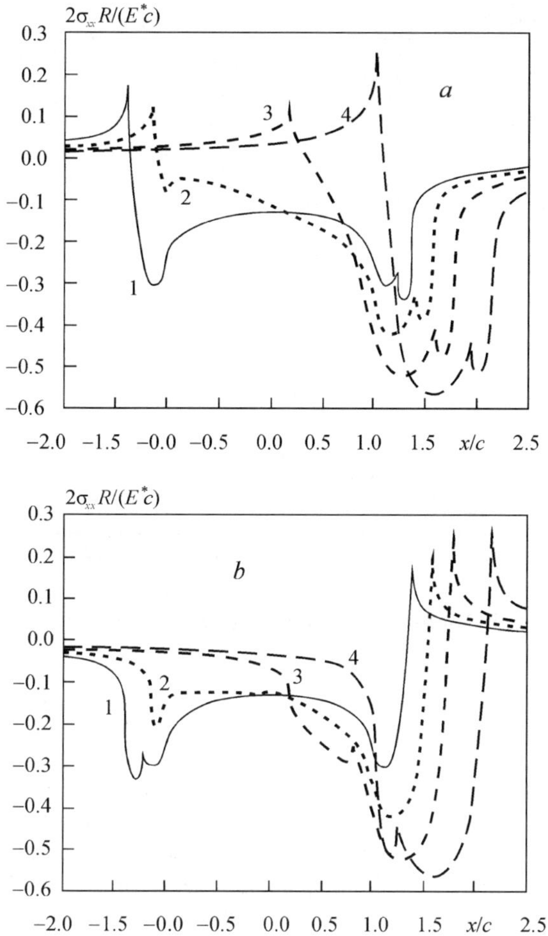

Fig. 11 Stress component σ_{xx} at $y = 0$ at the lower (a) and upper (b) half-spaces for $\bar{\alpha} = 0$ (1), $\bar{\alpha} = 0.1$ (2), $\bar{\alpha} = 0.25$ (3) and $\bar{\alpha} = 0.6$ (4).

for two values of the dimensionless normal load $\bar{P}$ is presented in Figure 12. If $\alpha \leq \alpha_1$ the dependence is close to a linear function.

Based on the analysis we can evaluate the stress field for the given external conditions, and also to provide the appropriate contact stresses by choosing the optimal external force application.

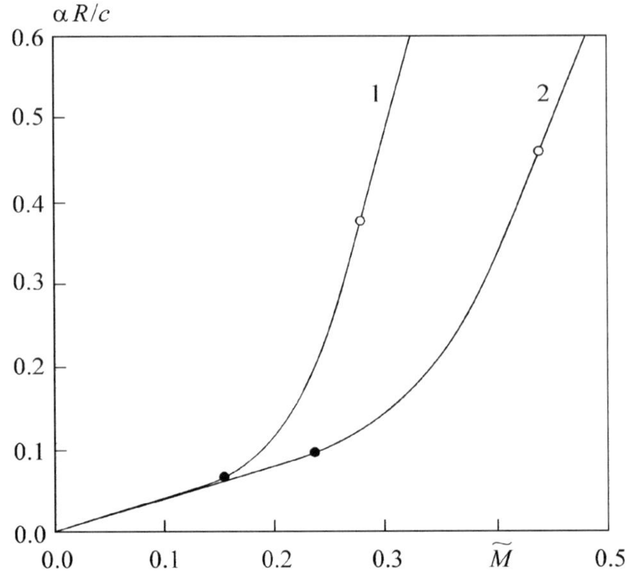

Fig. 12 Variation of $\bar{\alpha}$ with the moment for $\bar{P} = 0.2$ (1) and $\bar{P} = 0.3$ (2).

2.6 Conclusions

The solution of the contact problem with partial slip for the inclined punch having flat base and rounded edges is presented in a closed form, i.e. the analytical expressions for distribution of stick and slip zones, the shear stress and pressure distribution, for the interfacial slip displacement and for σ_{xx} stress component are given. This solution makes it possible to analyze the influence of the inclination angle on the asymmetry of the contact stress and stress concentration zones.

It is shown that for the inclined punch with rounded edges and similar materials of contacting bodies only one stick zone exists within the contact region.

Some specific features of the stress distribution have been established for the inclined punch, such as

- one of the ends of the contact is located at the flat base of the punch for the inclined punch if $\alpha > \alpha_1$, while for the horizontal base ($\alpha = 0$) the contact region and the stick zone always include the flat base of the punch;
- the maximum contact pressure occurs near the end of the contact region where there is the larger indentation of the punch into foundation;
- the stick zone is located close to the same end of the contact; the size of the stick zone depends on the inclination angle and the tangential force;
- the size of the slip zone and the slip displacements are higher at the left-hand side of the contact region where the values of the pressure are smaller;

- the maximum tensile stress occurs near the end of the contact where the minimum indentation occurs. The value is less than that for the symmetrical case and the same tangential force.

The results obtained can be used to analyze crack nucleation in fretting when the punch is acted upon by an oscillating tangential force and moment. Note that the results illustrate that the tensile stresses parallel to the surface that causes the fretting fatigue cracks to nucleate and grow are strongly dependent on small inclinations. Thus, great care must be taken in conducting fretting fatigue experiments with geometry similar to dovetail joints in engine hardware as slight misalignments will change this inclination; hence fatigue life.

2.7 Appendix

The following integrals have been used to reduce Eqs. (51), (58), (75) and (79):

$$\int \frac{dt}{\sqrt{(t+a)(b-t)}} = \arcsin\frac{2t+a-b}{a+b},$$

$$\int_{x_1}^{x_2} \frac{dt}{(t-x)\sqrt{(t+a)(b-t)}} = \frac{2(1+y^2)}{(a+b)} \int_{y_1}^{y_2} \frac{d\tau}{(\tau-y)(1-\tau y)} =$$

$$= \frac{2(1+y^2)}{(a+b)(1-y^2)} \ln\left|\frac{(y_2-y)(1-yy_1)}{(y_1-y)(1-yy_2)}\right|.$$

The last integral has been calculated using the substitutions:

$$x = \frac{b-a}{2} + \frac{y}{1+y^2}(a+b),$$
$$t = \frac{b-a}{2} + \frac{\tau}{1+\tau^2}(a+b).$$

The following integrals have been used to reduce the relationships (53), (54), (59), (60), (77), (78), (80) and (81):

$$\int \frac{t\,dt}{\sqrt{(t+a)(b-t)}} = \frac{b-a}{2}\arcsin\frac{2t+a-b}{a+b} - \sqrt{(t+a)(b-t)},$$

$$\int \sqrt{\frac{b-t}{t+a}}\,dt = \sqrt{(t+a)(b-t)} + \frac{a+b}{2}\arcsin\frac{2t+a-b}{a+b},$$

$$\int t\sqrt{\frac{b-t}{t+a}}\,dt = \frac{2t-3a-b}{4}\sqrt{(t+a)(b-t)}$$
$$+ \frac{(a+b)(b-3a)}{8}\arcsin\frac{2t+a-b}{a+b}.$$

The relationships (56), (57) and (58) have been obtained based on the following transformations and integrals:

$$\int_{-a}^{b} \frac{\sqrt{(x+a)(b-x)}}{x-t} dx = \sqrt{(x+a)(b-x)} \Big|_{-a}^{b} +$$

$$[ab + t(b-a-t)] \int_{-a}^{b} \frac{dx}{(x-t)\sqrt{(x+a)(b-x)}} +$$

$$\left(\frac{b-a}{2} - t\right) \int_{-a}^{b} \frac{dx}{\sqrt{(x+a)(b-x)}} = \left(\frac{b-a}{2} - t\right) \pi,$$

$$\int \frac{t^2 dt}{\sqrt{(t+a)(b-t)}} = \frac{3a^2 - 2ab + 3b^2}{8} \arcsin \frac{2t+a-b}{a+b} - \frac{2t+3b-3a}{4} \sqrt{(t+a)(b-t)},$$

$$\int \frac{t^3 dt}{\sqrt{(t+a)(b-t)}} = \frac{(b-a)(5a^2 + 2ab + 5b^2)}{16} \arcsin \frac{2t+a-b}{a+b} - \frac{8t^2 + 10(b-a)t + 15(b-a)^2 + 16ab}{24} \sqrt{(t+a)(b-t)}.$$

3 Three-Dimensional Sliding Contact of Elastic Bodies

We investigate three-dimensional contact problems under the assumption that friction forces are parallel to the motion direction. This case holds if the punch slides along the boundary of an elastic half-space with anisotropic friction. The friction depends in magnitude and direction on the direction of sliding. The description of the anisotropic friction has been made by Vantorin (1962) and Zmitrovicz (1990). This friction occurs, for example, in sliding of monocrystals, which have properties in different directions which depend on the orientation of the crystal. Seal (1957) investigated friction between two diamond samples, and showed that the friction coefficient changes from 0.07 to 0.21, depending on the mutual orientation of the samples. A similar phenomenon was observed by Tabor and Wynne-Williams (1961) in experiments on polymers, where polymeric chains at the surface have special orientations.

For arbitrary surfaces, the assumption that friction forces are parallel to the motion direction is satisfied approximately.

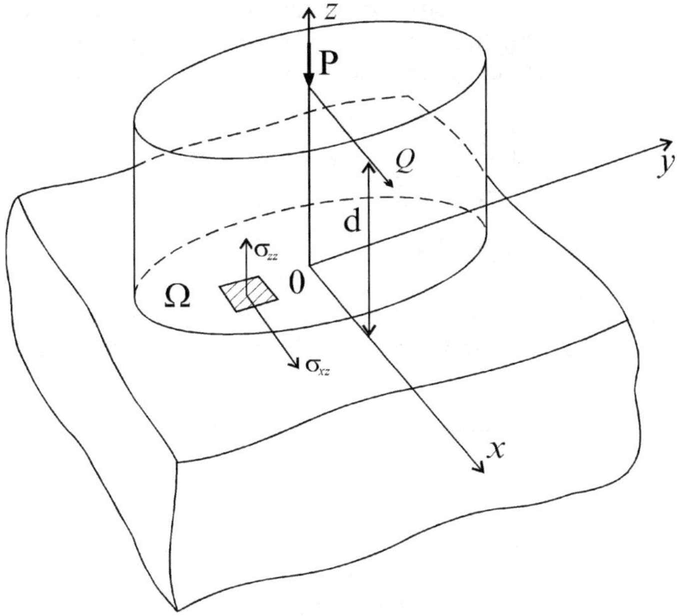

Fig. 13 Sliding contact of a punch and an elastic half-space.

This section reproduces in part the paper by Galin L.A. and Goryacheva I.G. (1983) published in the *Journal of Applied Mathematics and Mechanics* after Galin's death.

3.1 The Friction Law has the Form $\sigma_{xz} = \rho p$

We consider the contact of a punch sliding along the surface of an elastic half-space. We assume the problem to be quasistatic, which imposes a definite restriction on the sliding velocity, and we introduce a coordinate system (x, y, z) connected with the moving punch (Figure 13). The shear stresses within the contact region Ω are assumed to be directed along the x-axis, and $\sigma_{xz} = \rho p(x, y)$, where $p(x, y) = -\sigma_{zz}(x, y, 0)$ is the contact pressure ($p(x, y) \geq 0$). The boundary conditions have the form

$$\begin{aligned} &w = f(x, y) - D, \quad \sigma_{xz} = -\rho\sigma_{zz}, \quad \sigma_{yz} = 0, \quad && x, y \in \Omega, \\ &\sigma_{zz} = \sigma_{xz} = \sigma_{yz} = 0, && x, y \notin \Omega. \end{aligned} \tag{98}$$

Here $f(x, y)$ is the shape of the punch, and D is its displacement along the z-axis.

The displacement w of the half-space boundary in the direction of the z-axis can be represented as the superposition of the displacements caused by the normal pressure $p(x, y)$ and the shear stress σ_{xz} within the contact zone. The solution of the problem for the elastic half-space loaded by a concentrated force at the origin with components Q_x, Q_z along the x- and z-axis, gives the vertical displacement w on the plane $z = 0$ as

$$w = \frac{1-\nu^2}{\pi E} \cdot \frac{Q_z}{R} + \frac{(1+\nu)(1-2\nu)}{2\pi E} \cdot \frac{x Q_x}{R^2}, \qquad \left(R = \sqrt{x^2+y^2}\right). \tag{99}$$

Integrating (99) over the contact area Ω and taking into account conditions (98), we obtain the following integral equation to determine the contact pressure $p(x, y)$

$$\iint\limits_{\Omega} \left[\frac{1}{\sqrt{(x-x')^2+(y-y')^2}} + \frac{\rho\beta(x-x')}{(x-x')^2+(y-y')^2} \right] p(x', y')\, dx'dy' = \frac{\pi E}{1-\nu^2}\left[D - f(x, y)\right], \qquad \beta = \frac{1-2\nu}{2-2\nu}. \tag{100}$$

The coefficient β is equal to zero when $\nu = 0.5$, i.e. the elastic body is incompressible; in this case, friction forces do not affect the magnitude of the normal pressure. For real bodies, Poisson's ratio ν satisfies the inequality $0 < \nu < 0.5$, hence the coefficient β varies between the limits $0.5 > \beta > 0$; for example, $\beta = 0.286$ for $\nu = 0.3$. Moreover, it should be remembered that the magnitude of the friction coefficient ρ is also small. For dry friction of steel on steel, $\rho = 0.2$. In the case $\nu = 0.3$, $\rho\beta \approx 0.057$. For lubricated surfaces, the coefficient $\rho\beta$ takes a still smaller value.

We investigate Eq. (100), assuming the parameter $\rho\beta = \varepsilon$ to be small, and use the notation $p_0(x, y)$ for the solution of the integral equation (100) in the case $\rho\beta = 0$. We represent the function $p(x, y)$ in the form of the series

$$p(x, y) = p_0(x, y) + \varepsilon p_1(x, y) + \cdots + \varepsilon^n p_n(x, y) + \cdots. \tag{101}$$

Substituting the series (101) into the integral equation (100), we obtain a recurrent system of equations for the unknown functions $p_n(x, y)$

$$A[p_n(x, y)] = B[p_{n-1}(x, y)], \quad n = 1, 2, \ldots. \tag{102}$$

Here the following notations are introduced for operators

$$\begin{aligned} A[\omega] &= \iint\limits_{\Omega} \frac{\omega(x', y')dx'dy'}{\sqrt{(x-x')^2+(y-y')^2}}, \\ B[\omega] &= -\iint\limits_{\Omega} \frac{\omega(x', y')(x-x')dx'dy'}{(x-x')^2+(y-y')^2}. \end{aligned} \tag{103}$$

The convergence of the series (101) was proved (Galin and Goryacheva, 1983) for the case of a bounded function ω.

As an illustration, let us consider sliding contact of an axisymmetric punch of circular planform, $f(r) = r^2/2R$, ($r \le a$, a is the radius of the contact region Ω, R is the radius of curvature of the punch surface). We introduce the polar coordinates (r, θ), i.e.

$$x = r\cos\theta, \quad y = r\sin\theta.$$

As is known (see, for example, Galin, 1953, Chapter 3 of the text, or Johnson, 1987), in this case the function $p_0(x, y) = p_0(r)$ is

$$p_0(r) = \frac{4}{R\pi^2 K}\sqrt{a^2 - r^2}.$$

where $K = 2(1 - \nu^2)/(\pi E)$.

To find the next term $p_1(r, \theta)$ in the series (101), first we find $B[p_0(r)]$, which is the result of integration

$$B\left[p_0(r)\right] = b(r)\cos\theta,$$

$$b(r) = \frac{8}{3R\pi K r}\left[\left(a^2 - r^2\right)^{3/2} - a^3\right].$$

Then we solve the equation

$$A[p_1(r, \theta)] = b(r)\cos\theta. \tag{104}$$

We will seek the solution of the equation (104) in the form

$$p_1(r, \theta) = q(r)\cos\theta.$$

Changing to polar coordinates in Eq. (103) we obtain

$$A[p_1(r, \theta)] = \int_0^a \int_0^{2\pi} \frac{q(r')\cos(\theta')r'\,dr'd\theta'}{\sqrt{r^2 + r'^2 - 2rr'\cos(\theta - \theta')}}.$$

Using tables of Gradshteyn and Ryzhik (1963, 3.674), we calculate the integral

$$\int_0^{2\pi} \frac{\cos(\theta')\,d\theta'}{\sqrt{r^2 + r'^2 - 2rr'\cos(\theta - \theta')}} = Q(r, r')\cos\theta,$$

where

$$Q(r, r') = \begin{cases} \dfrac{4}{r'}\left[K\left(\dfrac{r'}{r}\right) - E\left(\dfrac{r'}{r}\right)\right], & r' < r, \\ \dfrac{4}{r}\left[K\left(\dfrac{r}{r'}\right) - E\left(\dfrac{r}{r'}\right)\right], & r' > r. \end{cases}$$

$K(x)$ and $E(x)$ are the complete elliptic integrals of the first and second kinds, respectively. So Eq. (104) reduces to the equation for determining the function $q(r)$

$$\int_0^r \left[K\left(\frac{r'}{r}\right) - E\left(\frac{r'}{r}\right)\right] q(r')dr' +$$

$$+\frac{1}{r}\int_r^a \left[K\left(\frac{r}{r'}\right) - E\left(\frac{r}{r'}\right)\right] r'q(r')dr' = \frac{1}{4}b(r).$$

The other terms in the series (101) have the form (Galin and Goryacheva, 1983)

$$p_n(r,\theta) = \sum_{k=1}^{n} q_{nk}(r)\cos k\theta.$$

So in the case of sliding contact with friction, the contact pressure has the form $p(r,\theta) = p_0(r) + \varepsilon q(r)\cos\theta + O\left(\varepsilon^2\right)$ which indicates, in particular, that the contact pressure is distributed nonsymmetrically, so that there is an additional moment M_y with respect to the y-axis:

$$M_y = \int_0^a \int_0^{2\pi} p(r,\theta) r^2 \cos\theta dr\, d\theta = \varepsilon\pi \int_0^a q(r) r^2 dr + O\left(\varepsilon^2\right).$$

It follows from the equilibrium condition that the force Q directed along the x-axis that causes the punch motion, should be applied at a distance $d = \left|M_y/\rho P\right|$ from the base. When this is not satisfied, the punch has an inclined base, which implies a change of the boundary conditions (98).

For a punch with a flat circular base, the contact pressure can be presented in the form (see Galin and Goryacheva, 1983):

$$p(r,\theta) = \frac{\psi(r,\theta)}{(a-r)^{1/2+\eta}},$$

where $\eta = \frac{1}{\pi}\arctan(\varepsilon\cos\theta)$, and $\psi(r,\theta)$ is a bounded and continuous function. To obtain this function, we again use the method of series-expansion with respect to the small parameter ε.

For the flat punch, the function $w(r,\theta)$ in (98) has the form $w(r,\theta) = \alpha r\cos\theta - D$. The unknown coefficient α governing the inclination of the punch can be found from the equilibrium condition for the moments acting on the punch (see Section 1).

3.2 The Friction Law Has the Form $\sigma_{xz} = \tau_0 + \rho p$

Consider the sliding contact of the punch and an elastic half-space, and assume that shear stresses within the contact region are directed along the x-axis and satisfy the two-terms friction law by Coulomb. Based on Eq. (99), we obtain the following integral equation for the contact pressure $p(x, y)$

$$\iint\limits_{\Omega}\left[\frac{1}{\sqrt{(x-x')^2+(y-y')^2}}+\frac{\rho\beta(x-x')}{(x-x')^2+(y-y')^2}\right]p(x',y')\,dx'dy'+$$
$$+\beta\tau_0\iint\limits_{\Omega}\frac{x-x'}{(x-x')^2+(y-y')^2}\,dx'dy'=\frac{2}{K}\left[D-f(x,y)\right]. \tag{105}$$

The second integral in the left-hand part of Eq. (105) can be calculated if the contact domain Ω is given. For example, if Ω is a circle of radius a, we may change to polar coordinates, and find

$$\iint\limits_{x'^2+y'^2\le a^2}\frac{(x-x')\,dx'dy'}{(x-x')^2+(y-y')^2}=\int\limits_0^a\int\limits_0^{2\pi}\frac{(r\cos\theta-r'\cos\theta')r'dr'd\theta'}{r^2+r'^2-2rr'\cos(\theta-\theta')}.$$

Using the relationship

$$\int\limits_0^{2\pi}\frac{(r\cos\theta-r'\cos\theta')\,d\theta'}{r^2+r'^2-2rr'\cos(\theta-\theta')}=\begin{cases}\dfrac{2\pi}{r}\cos\theta, & \left|\dfrac{r'}{r}\right|<1,\\[2ex] 0, & \left|\dfrac{r'}{r}\right|>1,\end{cases}$$

and the result of integration

$$\int\limits_0^r\frac{2\pi}{r}\cos\theta r'\,dr'=\pi r\cos\theta=\pi x,$$

we reduce Eq. (105) to

$$\iint\limits_{\Omega}p(x',y')\left[\frac{1}{\sqrt{(x-x')^2+(y-y')^2}}+\frac{\rho\beta(x-x')}{(x-x')^2+(y-y')^2}\right]dx'dy'=$$
$$=\frac{2}{K}\left[D-f(x,y)\right]-\pi\beta\tau_0x. \tag{106}$$

Eq. (106) differs from Eq. (100) only by the right side. The method of expansion with respect to the small parameter $\varepsilon=\rho\beta$ can again be used to solve Eq. (106).

Let us analyze the influence of the parameter τ_0/E on the solution of Eq. (106). At first, we consider a smooth punch with surface described by the function $f(x,y)=(x^2+y^2)/2R$. Then the right-hand side of Eq. (106) can be rewritten in the form

$$\frac{2}{K}\left[D-f(x,y)\right]-\pi\beta\tau_0x=\frac{2}{K}\left[D_1-\frac{(x+e)^2+y^2}{2R}\right], \tag{107}$$

where

$$e = \frac{\beta\tau_0 R\pi K}{2}, \quad D_1 = D + \frac{R\beta^2\tau_0^2\pi^2 K^2}{8}. \tag{108}$$

The relationships (108) indicate that the shift of the contact region e and the indentation of the punch D depend on the value of τ_0/E.

Then let us consider the sliding contact of a punch with a flat base ($f(x, y) = 0$, $x^2 + y^2 \leq a^2$). In this case the right-hand side of Eq. (106) has the form

$$\frac{2}{K}\left[D - f(x, y)\right] - \pi\beta\tau_0 x = \frac{2}{K}D - \pi\beta\tau_0 x.$$

So the contact pressure distribution corresponds to the solution of Eq. (100) for the punch with inclined flat base; the angle of inclination is proportional to $\pi\beta\tau_0$.

This conclusion about the influence of τ_0 on the contact characteristics is in a good agreement with that made in the two-dimensional problem (see Section 1).

4 Periodic Contact Problem

The solution of the contact problem with the additional loading applied outside the contact region obtained by Galin (see Section 5.7 of the text) has been used in Goryacheva (1987, 1998a) to solve the periodic contact problem for the elastic half-space. The periodic contact problem solutions are applied for analysis of the real contact characteristics for surfaces with regular microgeometry (for example, wavy surfaces).

The 2-D periodic contact problem for elastic bodies in the absence of friction was investigated by Westergaard (1939) and Staierman (1949). Kuznetsov and Gorokhovsky (1978a, 1978b, 1980) obtained the solution of a 2-D periodic contact problem with friction force, and analysed the stress-strain state of the surface layer for different parameters characterizing the surface shape. Johnson, Greenwood and Higginson (1985) developed a method of analysis of a periodic contact problem for an elastic body, the surface of which in two mutually perpendicular directions was described by two sinusoidal functions; the counter body had a smooth surface.

The approach based on the periodic contact problem solution has then been developed to analyse the real and nominal contact characteristics in normal contact of the bodies with given macro- and micro-geometries (see Goryacheva, 1998a, 1999, 2006).

In what follows the solution of the 3-D periodic contact problem for a system of asperities of regular shape and an elastic half-space is presented. This part reproduces the results obtained by Goryacheva (1987, 1998a).

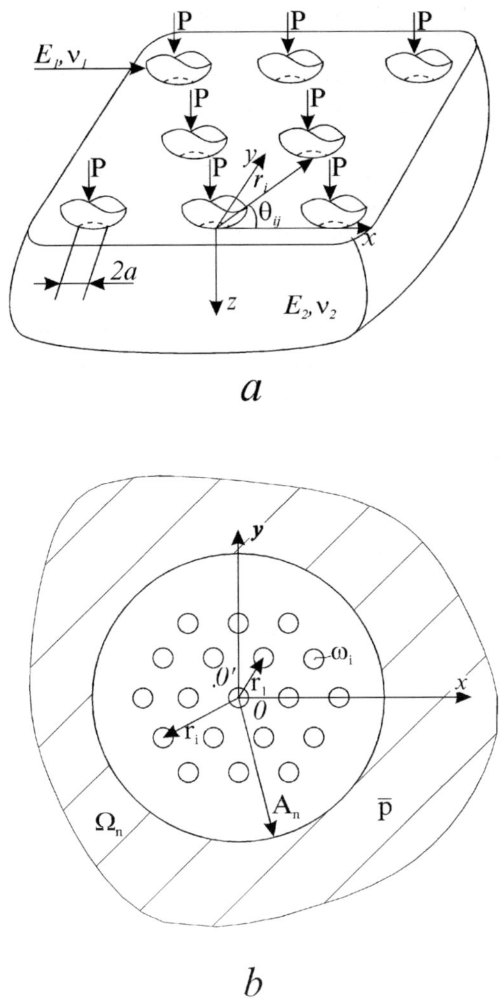

Fig. 14 Scheme of contact of a periodic system of indenters and an elastic half-space (a) and representation of the contact region based on the principle of localization (b) (the nominal pressure $\bar{p}$ is applied to the shaded region).

4.1 One-Level Model

We consider a system of identical axisymmetric elastic indenters ($z = f(r)$) of the same height (one-level model), interacting with an elastic half-space (Figure 14). The axes of the indenters are perpendicular to the half-space surface $z = 0$ and intersect this surface at points which are distributed uniformly over the plane $z = 0$.

As an example of such a system we can consider indenters located at the sites of a rectangular or hexagonal lattice.

Let us fix an arbitrary indenter and locate the origin O of a polar system of coordinates (r, θ) in the plane $z = 0$ at the point of intersection of the axis of this indenter with the plane $z = 0$ (see Figure 14a). The tops of the indenters have the coordinates (r_i, θ_{ij}) ($i = 1, 2, \ldots$; $j = 1, 2, \ldots, m_i$, where m_i is the number of indenters located at the circumference of the radius r_i, $r_i < r_{i+1}$).

Due to the periodicity of the problem, each contact occurs under the same conditions. We assume that contact spots are circles of radius a, and that only normal pressure $p(r, \theta)$ acts at each contact spot ($r \leq a$) (the shear stress is negligibly small). To determine the pressure $p(r, \theta)$ acting at an arbitrary contact spot with a center O, we use the solution of a contact problem for an axisymmetric indenter ($z = f(r)$) and an elastic half-space subjected to the pressure $q(r, \theta)$, distributed outside the contact region (see Section 5.7 of the text). The contact pressure $p(r, \theta)$ ($r \leq a$) is determined by the formula

$$p(r, \theta) = G(r) + \frac{c(\theta)}{\sqrt{a^2 - r^2}} - \frac{1}{\sqrt{a^2 - r^2}} \int_a^{+\infty} \int_0^{2\pi} q(r', \theta') H_2(r, \theta, r', \theta') r' dr' d\theta', \tag{109}$$

where

$$G(r) = \frac{E^*}{4\pi^2} \int_0^a \Delta f(r') H_1(r, r')\, dr', \tag{110}$$

$$H_1(r, r') = \int_0^{2\pi} \frac{2r'}{\sqrt{r^2 - 2rr' \cos\theta' + r'^2}} \arctan \frac{\sqrt{a^2 - r^2}\sqrt{a^2 - r'^2}}{a\sqrt{r^2 - 2rr' \cos\theta' + r'^2}} d\theta', \tag{111}$$

$$H_2(r, \theta, r', \theta') = \frac{\sqrt{r'^2 - a^2}}{\pi^2 \left[r^2 + r'^2 - 2rr' \cos(\theta - \theta')\right]}, \tag{112}$$

$$E^* = \left(\frac{1 - \nu_1^2}{E_1} + \frac{1 - \nu_2^2}{E_2}\right)^{-1}. \tag{113}$$

Here E_1, ν_1 and E_2, ν_2 are the moduli of elasticity of the indenters and the half-space, respectively. The function $c(\theta)$ depends on a shape of the indenter $f(r)$. For example, if the indenter is smooth (the function $f'(r)$ is continuous at $r = a$), then the contact pressure is zero at $r = a$, i.e. $p(a, \theta) = 0$, and the function $c(\theta)$ has the form

$$c(\theta) = \int_a^{+\infty} \int_0^{2\pi} q(r', \theta') H_2(a, \theta, r', \theta') r' dr' d\theta'. \tag{114}$$

The first term in Eq. (109) means the pressure that occurs under a single axisymmetric indenter of the shape function $f(r)$ penetrating an elastic half-space, the last two terms are the additional contact pressure occurring due to the pressure $q(r,t)$ distributed outside the contact region.

For the periodic contact problem, the function $q(r,\theta)$ coincides with the pressure $p(r,\theta)$ at each contact spot located at (r_i,θ_{ij}) $(r_i > a)$, and is zero outside contact spots. So we obtain the following integral equation from Eq. (109), on the assumption that $f'(r)$ is a continuous function $(p(a,\theta)=0)$:

$$p(r,\theta) - \int_0^a \int_0^{2\pi} K\left(r,\theta,r',\theta'\right) p(r',\theta')r'dr'd\theta' = G(r), \tag{115}$$

where

$$K\left(r,\theta,r',\theta'\right) = \sum_{i=1}^{\infty} K_i\left(r,\theta,r',\theta'\right), \tag{116}$$

$$K_i\left(r,\theta,r',\theta'\right) = \frac{1}{\pi^2\sqrt{a^2-r^2}} \sum_{j=1}^{m_i} \left[K_{ij}\left(a,\theta,r',\theta'\right) - K_{ij}\left(r,\theta,r',\theta'\right)\right], \tag{117}$$

$$K_{ij}\left(r,\theta,r',\theta'\right) = \frac{\sqrt{r_i^2 + r'^2 + 2r_i r' \cos\left(\theta_{ij}-\theta'\right) - a^2}}{\left(r\cos\theta - r'\cos\theta' - r_i\cos\theta_{ij}\right)^2 + \left(r\sin\theta - r'\sin\theta' - r_i\sin\theta_{ij}\right)^2}. \tag{118}$$

It is worth noting that similar reasoning can be used to obtain the integral equation for the system of punches with a given contact region (for example, cylindrical punches with a flat base); the equation will have the same structure as Eq. (115).

The kernel $K\left(r,\theta,r',\theta'\right)$ of Eq. (115) is represented as a series (116). A general term (117) of this series can be transformed to the form:

$$K_i(r,\theta,r',\theta') = \frac{1}{\pi^2\sqrt{a^2-r^2}} \sum_{j=1}^{m_i} \left\{ \frac{2(a-r)\cos\left(\theta_{ij}-\theta\right)}{r_i^2} + \right.$$
$$+ \frac{(a-r)\left[-a-r-6r'\cos\left(\theta_{ij}-\theta'\right)\cos\left(\theta_{ij}-\theta\right) + 2r'\cos\left(\theta'-\theta\right)\right]}{r_i^3} +$$
$$\left. + O\left(\frac{1}{r_i^4}\right)\right\}. \tag{119}$$

We assume that for the periodic system of indenters under consideration, each contact spot with center $\left(r_i;\theta_{ij}\right)$ has a partner with center at the point $(r_i;\pi+\theta_{ij})$. So the sum on the first line of Eq. (119) is zero. Hence, the general term of the series (116) has order $O(1/r_i^2)$, since $m_i \sim r_i$, and the series converges.

4.2 Principle of Localization

In parallel with Eq. (115) we consider the following equation

$$p(r,\theta) - \int_0^a \int_0^{2\pi} \sum_{i=1}^{n} K_i\left(r,\theta,r',\theta'\right) p(r',\theta')r'dr'd\theta' = \\ = G(r) + \frac{2\bar{N}P}{\pi}\arctan\frac{\sqrt{a^2-r^2}}{\sqrt{A_n^2-a^2}}, \qquad (120)$$

where P is a load applied to each contact spot. This load satisfies the equilibrium equation

$$P = \int_0^a \int_0^{2\pi} p(r,\varphi) r\, dr d\varphi. \qquad (121)$$

To obtain Eq. (120) we substitute integration over region Ω_n ($\Omega_n : r \geq A_n, 0 \leq \theta \leq 2\pi$) for summation over $i > n$ in Eq. (116), taking into account that the centers of contact spots are distributed uniformly over the plane $z = 0$ and their number per unit area is characterized by the value $\bar{N}$. Actually, the following transformation demonstrates the derivation Eq. (120)

$$J_n = \sum_{i=n+1}^{\infty} K_i\left(r,\theta,r',\theta'\right) \approx \bar{N} \int_{A_n}^{+\infty}\int_0^{2\pi} \frac{\sqrt{x^2+r'^2+2xr'\cos(\phi-\theta')-a^2}}{\pi^2\sqrt{a^2-r^2}} \times \\ \left[\frac{1}{(a\cos\theta - r'\cos\theta' - x\cos\phi)^2 + (a\sin\theta - r'\sin\theta' - x\sin\phi)^2} - \right. \\ \left. - \frac{1}{(r\cos\theta - r'\cos\theta' - x\cos\phi)^2 + (r\sin\theta - r'\sin\theta' - x\sin\phi)^2}\right] x\, dx d\phi.$$

Changing the variables $y\cos\varphi = x\cos\phi + r'\cos\theta'$, $y\sin\varphi = x\sin\phi + r'\sin\theta'$ and taking into account that $r' \leq a \ll A_n$, we finally obtain

$$J_n \approx \frac{\bar{N}}{\pi^2\sqrt{a^2-r^2}} \times \\ \int_{A_n}^{+\infty}\int_0^{2\pi} \sqrt{y^2-a^2}\left[\frac{1}{a^2+y^2-2ay\cos\varphi} - \frac{1}{r^2+y^2-2ry\cos\varphi}\right] y\, dy d\varphi = \\ = \frac{2\bar{N}}{\pi}\arctan\frac{\sqrt{a^2-r^2}}{\sqrt{A_n^2-a^2}},$$

where A_n is the radius of a circle in which there are $\sum_{i=1}^{n} m_i + 1$ central indenters. It is apparent that

$$A_n^2 = \frac{1}{\pi \bar{N}} \left(\sum_{i=1}^{n} m_i + 1 \right). \tag{122}$$

We note that the solution of Eq. (120) tends to the solution of Eq. (115) if $n \to \infty$.

Let us analyze the structure of Eq. (120). The integral term on the left side of Eq. (120) governs the influence of the real pressure distribution at the neighboring contact spots ($r_i < A_n$), on the pressure at the fixed contact spot with center $(0, 0)$ (local effect). The effect of the pressure distribution at the remaining contact spots which have centers (r_i, θ_{ij}), $r_i > A_n$, is taken into account by the second term in the right side of Eq. (120). This term describes the additional pressure $p_a(r)$ which arises within a contact spot ($r < a$) from the nominal pressure $\bar{p} = P\bar{N}$ in the region Ω_n ($r > A_n$). Indeed, from Eqs. (109) and (114) it follows that the additional pressure $p_a(r)$ within the contact spot ($r \leq a$) arising from the pressure $q(r, \theta) = \bar{p}$ distributed uniformly in the region Ω_n has the form

$$p_a(r) = \frac{\bar{p}}{\pi^2 \sqrt{a^2 - r^2}} \times$$

$$\times \int_{A_n}^{+\infty} \int_{0}^{2\pi} \sqrt{r'^2 - a^2} \left[\frac{1}{a^2 + r'^2 - 2ar'cos\theta} - \frac{1}{r^2 + r'^2 - 2rr'cos\theta} \right] r'dr'd\theta =$$

$$= \frac{2\bar{p}}{\pi} \arctan \frac{\sqrt{a^2 - r^2}}{\sqrt{A_n^2 - a^2}}.$$

Thus, the effect of the real contact pressure distribution over the contact spots ω_i far away from the contact spot under consideration ($\omega_i \in \Omega_n$) can be taken into account to sufficient accuracy by the nominal pressure $\bar{p}$ distributed over the region Ω_n (Figure 14b).

This conclusion stated for the periodic contact problem is a particular case of a general principle which we call *a principle of localization*: in conditions of discrete contact, the stress-strain state near one contact spot can be calculated to sufficient accuracy by taking into account the real contact conditions (real pressure, shape of bodies, etc.) at this contact spot and at the nearby contact spots (in the local vicinity of the fixed contact), and the averaged (nominal) pressure over the remaining part of the region of interaction (nominal contact region). This principle was supported by results of investigation of some particular problems considered by Goryacheva et al. (1991, 1998a).

Eqs. (120) and (121) are used to determine the contact pressure $p(r, \theta)$ and the radius a of each spot. The stress distribution in the subsurface region ($z > 0$) arising from the real contact pressure distribution at the surface $z = 0$ can then be found by superposition, using the potentials of Boussinesq (1885) or the particular solution of the axisymmetric problem given by Timoshenko and Goodier (1951).

To simplify the procedure, we can use the principle of localization to determine of internal stresses, substituting the real contact pressure at distant contact spots by the nominal contact pressure. We give here the analytical expressions for the additional stresses which occur on the axis of symmetry of any fixed contact spot from the action of the nominal pressure $\bar{p}$ within the region $\Omega_n(r > A_n)$.

$$\sigma_{zz} = -\frac{\bar{p}z^3}{\left(A_n^2 + z^2\right)^{3/2}},$$

$$\sigma_{rr} = \sigma_{\theta\theta} = \frac{\bar{p}z}{\sqrt{A_n^2 + z^2}}\left[\frac{z^2}{2\left(A_n^2 + z^2\right)} - (1+\nu)\right], \tag{123}$$

$$\sigma_{rz} = \sigma_{\theta z} = \sigma_{r\theta} = 0.$$

4.3 System of Indenters of Various Heights

The method described above is used to determine the real pressure distribution in contact interaction between a periodic system of elastic indenters of the various heights, and an elastic half-space. We assume that the shape of an indenter is described by a continuously differentiable function $z = f_m(r) + h_m$, where h_m is a height of indenters of a given level m ($m = 1, 2, \ldots, k$), k is the number of levels. An example of positions of indenters of each level for $k = 3$ for a hexagonal lattice is shown in Figure 15a. We assume also that the contact spot of the m-th level is a circle of radius a_m.

Let us fix any indenter of the m-th level and place the origin of the polar system of coordinates at the center of its contact spot (Figure 15b). Using the principle of localization, we take into account the real pressure $p_j(r, \theta)$ ($j = 1, 2, \ldots, k$) at the contact spots which are inside the region $\bar{\Omega}_m$ which is a circle of radius A_m ($\bar{\Omega}_m : r \leq A_m$):

$$A_m^2 = \frac{1}{\pi}\left(\sum_{j=1}^{k}\frac{k_{jm}}{\bar{N}_j} + \frac{1}{\bar{N}_m}\right),$$

where k_{jm} is the number of indenters of the j-th level inside the region $\bar{\Omega}_m$, $\bar{N}_j$ is the density of indenters of the j-th level, which is the number of indenters at the j-th level for the unit area. It must be noted that the number of indenters of the m-th level ($j = m$) inside the region $\bar{\Omega}_m$ is $k_{mm} + 1$. Replacing the real contact pressure at the removed contact spots ($r_i > A_m$) by the nominal pressure $\bar{p}$ acting within the region ($r > A_m$)

$$\bar{p} = \sum_{j=1}^{k}\bar{N}_j\int_0^{a_j}\int_0^{2\pi} p_j(r', \theta')r'dr'd\theta',$$

we obtain the following relationship similar to Eq. (120)

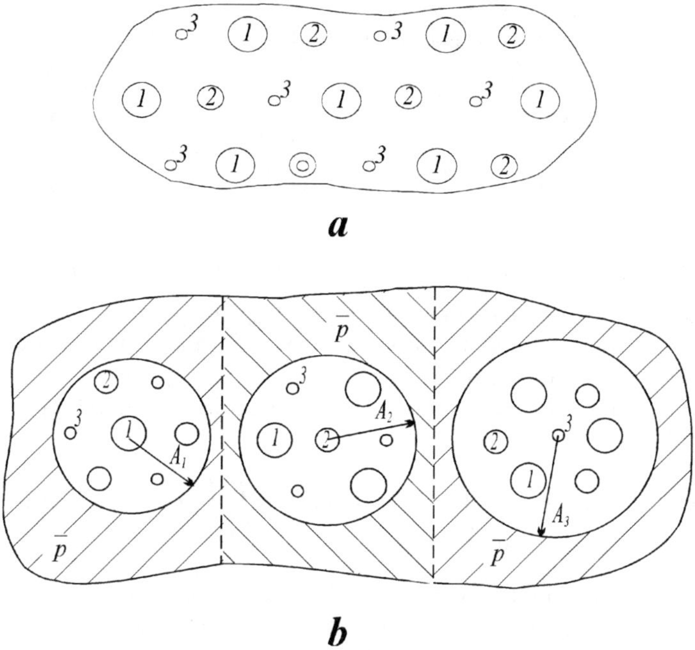

Fig. 15 The location of indenters of each level in the model ($k = 3$) (a) and scheme of calculations based on Eqs. (124)–(126) for $n = 1$ (b).

$$
\begin{aligned}
p_m(r,\theta) - \sum_{j=1}^{k} \int_0^{a_j} \int_0^{2\pi} K_n\left(a_m, r, \theta, r', \theta'\right) p_j(r',\theta') r' dr' d\theta' = \\
= G_m(r) + \frac{2\bar{p}}{\pi} \arctan \frac{\sqrt{a_m^2 - r^2}}{\sqrt{A_m^2 - a_m^2}} .
\end{aligned}
\tag{124}
$$

The kernel of Eq. (124) has the form

$$
K_n\left(a_m, r, \theta, r', \theta'\right) = \sum_{i=1}^{n} K_i\left(a_m, r, \theta, r', \theta'\right),
$$

where functions $K_i\left(a_m, r, \theta, r', \theta'\right)$ are determined by Eqs. (117) and (118), in which we must put $a = a_m$. The function $G_m(r)$ is determined by Eq. (110), where $a = a_m$ and $f(r) = f_m(r)$.

Repeating the same procedure for indenters of each level (see Figure 15b), we obtain the system of k integral equations (124) ($m = 1, 2, \ldots, k$) for determination of the pressure $p_m(r,\theta)$ within the contact spot ($r \leq a_m$) of each level.

Usually the radius of a contact spot a_m is unknown. If an origin of a polar system of coordinates is placed in the center O_m of the m-th level contact spot, we can write

$$h_m = \frac{1}{\pi E^*}\Bigg[\int_0^{a_m}\int_0^{2\pi} p_m(r,\theta)drd\theta + 2\pi\bar{p}\,(A_\infty - A_m) + \\ + \sum_{j=1}^{k}\sum_{i=1}^{k_{jm}}\int_0^{a_j}\int_0^{2\pi}\frac{p_j(r,\theta)rdrd\theta}{\sqrt{r^2 - 2rr_{ij}^{(m)}\cos(\theta-\theta_{ij}^{(m)}) + r_{ij}^{(m)2}}}\Bigg], \tag{125}$$

where $r_{ij}^{(m)}$, $\theta_{ij}^{(m)}$ are the coordinates with respect to the system $(O_m r\theta)$ of the centers of contact spots located within the region Ω_m ($a_m < r_{ij}^{(m)} < A_m$, $0 < \theta_{ij}^{(m)} < 2\pi$), A_∞ is a constant which can be excluded from the system of Eqs. (125) by consideration of differences of heights $h_1 - h_m$, where h_1 is the largest height. The system of equations is completed if we add the equilibrium condition

$$\bar{p}\pi A_m^2 = \sum_{j=1}^{k} k_{jm}\int_0^{a_j}\int_0^{2\pi} p_j(r,\theta)rdrd\theta + \int_0^{a_m}\int_0^{2\pi} p_m(r,\theta)rdrd\theta. \tag{126}$$

It should be remarked that for given height distribution h_m all indenters enter into contact only if the nominal pressure reaches the definite value $\bar{p}^*$. For $\bar{p} < \bar{p}^*$ there are less than k levels of indenters in contact.

4.4 Stress Field Analysis

We use the relationships obtained in Sections 4.1–4.3 to analyze a real contact pressure distribution and the internal stresses in a periodic contact problem for a system of indenters and the elastic half-space. Particular emphasis will be placed upon the influence of the geometric parameter which describes the density of indenter location, on the stress-strain state. This will allow us to determine the range of parameter variations in which it is possible to use the simplified theories which neglect the interaction between contact spots (the integral term in Eq. (115)) or the local effect of the influence of the real pressure distribution at the neighboring contact spots on the pressure at the fixed spot (the integral term in Eqs. (120)).

Numerical results are presented here for a system of spherical indenters, ($f(r) = r^2/2R$, R is a radius of curvature), located on a hexagonal lattice with a constant pitch l. Figure 15a shows the location of indenters of different levels at the plane $z = 0$ for a three-level model ($k = 3$). We introduce the following dimensionless parameters and functions

$$\rho^1 = \frac{r}{R}, \quad A_n^1 = \frac{A_n}{R}, \quad a^1 = \frac{a}{R}, \quad l^1 = \frac{l}{R}, \\ p^1(\rho^1,\theta) = \frac{\pi p(\rho^1 R,\theta)}{2E^*}, \quad P^1 = \frac{\pi P}{2E^* R^2}. \tag{127}$$

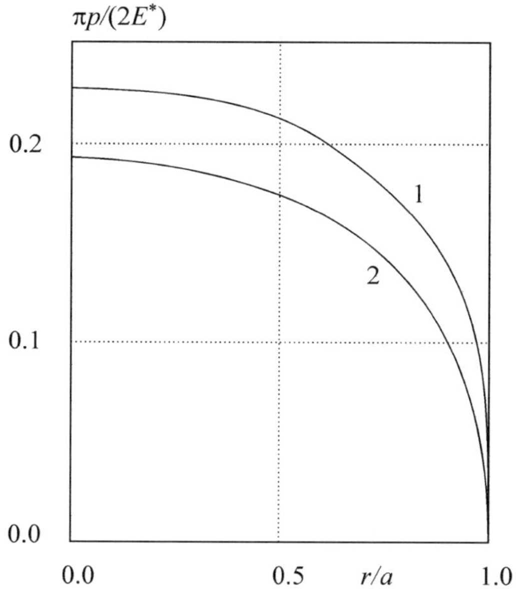

Fig. 16 Pressure distribution within a contact spot, calculated from Eq. (120) for $n = 0$ (curve 1), $n = 1$ and $n = 2$ (curve 2) and $a/R = 0.1$, $l/R = 0.2$ (one-level model).

The systems of Eqs. (120) and (121) for the one-level model and of Eqs. (124)-(126) for the three-level model are solved by iteration. The density $\bar{N}_j$ of arrangement of indenters in the three-level model under consideration is determined by the formula

$$\bar{N}_j = \frac{2}{3l^2\sqrt{3}}, \quad (j = 1, 2, 3). \tag{128}$$

For the one-level model $\bar{N} = 3\bar{N}_j = 2/l^2\sqrt{3}$.

For determination of the radius A_n of the circle ($r \le A_n$) where the real pressure distribution within a nearby contact spots is taken into account (local effect) and the corresponding value of n which gives an appropriate accuracy of the solution of Eq. (120), we calculated the contact pressure $p^1(\rho^1, \theta)$ from Eqs. (120) and (121) for $n = 0$, $n = 1$, $n = 2$ and so on. For $n = 0$, the integral term on the left of Eq. (120) is zero, so that the effect of the remaining contact spots surrounding the fixed one (with the center at the origin of coordinate system O) is taken into account by a nominal pressure distributed outside the circle of radius A_0 (the second term in the right side of Eq. (120)), where A_0 is determined by Eq. (122). For $n = 1$ we take into account the real pressure within 6 contact spots located at the distance l from the fixed one, for $n = 2$ they are 12 contact spots, six located at the distance l and the another six at a distance $l\sqrt{3}$, and so on. Figure 16 illustrates the results calculated for $a^1 = 0.1$ and $l^1 = 0.2$, i.e. $a/l = 0.5$, this case corresponds to the limiting value of contact density. The results show that the contact pressure calculated for

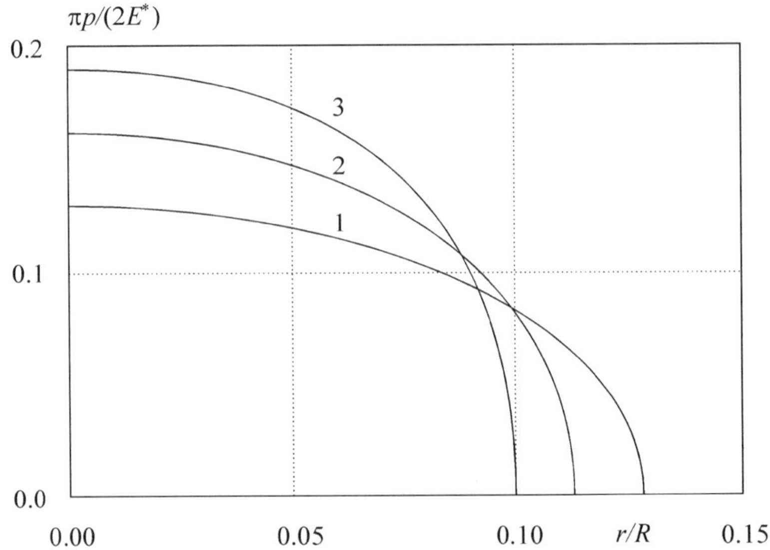

Fig. 17 Pressure distribution under an indenter acted on by the force $P^1 = 0.0044$ for the one-level model characterized by the various distances between indenters: $l/R = 0.2$ (curve 1), $l/R = 0.25$ (curve 2), $l/R = 1$ (curve 3).

$n = 1$ and $n = 2$ differ from one another less than 0.1%. If contact density decreases (a/l decreases) this difference also decreases. Based on this estimation, we will take $n = 1$ in subsequent analysis.

We first analyze the effect of interaction between contact spots and pressure distribution. Figure 17 illustrates the contact pressure under some indenter of the one-level system for different values of the parameter l^1 characterizing the distance between indenters. In all cases, the normal load $P^1 = 0.0044$ is applied to each indenter. The results show that the radius of the contact spot decreases and the maximum contact pressure increases if the distance l between indenters decreases; the contact density characterized by the parameter a/l also increases ($a/l = 0.128$ (curve 3), $a/l = 0.45$ (curve 2), $a/l = 0.5$ (curve 1)). The curve 3 practically coincides with the contact pressure distribution calculated from Hertz theory which neglects the influence of contact spots surrounding the fixed one. So, for small values of parameter a/l, it is possible to neglect the interaction between contact spots for determination of the contact pressure.

The dependencies of the radius of a contact spot on the dimensionless nominal pressure $\bar{p}^1 = \bar{p}\pi/2E^*$ calculated for different values of parameter l^1 and a one-level model are shown in Figure 18 (curves 1, 2, 3). The results of calculation based on the Hertz theory are added for comparison (curves 1′, 2′, 3′). The results show that under a constant nominal pressure $\bar{p}$ the radius of each contact spot and, hence the real contact area, decreases if the relative distance l/R between contact spots decreases. The comparison of these results with the curves calculated from Hertz

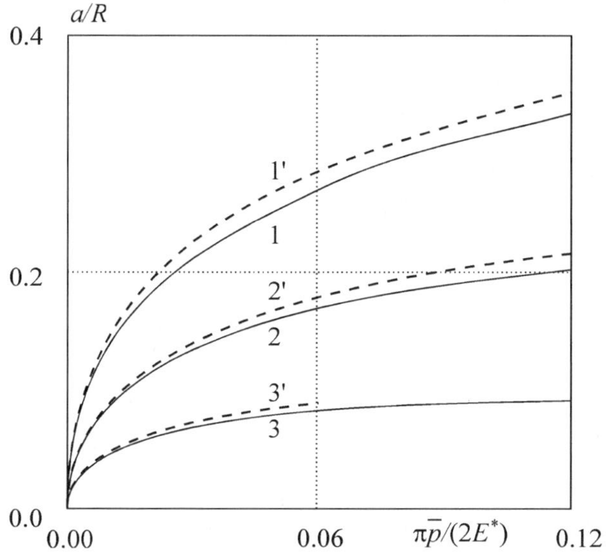

Fig. 18 Dependence of the radius of a contact spot on the nominal pressure for $l = 1$ (curves 1, 1′), $l = 0.5$ (curves 2, 2′), $l = 0.2$ (curves 3, 3′), calculated from Eq. (120) (1, 2, 3) and from Hertz theory (1′, 2′, 3′).

theory makes it possible to conclude that for $a/l < 0.25$ the discrepancy between the results predicted from the multiple contact theory and Hertz theory does not exceed 2.5%. For higher nominal pressure and, hence higher contact density, the discrepancy becomes serious. Thus, for $l = 0.5$ (curves 2, 2′) and $a/l = 0.44$ the calculation of the real contact area from Hertz theory gives an error of about 15%.

Investigation of contact characteristics in the three-level model is a subject of particular interest because this model is closer to the real contact situation than is the one-level model. The multiple contact model developed in this section takes into account the influence of the density of contact spots on the displacement of the surface between contact spots, and so the load, which must be applied to bring a new level of indenters into contact, depends not only on the height difference of the indenters, but also on the contact density. The calculations were made for a model with fixed height distribution: $(h_1 - h_2)/R = 0.014$ and $(h_1 - h_3)/R = 0.037$. Figure 19 illustrates the pressure distribution within the contact spots for each level if $P^1 = 0.059$ where P^1 is the load applied to 3 indenters ($P^1 = P_1^1 + P_2^1 + P_3^1$). The curves 1, 2, 3 and the curves 1′, 2′, 3′ correspond to the solutions of the periodic contact problem and to the Hertz problem, respectively. The results show that the smaller the height of the indenter, the greater is the difference between the contact pressure calculated from the multiple contact and Hertz theory.

We also investigated the internal stresses for the one-level periodic problem and compared them with the uniform stress field arising from the uniform loading by the

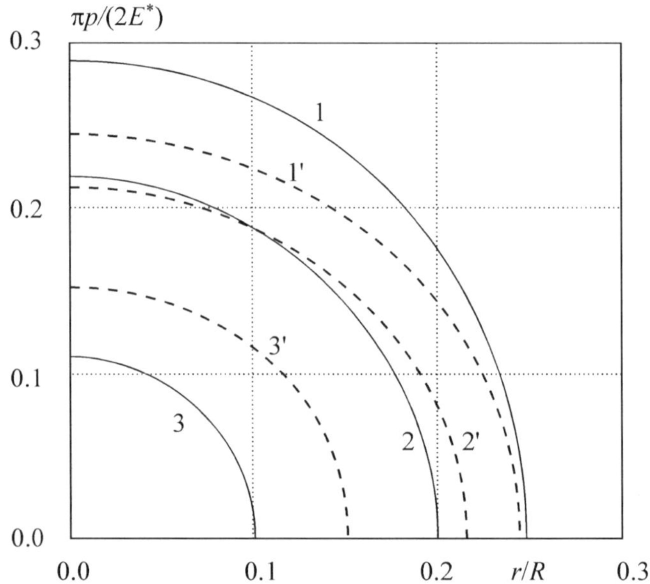

Fig. 19 Pressure distribution at the contact spots of indenters with the heights h_1 (curves 1, 1′), h_2 (curves 2, 2′) and h_3 (curves 3, 3′) for the three-level model ($(h_1 - h_2)/R = 0.014$, $(h_1 - h_3)/R = 0.037$, $P^1 = 0.059$) calculated from Eqs. (124)–(126) (1, 2, 3) and from Hertz theory (1′, 2′, 3′).

nominal pressure $\bar{p}_n$. It follows from the analysis that for periodic loading by the system of indenters, there is a nonuniform stress field in the subsurface layer, the thickness of which is comparable with the distance l between indenters. The stress field features depend essentially on the contact density parameter a/l. Figure 20 illustrates the principal shear stress $\tau_1/\bar{p}$ along the z-axis which coincides with the axis of symmetry of the indenter (curves 1, 2) and along the axis $O'z$ (curves 1′, 2′) equally spaced from the centers of the contact spots (see Figure 14). The results are calculated for the same nominal pressure $\bar{p}^1 = 0.12$, and the different distances l/R between the indenters: $l/R = 1$, $(a/R = 0.35)$ (curves 1, 1′) and $l/R = 0.5$ $(a/R = 0.21)$ (curves 2, 2′). The maximum value of the principal shear stress is related to the nominal pressure; the maximum difference of the principal shear stress at the fixed depth decreases as the parameter a/l increases. The maximum value of the principal shear stress occurs at the point $r = 0$, $z/a = 0.43$ for $a/l = 0.35$ (curve 1) and at the point $r = 0$, $z/a = 0.38$ for $a/l = 0.42$ (curve 2). At infinity the principal shear stresses depend only on the nominal stress $\bar{p}$. The results show that internal stresses differ noticeably from ones calculated from the Hertz model if the parameter a/l varies between the limits $0.25 < a/l \leq 0.5$.

Figure 21 illustrates contours of the function $\tau_1/\bar{p}$ at the plane $z/R = 0.08$, which is parallel to the plane Oxy. The principal shear stresses are close to the maximum values at the point $x = 0$, $y = 0$ of this plane. Contours are presented

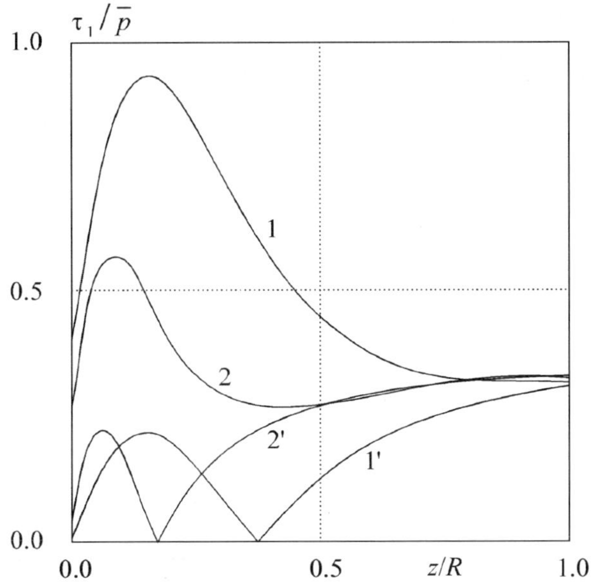

Fig. 20 The principal shear stress $\tau_1/\bar{p}$ along the axes Oz (curves 1, 2) and $O'z$ (curves 1′, 2′) for $l/R = 1$ (1, 1′), $l/R = 0.5$ (2, 2′), $\bar{p}^1 = 0.12$.

within the region

$$\left(-\frac{l^1}{2} < x < l^1, \frac{l^1\sqrt{3}}{4} < y < \frac{l^1\sqrt{3}}{2}\right)$$

for $a^1 = 0.2$ and $l^1 = 1$ (Figure 21a) and $l^1 = 0.44$ (Figure 21b). The results show that the principal shear stress at the fixed depth varies only slightly if the contact density parameter is close to 0.5. Similar conclusions follow for all the components of the stress tensor.

Thus, as a result of the nonuniform pressure distribution at the surface of the half-space (discrete contact), there is a nonuniform stress field dependent on the contact density parameter in the subsurface layer. The increase of stresses in some points of the layer may cause plastic flow or crack formation. The results obtained here coincide with the conclusions which follow from the analysis of the periodic contact problem for the sinusoidal punch and an elastic half-plane (2-D contact problem) in Kuznetsov and Gorokhovsky (1978a, 1978b).

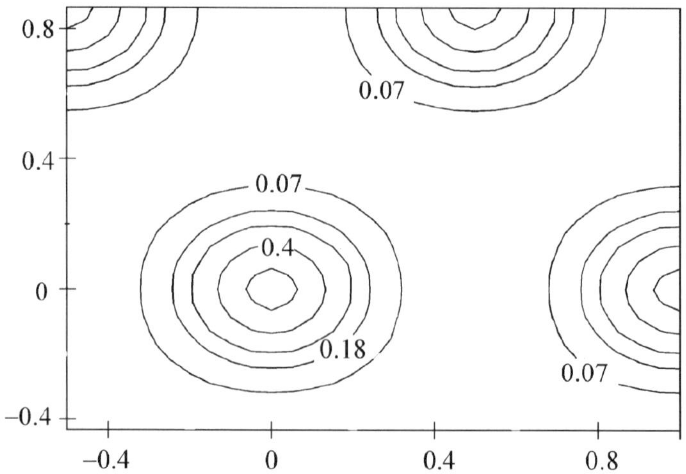

a

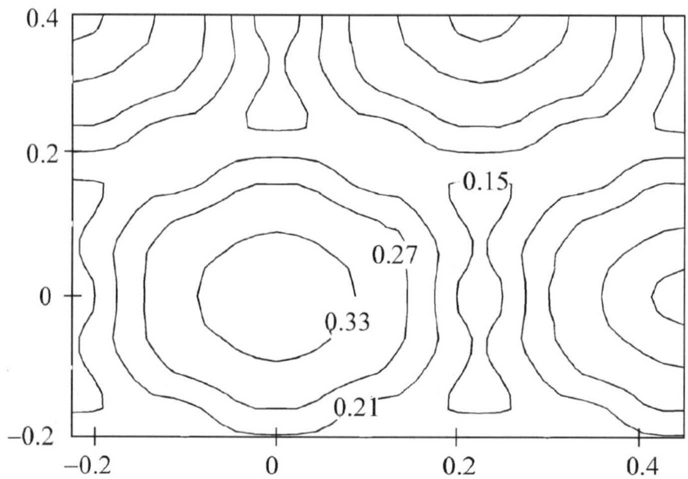

b

Fig. 21 Contours of the function $\tau_1/\bar{p}$ at the plane $z/R = 0.08$ for $l^1 = 1$ (a) and $l^1 = 0.44$ (b); $a^1 = 0.2$.

References

Amontons G. (1699) De la résistance cause dans les machines, *Mémoire de l'Académie Royale*, **A**, 275–282.

Boussinesq J. (1885) *Application des Potentiels à l'Étude de l'Équilibre et du Mouvement des Solides Élastique*, Gauthier-Villars, Paris.

Cattaneo C. (1938) Sul contatto di due corpi elastici: Distribuzion locale degli sforzi. *Reconditi Accad. Naz. Lincei*, **27**, 342–348, 434–436, 474–478.

Ciavarella M., Hills D.A., Monno G. (1998) The influence of rounded edges on indentation by a flat punch. *Proceedings of the Institution of Mechanical Engineers, Part C, Journal of Mechanical Engineering Sciences*, **212**(4), 319–328.

Coulomb Ch.A. (1785) Théorie des machines simples, *Mémoire de Mathématiques et de Physique de l'Académie Royale*, 161–331.

Galin L.A. (1945) Pressure of punch with friction and adhesion domains, *J. Appl. Math. Mech.*, **9**, 5, 413–424.

Galin L.A. (1953) *Contact Problem of the Theory of Elasticity*, Gostekhizdat, Moscow [in Russian].

Galin L.A., Goryacheva I.G. (1983) Three-dimensional contact problem of the motion of a stamp with friction, *J. Appl. Math. Mech.*, **46**, 6, 819–824.

Goryacheva I.G. (1987) Wear contact problem for the system of punches, *Izv. AN SSSR, Mechanics of Solids*, **6**, 62–68.

Goryacheva I.G. (1998a) Periodic contact problem for the elastic half-space, *J. Appl. Math. Mech.*, **62**, 6, 1036–1044.

Goryacheva I.G. (1998b) *Contact Mechanics in Tribology*, Kluwer Academic Publishers, Dordrecht/Boston/London.

Goryacheva I.G. (1999) Calculation of contact characteristics taking into account surface macro- and micro-shape. *Friction and Wear*, **20**, 3.

Goryacheva I.G. (2006) Mechanics of Discrete Contact. *Tribology International*, **39**, 381–386.

Goryacheva I.G., Dobychin M.N. (1991) Multiple contact model in the problems of tribomechanics. *Tribology International*, **24**, 1, 29–35.

Goryacheva I.G., Murthy H., Farris T. (2002). Contact problem with partial slip for the inclined punch with rounded edges. *Int. J. of Fatigue*, **24**, 1191–1201.

Gradshteyn I.S., Ryzhik I.M. (1963) *Tables of Integrals, Series and Products*, 4th Edn., Nauka, Moscow [in Russian].

Hills D.A., Sosa G. Urriolagoitia (1999) Origins of partial slip in fretting – A review of known and potential solutions. *J. of Strain Analysis*, **34**, 3, 175–181.

Johnson, K.L. (1987) *Contact Mechanics*, Cambridge University Press, Cambridge.

Johnson K.L., Greenwood J.A., Higginson J.G. (1985) The contact of elastic wavy surfaces, *Internat. J. Mech. Sci.*, **27**, 6, 383–396.

Kuznetsov E.A., Gorokhovsky G.A. (1978a) Stress distribution in a polymeric material subjected to the action of a rough-surface indenter, *Wear*, **51**, 299–308.

Kuznetsov E.A., Gorokhovsky G.A. (1978b) Influence of roughness on stress state of bodies in friction contact, *Prikl. Mekhanika*, **14**, 9, 62–68 [in Russian].

Kuznetsov E.A., Gorokhovsky G.A. (1980) Friction interaction of rough bodies from the point of view of mechanics of solids, *Soviet Journal of Friction and Wear*, **4**, 638–649.

McVeigh P.A., Harish G., Farris T.N., Szolvinski M.P. (1999) Modelling interfacial conditions in nominally at contacts for application to fretting fatigue of turbine engine components. *Int. J. Fatigue*, **21**, 157–165.

Mindlin R.D. (1949) Compliance of elastic bodies in contact. *J. Appl. Mech.*, **16**, 259–268.

Muskhelishvili N.I. (1949) *Some Basic Problems of the Mathematical Theory of Elasticity*, 3rd Edn., Nauka, Moscow [in Russian].

Plemelj J. (1908) Ein Ergänzungssatz zur Cauchyschen Integraldarstellung analytischer Funktionen, Randwerte betreffend, *Monatsh. Math. Phys.*, **19**, 205–210.

Seal M. (1957) Friction and wear of diamond, *Conf. on Lubr. and Wear*, **12**, 252–256.

Staierman I.Ya. (1949) *Contact Problem of the Elasticity Theory*, Gostekhizdat, Moscow [in Russian].

Tabor D., Wynne-Williams (1961) The effect of orientation on friction of polytetrafluoroethylene, *Wear*, **4**, 5, 391–400.

Timoshenko S.P., Goodier J.N. (1951) *Theory of Elasticity*, 3rd Edn., McGraw-Hill, New York/London.

Vantorin V.D. (1962) Motion along a plane with anisotropic friction, in *Friction and Wear in Machines*, **16**. Izdat. Akad. Nauk, Moscow, pp. 81–120 [in Russian].

Westergaard H.M. (1939) Bearing pressure and cracks, *ASME, Ser. E, J. Appl. Mech.*, **6**, 1, 49–53.

Zmitrovicz A. (1990) Liniowe i nieliniove równania konstytutywne tarcia anizotropowego, *Zeszyty Naukowe Akademii Gorniczo-Hutniczei, Mechanica*, **9**, 2, 141–154.

Hertz Type Contact Problems for Power-Law Shaped Bodies

Feodor M. Borodich

1 Introduction

1.1 Hertz Type Contact Problems for Rigid Indenters

In January 1881 H. Hertz submitted his famous paper on contact theory to the journal *Reine und angewandte Mathematik.* The paper was published in 1882 (Hertz, 1882a). He considered three-dimensional (3D) frictionless contact of two isotropic, linear elastic solids. It is well known (see, e.g. Galin, 1953; Gladwell, 1980; Johnson, 1985) that the main feature of this type of contact problems is that the region of contact is not known *a priori*. Hence, the contact problem for even linear elastic solids is non-linear. This is the main difficulty of solving these contact problems.

Hertz made three basic assumptions in his treatment of the contact between two elastic bodies: initially there is just one point of contact; the linear dimensions of the contact region are small compared to the radii of curvature of the bodies in the contact region; and each body may be considered to be an elastic half-space with a shape function f defining its shape in the contact region. We may specify the elastic properties of a homogeneous isotropic elastic body by its shear modulus μ and its Poisson's ratio ν. Then we define ϑ by $\vartheta = (1-\nu)/\mu$. As shown in Section 5.14, the contact problem for two such elastic bodies with shape functions f_1 and f_2, and with elastic constants ϑ_1 and ϑ_2 respectively, is equivalent to the contact problem for a rigid punch with shape function $f = f_1 + f_2$ indenting an elastic body with elastic constant $\vartheta = \vartheta_1 + \vartheta_2$. For this reason we may, and will, confine our attention to the contact problem for a rigid punch indenting an elastic body.

We use Cartesian coordinates $x_1 \equiv x, x_2 \equiv y, x_3 \equiv z$ with the origin (O) at the point of initial contact between the punch and the half-space $x_3 \geq 0$. The boundary plane $x_3 = 0$ is denoted by $\mathbb{R}^2$. Hence, the equation of the surface given by a function f, can be written as $x_3 = -f(x_1, x_2)$, $f \geq 0$.

After the punch contacts the half-space, displacements u_i and stresses σ_{ij} are generated. If the material properties are time independent, then the current state of the contact process can be completely characterized by an external parameter ($\mathcal{P}$), e.g., the compressive force (P), the relative approach of the bodies (δ) (for a rigid indenter, δ is the depth of indentation) and the size of the contact region (l). For axi-symmetric problems l is equal to the contact radius a. It should be noted that (l) can be used as the external parameter of the problem only for convex punches (Borodich and Galanov, 2002).

Thus, it is supposed that the shape of the punch and the external parameter $\mathcal{P}$ are given, and one has to find the bounded region G on the boundary plane $x_3 = 0$ of the half-space at the points where the punch and the medium are in mutual contact, displacements u_i, and stresses σ_{ij}. If the compressive force P is taken as the external parameter $\mathcal{P}$, then one has to find the depth of indentation δ and the size of the contact region l. If δ is taken as $\mathcal{P}$ then one has to find P and l.

1.2 Formulation of a Hertz Type Contact Problem

In the general case of a 3D Hertz type contact problem, it is not assumed that the punch shape is described by an elliptic paraboloid and the contact region is an ellipse as Hertz did. However, the formulation of the problem is geometrically linear, the contact region is unknown and is to be found; only vertical displacements of the boundary are taken into account, and the problem has the same boundary conditions within and outside the contact region as in the original Hertz problem.

In the problem the quantities sought satisfy the following equations

$$\begin{aligned} \sigma_{ji,j} &= 0, \quad i,j = 1,2,3; \\ \sigma_{ij} &= \mathcal{F}(\epsilon_{ij}), \quad \epsilon_{ij} = (u_{i,j} + u_{j,i})/2; \\ &\iint_{\mathbb{R}^2} \sigma_{33}(\mathbf{x}, \mathcal{P}) d\mathbf{x} = -P, \end{aligned} \tag{1}$$

in which ϵ_{ij} are the components of the strain tensor and $\mathcal{F}$ is the operator of constitutive relations for the material. Here and henceforth, a comma before the subscript denotes the derivative with respect to the corresponding coordinate; and summation from 1 to 3 is assumed over repeated Latin subscripts. The material behavior of the medium may be linear or non-linear, anisotropic or isotropic, depending on the form of the operator $\mathcal{F}$. For anisotropic, linear elastic media, the constitutive relations have the form of Hooke's law (4.5.1).

The displacement vector $\mathbf{u}$ should satisfy the conditions at infinity

$$\mathbf{u}(\mathbf{x}) \to 0 \quad \text{when} \quad |\mathbf{x}| \to \infty. \tag{2}$$

The contact region G is defined as an open region such that if $\mathbf{x} \in G$ then the gap $(u_3 - g)$ between the punch and the half-space is equal to zero and surface stresses are non-positive, while for $\mathbf{x} \in \mathbb{R}^2 \setminus G$ the gap is positive and the stresses are zero. Thus, $\mathbf{u}$ and σ_{ij} should satisfy the following boundary conditions within and outside the contact region

$$
\begin{aligned}
&u_3(\mathbf{x};\mathcal{P}) = g(\mathbf{x};\mathcal{P}), \quad && \sigma_{33}(\mathbf{x};\mathcal{P}) \le 0, \quad && \mathbf{x} \in G(\mathcal{P}),\\
&u_3(\mathbf{x};\mathcal{P}) > g(\mathbf{x};\mathcal{P}), \quad && \sigma_{3i}(\mathbf{x};\mathcal{P}) = 0, \quad && \mathbf{x} \in \mathbb{R}^2 \setminus G(\mathcal{P}),
\end{aligned} \tag{3}
$$

In the problem of vertical indentation of an isotropic or a transversally-isotropic medium by an axi-symmetric punch, the contact region is always a circle. This fact simplifies analysis of the problem. The analysis of three-dimensional contact is usually more complicated.

For the general case of the problem of vertical pressure, one has

$$
g(\mathbf{x};\mathcal{P}) = \delta - f(x_1, x_2). \tag{4}
$$

If one considers the frictionless problem, then the following two conditions hold within the contact region

$$
\sigma_{31}(\mathbf{x};\mathcal{P}) = \sigma_{32}(\mathbf{x};\mathcal{P}) = 0, \quad \mathbf{x} \in G(\mathcal{P}) \subset \mathbb{R}^2. \tag{5}
$$

The frictional conditions within the contact region may be formulated in the following way. Let us denote the quantities referring to the body $x_3^+ \le 0$ by a superscript "plus" sign, and those referring to the second body by a superscript "minus" sign. In the adhesive contact problem, there is no relative slip between the bodies within the contact region. If the following values are introduced

$$
v_1(x_1, x_2) \equiv u_1^+(x_1, x_2, 0, P) - u_1^-(x_1, x_2, 0, P)
$$

and

$$
v_2(x_1, x_2) \equiv u_2^+(x_1, x_2, 0, P) - u_2^-(x_1, x_2, 0, P)
$$

then the condition within this region is that these values do not change with augmentation of the external parameter $\mathcal{P}$. These conditions can be expressed by

$$
\frac{\partial}{\partial \mathcal{P}} v_i(x_1, x_2, 0, \mathcal{P}) = 0, \quad d\mathcal{P} > 0. \tag{6}
$$

In the frictional contact problem, it is usually assumed (see Bryant and Keer, 1982) that the contact region consists of the following parts: in the inner part G_1 the interfacial friction must be sufficient to prevent any slip taking place between bodies, i.e., (6) holds; in the outer part $G \setminus G_1$ the friction must satisfy the Coulomb frictional law with the coefficient of friction ρ (see Section 3.1). Let us define the vector of tangential stresses $\boldsymbol{\tau}^{\pm}(x_1, x_2, 0, P) \equiv (\sigma_{31}^{\pm}(x_1, x_2, 0, P), \sigma_{32}^{\pm}(x_1, x_2, 0, P))$. Then the frictional contact conditions can be written as

$$
\begin{aligned}
&\frac{\partial}{\partial \mathcal{P}} v_i(x_1, x_2, 0, \mathcal{P}) = 0, \quad d\mathcal{P} > 0, \quad (x_1, x_2) \in G_1,\\
&\boldsymbol{\tau}^{\pm}(x_1, x_2, 0, P) = -\rho \sigma_{33}^{\pm}(x_1, x_2, 0, P) \left[\frac{\mathbf{v}(x_1, x_2, 0, P)}{|\mathbf{v}(x_1, x_2, 0, P)|} \right],\\
&\qquad (x_1, x_2) \in G \setminus G_1.
\end{aligned} \tag{7}
$$

2 Frictionless Axisymmetric Contact

2.1 The Galin Solution for Frictionless Axisymmetric Contact

Let us use Cartesian and cylindrical coordinate frames, namely $x_1 = x, x_2 = y, x_3 = z$ and r, θ, z, where $r = \sqrt{x^2 + y^2}$ and $x = r\cos\theta, y = r\sin\theta$ (see (5.5.22)).

In 1946 considering axisymmetric frictionless contact problems for an elastic isotropic half-space and applying results derived by Kochin (1940), Galin obtained expressions for the contacting force P, the depth of penetration δ and the pressure distribution under a convex, smooth in $\mathbb{R}^2\backslash\{0\}$ punch of arbitrary shape $x_3 = -f(r)$, $f(0) = 0$. In particular, he wrote (see (5.5.18) and (5.5.41))

$$P = 4\vartheta^{-1} \int_0^a \rho_1 \Delta f(\rho_1)\sqrt{a^2 - \rho_1^2}d\rho_1, \tag{8}$$

$$\delta = \int_0^a \rho_1 \Delta f(\rho_1) \tanh^{-1}(\sqrt{1 - \rho_1^2/a^2})d\rho_1. \tag{9}$$

Here a is the radius of contact and Δ denotes the two-dimensional Laplace operator (see (5.5.43))

$$\Delta = \partial^2/\partial x_1^2 + \partial^2/\partial x_2^2 = \partial^2/\partial r^2 + r^{-1}\partial/\partial r. \tag{10}$$

2.2 The Sneddon Representation of the Galin Solution

Sneddon (1965) gave another representation of the Galin solution (8) and (9); he presented the following formulae for the force and the indentation depth of a punch having contact radius $r = a$

$$P = \frac{4a}{\vartheta} \int_0^1 \frac{\xi^2 w'(\xi)d\xi}{\sqrt{1 - \xi^2}}, \tag{11}$$

$$\delta = \int_0^1 \frac{w'(\xi)d\xi}{\sqrt{1 - \xi^2}} \tag{12}$$

where $f(r) = w(r/a)$. Let us derive (11) and (12).

Substituting (10) into (8), one has

$$\frac{P\vartheta}{4} = I_1 + I_2,$$

$$I_1 = \int_0^a f''(\rho_1)\rho_1\sqrt{a^2 - \rho_1^2}d\rho_1, \quad I_2 = \int_0^a f'(\rho_1)\sqrt{a^2 - \rho_1^2}d\rho_1.$$

Integrating by parts, one obtains

$$I_1 = \left[f'(\rho_1)\rho_1\sqrt{a^2-\rho_1^2} \right]_0^a - \int_0^a f'(\rho_1) d\left(\rho_1\sqrt{a^2-\rho_1^2}\right)$$
$$= -I_2 + \int_0^a \frac{\rho_1^2 f'(\rho_1) d\rho_1}{\sqrt{a^2-\rho_1^2}}.$$

Hence,

$$\frac{P\vartheta}{4} = \int_0^a \frac{\rho_1^2 f'(\rho_1) d\rho_1}{\sqrt{a^2-\rho_1^2}} = \int_0^a \frac{\rho_1^2 df(\rho_1)}{\sqrt{a^2-\rho_1^2}}. \tag{13}$$

By making a substitution $\xi = \rho_1/a$, one obtains (11)

$$\frac{\vartheta}{4}P = a\int_0^1 \frac{\xi^2 dw(\xi)}{\sqrt{1-\xi^2}}.$$

Similarly, substituting (10) into (9), one has $\delta = I_3 + I_4$, where

$$I_3 = \int_0^a \rho_1 f''(\rho_1) \tanh^{-1}(\sqrt{1-\rho_1^2/a^2}) d\rho_1,$$
$$I_4 = \int_0^a f'(\rho_1) \tanh^{-1}(\sqrt{1-\rho_1^2/a^2}) d\rho_1.$$

Integrating by parts, one obtains

$$I_3 = \left[f'(\rho_1)\rho_1 \tanh^{-1}(\sqrt{1-\rho_1^2/a^2}) \right]_0^a$$
$$- \int_0^a f'(\rho_1) d\left(\rho_1 \tanh^{-1}(\sqrt{1-\rho_1^2/a^2})\right)$$
$$= -I_4 - \int_0^a \rho_1 f'(\rho_1) d\left(\tanh^{-1}(\sqrt{1-\rho_1^2/a^2})\right).$$

Hence, one has

$$\delta = -\int_0^a \rho_1 f'(\rho_1) d\left[\tanh^{-1}(\sqrt{1-\rho_1^2/a^2})\right]. \tag{14}$$

It is known that by definition (see, e.g., (4.6.3) and (4.6.22) in Abramovitz and Stegun, 1964)

$$\tanh^{-1} v = \int_0^v \frac{dt}{1-t^2} = \frac{1}{2}\ln\frac{1+v}{1-v}. \tag{15}$$

Using (15) and substituting $v = \sqrt{1-\rho_1^2/a^2}$ into (14), one obtains

$$d[\tanh^{-1} v] = \frac{dv}{1-v^2} = -\frac{1}{\rho_1\sqrt{1-\rho_1^2/a^2}}.$$

Using this formula, one obtains the following representation of Galin's formula

$$\delta(a) = \int_0^a \frac{f'(\rho_1)}{\sqrt{1-\rho_1^2/a^2}} d\rho_1. \tag{16}$$

As above, a substitution $\xi = \rho_1/a$ leads to the formula (12) for the depth of indentation of a axisymmetric punch having contact radius $r = a$

$$\delta = \int_0^a \frac{df(\rho_1)}{\sqrt{1-\rho_1^2/a^2}} = \int_0^1 \frac{dw(\xi)}{\sqrt{1-\xi^2}}.$$

2.3 The Galin Solution for Monomial Punches

Let us apply the general solution (8) and (9) to an axisymmetric punch with shape described by a power-law (monomial) function

$$f(r) = Ar^\lambda, \tag{17}$$

where A is a constant. If the shape function is described by (17) then $\Delta f(r) = A\lambda^2 r^{\lambda-2}$, and (8) and (9) lead to the following formulae (see, (5.5.49) and (5.5.58))

$$P = \frac{A}{\vartheta}\frac{\lambda^2}{\lambda+1}2^\lambda\frac{\Gamma^2(\lambda/2)}{\Gamma(\lambda)}a^{\lambda+1}, \quad \delta = A\lambda 2^{\lambda-2}\frac{\Gamma^2(\lambda/2)}{\Gamma(\lambda)}a^\lambda. \tag{18}$$

Using (18), Galin established the following relation between the force P and the displacement δ (see, (5.5.53) and (5.5.54))

$$P = 2\vartheta^{-1}\left[A^{-\frac{1}{\lambda}}2^{2/\lambda}\lambda^{\frac{\lambda-1}{\lambda}}\frac{1}{\lambda+1}[\Gamma(\lambda/2)]^{-\frac{2}{\lambda}}[\Gamma(\lambda)]^{\frac{1}{\lambda}}\right]\delta^{\frac{\lambda+1}{\lambda}}. \tag{19}$$

Here $\Gamma(\lambda)$ is the Euler gamma function.

Using the property of the Euler gamma functions $\Gamma(n+1) = n!$, we can see that the Shtaerman (1939) solution is a particular case $\lambda = 2n$ of the Galin solution (19). Here n is a natural number. In particular, one can obtain the Shtaerman (1939) formula

$$P = 8nA\vartheta^{-1}a^{2n+1}\frac{2\cdot 4\ldots 2n}{1\cdot 3\ldots 2n+1} = 8nA\vartheta^{-1}a^{2n+1}\frac{(2n)!!}{(2n+1)!!}. \tag{20}$$

It follows from (18) that

$$P = \frac{4\lambda}{\lambda+1}\cdot\vartheta^{-1}a\delta(a). \tag{21}$$

Taking a limit $\lambda \to \infty$ in (21), one obtains the Boussinesq relation (5.3.27) for a flat ended cylindrical indenter of radius a

$$P = 4\vartheta^{-1}a\delta(a). \tag{22}$$

2.4 A Solution for Polynomial Punches

Let us represent the shape function of a smooth body of revolution in the form of a power series with fractional exponents

$$f(r) = \sum_{k=1}^{\infty} A_k r^{\lambda_k}, \quad \lambda_k > 0. \tag{23}$$

Using the Galin solution (18) and (19), we can prove the following result (Borodich, 1990a).

Let the punch described by (23) be pressed into an elastic half-space by a force P. Then the contact radius a, the contact load P and the depth of indentation δ satisfy the following equations

$$P = 2\vartheta^{-1} \sum_{k=1}^{\infty} B(A_k, \lambda_k) a^{\lambda_k+1}, \quad B(A_k, \lambda_k) = A_k 2^{\lambda_k - 1} \frac{\lambda_k^2}{\lambda_k + 1} \frac{\Gamma^2(\lambda_k/2)}{\Gamma(\lambda_k)},$$

$$\delta = \sum_{k=1}^{\infty} A_k \lambda_k 2^{\lambda_k - 2} \frac{\Gamma^2(\lambda_k/2)}{\Gamma(\lambda_k)} a^{\lambda_k}. \tag{24}$$

Indeed, it follows from (18) that if a punch $f_k(r) = A_k r^{\lambda_k}$ is pressed by the force

$$P_k = 2\vartheta^{-1} B(A_k, \lambda_k) a^{\lambda_k+1}$$

then its contact radius with the half-space is equal to a.

If the punch described by (23) is pressed by the force

$$P_\Sigma = \sum_{k=1}^{\infty} P_k$$

then its contact radius is also a. This is because the Hertz type contact problems with identical contact regions (in our case contact regions having the same fixed contact radius a) can be superimposed on each other. Putting $P_\Sigma = P$ we obtain (24) as as a superposition of solutions to linear Hertz type contact problems. (Note Borodich, 1990a, erroneously multiplied by 2 the right hand expression in (24)).

3 Axisymmetric Contact Problems with Molecular Adhesion

In preceding problems we have assumed that there is no molecular adhesion between the contacting solids. Indeed, adhesion has usually a negligible effect on contact surface interactions at the macro-scale. However molecular adhesion becomes significant as the contact size decreases.

Apparently Bradley (1932) was the first who considered attraction of a rigid sphere of radius R to a flat surface, and obtained an expression for the attracting force F_a. However, he did not consider the elastic deformations of the contacting solids.

Derjaguin (1934) presented the first attempt to consider the problem of adhesion between an elastic sphere and a half-space. Although Derjaguin's basic thermodynamic argument was correct, the deformation he used was not exactly correct (Kendall, 2001). Later, Johnson (1958) made an attempt to solve the problem by adding two simple stress distributions, namely the Hertz stress field and a rigid punch tensile stress distribution. According to Kendall (2001), Johnson, Kendall, and Roberts (JKR) applied Derjaguin's method to Johnson's stress distribution, and created the the JKR theory of adhesive contact. Nowadays, two other theories of adhesion of elastic spheres are also in common use: the DMT theory (Derjaguin, Muller, and Toporov, 1975) and the Maugis theory (the JKR-DMT transition). The detailed description of the theories is given by Maugis (2000).

In the framework of the JKR theory, the following equation was obtained for adhesive contact between two spheres

$$P = 8a^3/(3R\vartheta) - \sqrt{16\pi w \vartheta^{-1} a^3} \tag{25}$$

where $w = 2\gamma$ is the work of adhesion of the spheres, γ is the surface energy, and R is the effective radius of the spheres ($1/R = 1/R_1 + 1/R_2$).

These theories were developed for the case when the initial distance between contacting solids is described as a paraboloid of revolution; $x_3 = r^2/(2R)$ is a good approximation to a sphere. However, solids in contact may often be described not only as spheres but also by more general shapes. Let us generalize the JKR theory of contact with molecular adhesion in order to describe contact between an indenter and an elastic sample, when the distances between the solids are described as monomial functions of arbitrary degrees (17).

Following Johnson et al. (1971), let us consider the total energy of the contact system U_T that is made up of three terms: the stored elastic energy U_E, the mechanical energy in the applied load U_M, and the surface energy U_S. It is assumed that the equilibrium at contact satisfies the equation

$$\frac{dU_T}{da} = 0, \qquad U_T = U_E + U_M + U_S. \tag{26}$$

If there were no surface forces, then the contact radius a_0 under the external load P_0 could be found from the Galin solution (18):

$$a_0 = \left(\frac{P_0\vartheta}{2C(\lambda)A}\right)^{1/(\lambda+1)}, \qquad C(\lambda) = \frac{\lambda^2}{\lambda+1} 2^{\lambda-1} \frac{\Gamma^2(\lambda/2)}{\Gamma(\lambda)}. \tag{27}$$

Surface forces (forces of molecular adhesion) tend to increase the contact radius in equilibrium to $a_1 > a_0$ under the same load P_0. It is assumed that the contact systeme has come to this state in two steps: (i) first it has attained the contact radius a_1 and depth of indentation δ_1 under some apparent Hertz load P_1,

$$a_1 = \left(\frac{P_1\vartheta}{2C(\lambda)A}\right)^{1/(\lambda+1)} \tag{28}$$

then (ii) it is unloaded from P_1 to P_0 keeping the contact radius a_1 constant. The Boussinesq solution (22) for contact between an elastic half-space and a flat punch of radius a_1 may be used on the latter step. In this case, one can calculate U_E as the difference between the stored elastic energies $(U_E)_1$ and $(U_E)_2$ on the loading and unloading branches respectively. It follows from (19) that

$$\delta = P^{\lambda/(\lambda+1)} \left[C(\lambda)A(\vartheta/2)^{\lambda}\right]^{\frac{1}{\lambda+1}} \left(\frac{\lambda+1}{2\lambda}\right)$$

and therefore,

$$(U_E)_1 = P_1\delta_1 - \int_0^{P_1} \delta dP = \frac{\lambda+1}{2(2\lambda+1)} P_1^{(2\lambda+1)/(\lambda+1)} \left[C(\lambda)A(\vartheta/2)^{\lambda}\right]^{\frac{1}{\lambda+1}}.$$

Using the Boussinesq solution (22) and (28), we obtain for the unloading branch

$$\begin{aligned}(U_E)_2 &= \int_{P_0}^{P_1} \frac{P\vartheta}{4a_1} dP = \frac{(P_1^2 - P_0^2)\vartheta}{8a_1} \\ &= \frac{[C(\lambda)A(\vartheta/2)^{\lambda}]^{1/(\lambda+1)}}{4} \left(P_1^{(2\lambda+1)/(\lambda+1)} - P_0^2 P_1^{-1/(\lambda+1)}\right).\end{aligned}$$

Thus, the stored elastic energy U_E is

$$\begin{aligned}U_E &= (U_E)_1 - (U_E)_2 \\ &= \frac{1}{4}\left[C(\lambda)A(\vartheta/2)^{\lambda}\right]^{1/(\lambda+1)} \left(\frac{1}{2\lambda+1} P_1^{(2\lambda+1)/(\lambda+1)} + P_0^2 P_1^{-1/(\lambda+1)}\right).\end{aligned} \quad (29)$$

The mechanical energy in the applied load $U_M = -P_0(\delta_1 - \Delta\delta)$ where $\Delta\delta$ is the change in the depth of penetration due to unloading. Taking into account that $\Delta\delta = (P_1 - P_0)\vartheta/(4a_1)$, one obtains

$$U_M = -P_0 \frac{\lambda+1}{2\lambda} [C(\lambda)A(\vartheta/2)^{\lambda}]^{1/(\lambda+1)} \left[\frac{P_1^{\lambda/(\lambda+1)}}{\lambda+1} + \frac{P_0 P_1^{-1/(\lambda+1)}\lambda}{\lambda+1}\right]. \quad (30)$$

The surface energy $U_S = -w\pi a_1^2$ or

$$U_S = -w\pi \left(\frac{P_1\vartheta}{2C(\lambda)A}\right)^{2/(\lambda+1)}. \quad (31)$$

Thus, the total energy U_T can be obtained by summation of (29), (30) and (31):

$$\begin{aligned}U_T = &\frac{1}{4}\left[C(\lambda)A(\vartheta/2)^{\lambda}\right]^{\frac{1}{\lambda+1}} \left[\frac{1}{2\lambda+1} P_1^{\frac{2\lambda+1}{\lambda+1}} - P_0^2 P_1^{\frac{-1}{\lambda+1}} - \frac{2}{\lambda} P_0 P_1^{\frac{\lambda}{\lambda+1}}\right] \\ &- w\pi \left(\frac{P_1\vartheta}{2C(\lambda)A}\right)^{\frac{2}{\lambda+1}}.\end{aligned} \quad (32)$$

The equilibrium equation (26) is equivalent to

$$\frac{dU_T}{dP_1} = 0 \tag{33}$$

and, hence, one may obtain from (32) and (33)

$$P_0^2 - 2P_0(2C(\lambda)A\vartheta^{-1})a_1^{\lambda+1} + (2C(\lambda)A\vartheta^{-1})^2 a_1^{2(\lambda+1)} - 16w\pi\vartheta^{-1}a_1^3 = 0.$$

Solving this equation with respect to P_0 and taking the stable solution, one obtains an exact formula giving relation between the load P and the radius of the contact region a

$$P = 2C(\lambda)\vartheta^{-1}Aa^{\lambda+1} - \sqrt{16\pi w\vartheta^{-1}a^3}. \tag{34}$$

In 1993 this solution was derived first by Galanov and later by Borodich (see, e.g. Galanov and Grigoriev, 1994). Evidently, (34) coincides with the classic JKR formula (25) when $\lambda = 2$. When λ is an even integer, (34) coincides with the solution obtained independently by Carpick et al. (1996).

It follows from the JKR model that the detachment of the spheres is accompanied by an abrupt change in the contact radius. Indeed, it follows from (34) that the radius a of the contact region at $P = 0$ is

$$a = [\sqrt{4\pi w\vartheta}/(C(\lambda)A)]^{2/(2\lambda-1)}.$$

4 Adhesive (No-Slip) Axisymmetric Contact Problems

Let us consider next axisymmetric Hertz type contact problems with adhesive (no slip) boundary conditions. If the external parameter of the problem $\mathcal{P}$ is gradually increased, then the surface displacements $u_r(r, 0, \mathcal{P})$ and $u_z(r, 0, \mathcal{P})$ will be functions of both r and the parameter $\mathcal{P}$. The adhesive (no slip) boundary conditions mean that once a point of the surface contacts the indenter, its radial displacement does not change further with $\mathcal{P}$. Hence, instead of the conditions (5), one can write the following no-slip condition within the contact region (compare with (6))

$$\frac{\partial u_r}{\partial \mathcal{P}}(r, 0, \mathcal{P}) = 0. \tag{35}$$

4.1 Two-Dimensional Problem for a Punch with Horizontal Base

An effective solution to the two-dimensional problem of adhesive contact between a punch with straight horizontal base and an elastic half-plane was first given by Abramov (1937) using Mellin integrals. Then Muskhelishvili (1949) gave another solution to the problem using Kolosov's (1914) complex potentials. It follows from the solution that the pressure $p(x)$ under the punch $-a \leq x \leq a$ loaded by a vertical force P_0 is determined by the following formula (see, e.g. equations (114.7a) and (114.8a) in Muskhelishvili (1949))

$$p(x) = \frac{2P_0(1-\nu)}{\pi\sqrt{a^2-x^2}\sqrt{3-4\nu}} \cos\left[\beta \ln \frac{a+x}{a-x}\right] \tag{36}$$

where

$$\beta = \frac{1}{2\pi}\ln(3-4\nu), \qquad \text{and} \qquad \chi(x,a) = \cos\left(\beta \ln \frac{a-x}{a+x}\right).$$

(Note the coefficient $1/2$ omitted by Muskhelishvili (1949) in his equations (114.7a) and (114.8a).)

4.2 Axisymmetric Adhesive Contact Problems for Curved Punches

The analysis of adhesive contact problems was first performed incrementally (Mossakovskii, 1954, 1963; Goodman 1962) for a growth in the contact radius a. Mossakovskii noted self-similarity of the problem for punches described by monomial functions (17). However, only Spence (1968) pointed out that the solution can be obtained directly without application of incremental techniques (see Johnson, 1985; Gladwell, 1980). Self-similarity of a general frictional Hertz type contact problem was shown later by Borodich (1993).

Mossakovskii (1954, 1963) considered only two particular examples of no-slip contact problems, namely those for a flat-ended cylinder and a parabolic punch. Spence (1968) introduced an alternative method for solution of the problems, corrected some misprints in the Mossakovskii examples and also presented the solution for a conical punch.

Following Mossakovskii and Spence, let us take the contact radius a as the external parameter $\mathcal{P}$.

4.3 The Mossakovskii Solution for Adhesive Contact

In 1954, Mossakovskii presented the solution to a mixed boundary value problem for an elastic half-space when the line separating the boundary conditions is a circle.

The Boussinesq relation for adhesive (no-slip) contact. Mossakovskii presented the following formula for the compressive normal stresses σ_{zz}^0 under a flat-ended punch of radius a in adhesive (no-slip) contact

$$\sigma_{zz}^0(r,0,a) = K\delta_0 \frac{1}{r}\frac{d}{dr}\int_0^r \sin\left(\beta \ln \frac{a-x}{a+x}\right)\frac{x}{\sqrt{r^2-x^2}}dx. \tag{37}$$

Here δ_0 is the depth of the punch and

$$K = \frac{8(1-\nu)^2}{\pi \vartheta (1-2\nu)\sqrt{3-4\nu}}.$$

The correctness of the formula (37) was later checked by Keer (1967) and Spence (1968). Speaking about the further calculations of the compressing stress by Mossakovskii, Spence (1968) made a remark that a factor of 2 was omitted throughout Mossakovskii's paper of 1963, beginning with Mossakovskii's equation (2.16). Indeed, Mossakovskii's papers have various misprints; for example, Mossakovskii's expression for the contact force obtained by integration of the pressure (37) over the contact region. The correct formula is (see Spence, 1968; Khadem and Keer, 1974)

$$P = 4\delta_0 a \frac{(1-\nu)\ln(3-4\nu)}{\vartheta(1-2\nu)}. \tag{38}$$

However, in this case his calculations were correct and Spence's comment was in error. The formula (38) was also presented with a misprint in Johnson's (1985) book (see his (3.105)). One can see that the solution differs from the frictionless Boussinesq solution (22).

Integrating (37) by parts, we obtain the following formula for the pressure under a circular plane punch with unit settlement

$$\sigma^0_{zz}(r,0,a) = -2\beta a K \int_0^r \frac{\chi(x,a)dx}{\sqrt{r^2-x^2(a^2-x^2)}},$$

$$\chi(x,a) = \cos\left(\beta \ln \frac{a-x}{a+x}\right). \tag{39}$$

Applying the incremental approach to the solution (39) with varying radius t of the punch, one can calculate the normal stress under a curved axisymmetric punch

$$\sigma_{zz}(r,0,a) = \int_r^a \delta'(t)\sigma^0_{zz}(r,0,t)dt \tag{40}$$

where $\delta'(t)$ is a derivative of the punch displacement.

Developing the Mossakovskii approach, Borodich and Keer (2004a) obtained the following formula for the contact force

$$P(a) = \frac{16(1-\nu)^2\ln(3-4\nu)}{\pi\vartheta(1-2\nu)\sqrt{3-4\nu}} I \int_0^a \delta'(t)t\,dt, \quad I = \int_0^a \frac{\chi(x,a)}{\sqrt{a^2-x^2}}dx. \tag{41}$$

The integral I can be calculated using the Abramov–Muskhelishvili solution to the two-dimensional problem of adhesive contact between a punch with straight horizontal base and an elastic half-plane. Using (36), one obtains

$$P_0 = \int_{-a}^{a} p(x)dx = 2\int_0^a p(x)dx = \frac{4P_0(1-\nu)}{\pi\sqrt{3-4\nu}} \int_0^a \frac{\chi(x,a)}{\sqrt{a^2-x^2}}dx.$$

Hence,

$$I = \int_0^a \frac{\chi(x,a)}{\sqrt{a^2-x^2}}dx = \frac{\pi}{4}\frac{\sqrt{3-4\nu}}{1-\nu}. \tag{42}$$

Thus, it follows from (41) that the general expression for the force acting on a curved axisymmetric punch at adhesive contact, is

$$P(a) = \frac{4(1-\nu)\ln(3-4\nu)}{\vartheta(1-2\nu)} \int_0^a \delta'(t)tdt. \tag{43}$$

4.4 Solution to the Problem for Punches of Monomial Shape

Let us consider in detail the adhesive contact for punches of monomial shape. In the adhesive contact problem, the relation between the derivative of the function $\delta'(t)$ and the shape function $x_3 = -f(r)$ is (Mossakovskii, 1963)

$$f(r) = \frac{2}{\pi}\int_0^r \frac{1}{\sqrt{r^2-x^2}} \left[\int_0^x \delta'(t)\cos\left(\beta \ln\frac{x-t}{x+t}\right)dt \right] dx \tag{44}$$

It follows from (44) that if $\delta'(t) = K_\lambda t^{\lambda-1}$ or $\delta(t) = K_\lambda t^\lambda/\lambda$ then $f(r) = Ar^\lambda$ where

$$A = K_\lambda C_\lambda, \quad C_\lambda = \frac{2}{\pi} I^*(\lambda) I^{**}(\lambda) \tag{45}$$

and

$$I^*(\lambda) = \int_0^1 t^{\lambda-1}\cos\left(\beta\ln\frac{1-t}{1+t}\right)dt, \quad I^{**}(\lambda) = \int_0^1 \frac{x^\lambda}{\sqrt{1-x^2}}dx.$$

Taking into account

$$I^{**}(\lambda) = \frac{1}{2}\frac{\Gamma\left(\frac{1}{2}\right)\Gamma\left(\frac{\lambda+1}{2}\right)}{\Gamma\left(\frac{\lambda+2}{2}\right)} = \frac{1}{\lambda}\frac{\sqrt{\pi}\Gamma\left(\frac{\lambda+1}{2}\right)}{\Gamma\left(\frac{\lambda}{2}\right)} = \frac{2^{1-\lambda}\pi}{\lambda}\frac{\Gamma(\lambda)}{\Gamma^2\left(\frac{\lambda}{2}\right)},$$

one obtains

$$C_\lambda = \frac{2^{2-\lambda}}{\lambda}\frac{\Gamma(\lambda)}{\Gamma^2\left(\frac{\lambda}{2}\right)} I^*(\lambda).$$

It follows from (43) that the force is

$$P(a) = \frac{4(1-\nu)\ln(3-4\nu)}{\vartheta(1-2\nu)} \cdot \frac{A}{C_\lambda} \cdot \frac{a^{\lambda+1}}{\lambda+1}. \quad (46)$$

Thus, for an axisymmetric punch with shape described by the monomial function (17), the relations between the force P and the contact radius a and between the displacement δ and a are given by the following exact formulae (Borodich and Keer, 2004b)

$$P = \frac{2(1-\nu)\ln(3-4\nu)}{\vartheta(1-2\nu)} A \frac{\lambda}{\lambda+1} 2^{\lambda-1} \frac{\Gamma^2(\lambda/2)}{\Gamma(\lambda)} \frac{1}{I^*(\lambda)} a^{\lambda+1},$$

$$\delta = A 2^{\lambda-2} \frac{\Gamma^2(\lambda/2)}{\Gamma(\lambda)} \frac{1}{I^*(\lambda)} a^{\lambda}. \quad (47)$$

Using (47), one can establish the following relation between the force P and the displacement δ for a monomial punch in adhesive contact

$$P = \frac{2(1-\nu)\ln(3-4\nu)}{\vartheta(1-2\nu)} \frac{\lambda}{\lambda+1} \left[\frac{4I^*(\lambda)}{A} \frac{\Gamma(\lambda)}{[\Gamma(\lambda/2)]^2} \right]^{1/\lambda} \delta^{\frac{\lambda+1}{\lambda}}. \quad (48)$$

In the case $\nu = 0.5$, one has

$$\lim_{\nu \to 0.5} \frac{2(1-\nu)\ln(3-4\nu)}{\vartheta(1-2\nu)} = \frac{8(1-\nu^2)}{3\vartheta},$$

$\beta = 0$ and $I^*(\lambda) = 1/\lambda$. Hence, the formulae (47) and (48) are identical with the corresponding formulae (18) and (19) obtained by Galin (1946) for frictionless contact.

Using the general solution for monomial punches, we can consider some particular cases.

Conical punch. For a cone of semi-vertical angle $\pi/2-\alpha$, $\lambda = 1$, $f(r) = A_1 r$, and $\delta'(a) = K_1$. For a linearized treatment to be possible, α must be small compared with 1 and $\tan\alpha = A_1 \approx \alpha$. It follows from (46) that the force is

$$P = \frac{2\mu \ln(3-4\nu)}{1-2\nu} \frac{A_1}{C_1} a^2.$$

Taking into account that $\Gamma\left(\frac{1}{2}\right) = \sqrt{\pi}$ and $\Gamma(1) = 1$, we obtain from (47)

$$P = \frac{\pi\mu \ln(3-4\nu)}{(1-2\nu)I^*(1)} A_1 a^2. \quad (49)$$

$I^*(1)$ can be represented as the following Fourier transform (see (4.6) by Spence, 1968)

$$I^*(1) = \int_0^1 \cos\left(\beta \ln \frac{1-t}{1+t}\right) dt = \int_0^\infty \cos(\beta\xi/2)\mathrm{sech}^2\xi d\xi,$$

$$\xi(t) = \frac{1}{2} \ln \frac{1+t}{1-t}$$

and using tables collected by Erdelyi (1954, p. 30)

$$I^*(1) = \pi\beta \operatorname{cosech}(\pi\beta) = \frac{2\pi\beta}{(e^{\pi\beta} - e^{-\pi\beta})} = \frac{\ln(3-4\nu)\sqrt{3-4\nu}}{2(1-2\nu)}. \tag{50}$$

Substituting (50) into (49), one obtains

$$P = \frac{2\mu\pi A_1 a^2}{\sqrt{3-4\nu}}.$$

This agrees with equation (4.26) of Spence (1968).

Spherical punch. For a sphere of radius R, $\lambda = 2$, $A_2 = 1/(2R)$, $f(r) = A_2 r^2$, and

$$\delta'(a) = K_2 a = a/(2RC_2). \tag{51}$$

It follows from (47) that

$$P = \frac{2\mu \ln(3-4\nu)}{3R(1-2\nu)} \frac{a^3}{C_2} = \frac{4\mu \ln(3-4\nu)}{3R(1-2\nu)I^*(2)} a^3.$$

The adhesive problem for a sphere was first considered by Mossakovskii (1963) and Spence (1968). Our constant C_2 is λ_1 in Mossakovskii's notation and $\gamma(\kappa)/4$ in Spence's notation. Their results are identical with the above, except for factors 2 and $\gamma(\kappa)$ which were respectively omitted by Mossakovskii in his equation (5.6) (this is because he omitted this factor earlier in his equation (5.2) which is our (51)) and by Spence in his equation (4.20).

5 Self-Similar Contact Problems

Let us consider three-dimensional Hertz type contact problems. Now the punch shape is no longer described by an axi-symmetric function. Although the classic Hertzian contact problem is non-linear, it is self-similar. For the classical self-similar problem, using the solution to the problem for only one value of the parameter $\mathcal{P}$, it is possible to obtain the solutions for any other parameter values by simple renormalization of the known solution. This was shown independently by Galanov (1981a) and Borodich (1983) using two different approaches.

The former approach is based on the use of the explicit form of the solution to the Boussinesq problem for a concentrated load, while the latter approach deals directly with the equations of elasticity and the general formulation of the problem. The advantage of the former approach is its suitability for effective numerical calculations of the stress fields in the contact problems. However, it can be applied directly only to frictionless contact problem for solids made of isotropic materials, for example elastic, viscoelastic, plastic or creeping materials (Galanov 1981a, 1981b, 1982; Galanov and Kravchenko, 1986). Using the latter approach,

Borodich attacked the problem for anisotropic materials. The conditions under which frictionless Hertz type contact problems possess classical self-similarity, are as follows (Borodich 1988):

The constitutive relationships are homogeneous with respect to the strains or the stresses. The shape of the indenter is described by a homogeneous function with degree greater than or equal to unity. During the process of the contact, the loading at any point is progressive.

This means that the shape function f describing the indenter should satisfy the identity $f(\Lambda x_1, \Lambda x_2) = \Lambda^\lambda f(x_1, x_2)$, for arbitrary positive Λ. Here λ is the degree of the homogeneous function f, in particular, $\lambda = 2$ for the elliptic paraboloid considered by Hertz. Additionally, the operators of the constitutive relations $\mathcal{F}$ for the materials of the contacting bodies should be homogeneous functions of degree κ with respect to the components of the strain tensor e_{ij}, i.e.,

$$\mathcal{F}(\Lambda e_{ij}) = \Lambda^\kappa \mathcal{F}(e_{ij}). \tag{52}$$

The theoretical analysis of Hertz type contact problems based on similarity transformations of the 3D contact problems is valid for anisotropic materials (Borodich, 1989, 1993a). Hooke's law is an example of a linear ($\kappa = 1$) homogeneous constitutive relationship. Another example is the set of constitutive relationships of a plastic isotropic non-compressible material of the form

$$\sigma_{ij}^D = K\Gamma^{\kappa-1}\epsilon_{ij}, \tag{53}$$

$$\sigma_{ij}^D = \sigma_{ij} - \delta_{ij}\sigma, \quad \sigma = \sigma_{ii}/3, \quad \Gamma = \sqrt{\epsilon_{ij}^D \epsilon_{ij}^D/2}$$

where δ_{ij} is the Kronecker delta, σ_{ij}^D and ϵ_{ij}^D are the components of stress and shear strain deviators respectively. Here Γ is the intensity of shear strains, and K and κ are material constants. The constitutive relationships of a plastic anisotropic material given by Pobedrya (1984) are homogeneous. These relationships are often referred to as the power law of material hardening.

The following 3D self-similar contact problems have been considered: plastic non-compressible materials with power-law hardening (Galanov, 1981b; Borodich, 1988, 1990b); creeping materials in both deformation type (Galanov, 1982; Borodich, 1990b), and incremental formulations (see, e.g., Borodich, 1988, 1990d), and other materials (Borodich, 1989, 1990a–d). It has been shown that the self-similarity can also be applied to problems of elastic or elasto-plastic collision (Borodich, 1983, 1989, 1990c). These results were discussed in detail in Borodich's DSc thesis (Borodich, 1990c). Then the similarity approach was developed and applied to problems with friction and adhesion (Borodich, 1993a). There is still considerable interest in solving such problems using similarity.

5.1 Similarity Transformations for Hertz Type Contact Problems

It was shown (Borodich, 1989, 1990c, 1993a) that one solution of the contact problem can be transformed into another solution by the following transformations when one punch is replaced by another.

Transformation A. The function of the shape of the punch is transformed by homogeneous dilations Λ along all axes, i.e.,

$$x_i \to \Lambda x_i, \qquad i = 1, 2, 3.$$

In this case, the punch is replaced by its scaled copy.

Transformation B. The function of the shape of the punch is transformed by dilation Λ along only x_3-axis, i.e.,

$$x_1 \to x_1, \quad x_2 \to x_2, \quad x_3 \to \Lambda x_3.$$

In this case, the punch is replaced by another one whose shape is just an extension along x_3-axis of the original shape.

Theorem 1. (Similarity I.) *Let f_1 be an arbitrary positive function of the blunt punch shape.*

Let the punch be pressed into a continuous half-space, whose operator of constitutive relations $\mathcal{F}$ is arbitrary.

Let the functions $u_i^*(\mathbf{x}, t, P_0)$, $\sigma_{ij}^*(\mathbf{x}, t, P_0)$, the quantity $\delta^*(t, P_0)$ and the region $G^*(t, P_0)$ give the solution to the contact problem (1)–(4) with one of conditions (5)–(7) for this punch and a pressing force P_0.

Then, the solution to this problem for another punch obtained by Transformation A from the original punch, i.e. its shape is described by the function f

$$f_1(x_1, x_2) = \Lambda f(\Lambda^{-1}x_1, \Lambda^{-1}x_2),$$

and pressed in the half-space by the force

$$P = \Lambda^2 P_0,$$

will be given by $u_i(\mathbf{x}, t, P)$, $\sigma_{ij}(\mathbf{x}, t, P)$, and $\delta(t, P)$, namely

$$u_i(\mathbf{x}, t, P) = \Lambda u_i^*(\Lambda^{-1}\mathbf{x}, t, P_0),$$

$$\sigma_{ij}(\mathbf{x}, t, P) = \sigma_{ij}^*(\Lambda^{-1}\mathbf{x}, t, P_0),$$

$$\delta(t, P) = \Lambda\delta^*(t, P_0),$$

and the contact region $G(t, P)$ changes according to the transformation of homothety, i.e.,

$$[(x_1, x_2) \in G(t, P)] \Longleftrightarrow [(\Lambda^{-1}x_1, \Lambda^{-1}x_2) \in G^*(t, P_0)].$$

Remarks. The punch shape that is invariant under transformation A is a cone.

The time t is included to cover viscoelastic and other materials with constitutive relations depend on time.

A blunt punch means that the slope to the shape function is small within the contact region. Hence, we can apply the linearised formulation of the contact problem (the Hertzian approach).

Theorem 2. (Similarity II.) *Let f_1 be an arbitrary positive function of the blunt punch shape.*

Let the punch be pressed into a non-linear half-space with the operator $\mathcal{F}$ satisfying (52). Let the functions $u_i^*(\mathbf{x}, t, P_0)$, $\sigma_{ij}^*(\mathbf{x}, t, P_0)$, the quantity $\delta^*(t, P_0)$ and the region $G^*(t, P_0)$ give the solution to the contact problem (1)-(4) with one of conditions (5)–(7) for this punch and a pressing force P_0.

Then, the solution to this problem for another punch obtained by Transformation B from the original punch, i.e. its shape is described by the function f

$$f(x_1, x_2) = \Lambda f_1(x_1, x_2),$$

and pressed in the half-space by the force

$$P = \Lambda^{\mu} P_0,$$

will be given by $u_i(\mathbf{x}, t, P)$, $\sigma_{ij}(\mathbf{x}, t, P)$, and $\delta(t, P)$, namely

$$u_i(\mathbf{x}, t, P) = \Lambda u_i^*(\mathbf{x}, t, P_0), \qquad i = 1, 2, 3,$$

$$\sigma_{ij}(\mathbf{x}, t, P) = \Lambda^{\mu} \sigma_{ij}^*(\mathbf{x}, t, P_0),$$

$$\delta(t, P) = \Lambda \delta^*(t, P_0),$$

and the contact region $G(t, P)$ changes according to the homothety transformation, i.e.,

$$[(x_1, x_2) \in G(t, P)] \Longleftrightarrow [(x_1, x_2) \in G^*(t, P_0)].$$

The contact region is not changed under transformation B. Let us unify these two transformations. Suppose the shape function f_1 of the punch is transformed by dilation Λ_1 along the x_1 and x_2 axes and by Λ_2 along the x_3 axis, i.e., the shape function f of the transformed punch is given by

$$f(x_1, x_2) = \Lambda_2 f_1(\Lambda_1^{-1} x_1, \Lambda_1^{-1} x_2). \tag{54}$$

Then the following similarity theorem holds.

Theorem 3. *Let f_1 be an arbitrary positive function of the blunt punch shape. Let the punch be pressed in a non-linear half-space with the operator $\mathcal{F}$ satisfying (52). Let the functions $u_i^*(\mathbf{x}, t, P_0)$, $\sigma_{ij}^*(\mathbf{x}, t, P_0)$, the quantity $\delta^*(P_0)$ and the region $G^*(P_0)$*

give the solution of the contact problem (1)-(4) with one of conditions (5)–(7) for this punch and a pressing force P_0.

Then, the solution of this problem for another punch, with shape function satisfying (54) and pressed into the half-space by the force

$$P = \Lambda_1^{(2-\mu)} \Lambda_2^{\mu} P_0,$$

will be given by $u_i(\mathbf{x}, P)$, $\sigma_{ij}(\mathbf{x}, t, P)$, *and* $\delta(t, P)$, *namely*

$$u_i(\mathbf{x}, t, P) = \Lambda_2 u_i^*(\Lambda_1^{-1}\mathbf{x}, t, P_0),$$

$$\sigma_{ij}(\mathbf{x}, t, P) = (\Lambda_2/\Lambda_1)^{\mu} \sigma_{ij}^*(\Lambda_1^{-1}\mathbf{x}, t, P_0),$$

$$\delta(t, P) = \Lambda_2 \delta^*(t, P_0),$$

and the contact region $G(t, P)$ *changes according to the homothety transformation, i.e.,*

$$[(x_1, x_2) \in G(t, P)] \Longleftrightarrow [(\Lambda_1^{-1}x_1, \Lambda_1^{-1}x_2) \in G^*(t, P_0)].$$

Remark. Formally, Theorem 3 would be valid for other conditions of friction within the contact region, in particular conditions when regions of stick and slip are alternating. However, these conditions look rather unreal.

5.2 Punches Described by Homogeneous Functions

Theorem 3 shows that there is a two-parameter transformation group of coordinate dilations that transforms one solution into another one. It gives also a mathematical explanation of existence of the similarity which was considered in papers where distance between contacting bodies was determined by a positive, homogeneous function of degree $\lambda \geq 1$. Indeed, if the punch shape is described by a homogeneous function H_λ of degree λ then the two-parameter transformation group becomes a one-parameter transformation group, and the punch is transformed into itself. Thus, taking $\Lambda_1 = s^{-1}$, and $\Lambda_2 = s^{-\lambda}$, one obtains the following similarity theorem.

Theorem 4. *Let the shape of a blunt punch be determined by a positive, homogeneous function of degree* $\lambda > 0$. *In addition let the operator of the constitutive relations* $\mathcal{F}$ *satisfy (52).*

Assume further that for a value of the compressing force P_0 *the solution of the contact problem (1)-(4) with one of conditions (5)–(7) is given by the functions* $\sigma_{ij}(\mathbf{x}, t, P_0)$, $\epsilon_{ij}(\mathbf{x}, t, P_0)$, $u_i(\mathbf{x}, t, P_0)$, *quantity* $\delta(t, P_0)$ *and region* $G(t, P_0)$ *and* $G_1(t, P_0)$.

Then, for any positive force P, *the solution of the contact boundary-value problem will be given by*

$$u_i(\mathbf{x}, t, P) = s^{-\lambda} u_i(s\mathbf{x}, t, P_0),$$

$$\epsilon_{ij}(\mathbf{x}, t, P) = s^{(1-\lambda)} \epsilon_{ij}(s\mathbf{x}, t, P_0),$$

$$\sigma_{ij}(\mathbf{x}, t, P) = s^{\kappa(1-\lambda)}\sigma_{ij}(s\mathbf{x}, t, P_0)$$
$$\delta(t, P) = s^{-\lambda}\delta(t, P_0),$$

where $s = (P_0/P)^{1/\alpha}$, *where* $\alpha = 2+\kappa(\lambda-1)$, *i.e.,* $P_0 = s^\alpha P$. *The contact regions* $G(t, P_0)$ *and* $G_1(t, P_0)$ *change according to the homothety transformation, i.e.,*

$$[(x_1, x_2) \in G(t, P)] \Longleftrightarrow [(sx_1, sx_2) \in G(t, P_0)].$$

The similarity properties of all Hertz contact problems follows from this theorem.

Corollary 1. *It follows from Theorem 3 that if for a punch loaded by the force* P_1 *and having shape function described by a homogeneous function* H_λ *of degree* λ, *the size of contact region is known and equal to* $l(1, t, P_1)$, *and the depth of indentation of the punch is equal to* $\delta(1, t, P_1)$, *then for a punch loaded by some force* P *and whose shape is described by the function* cH_λ, $c > 0$, *the size of contact region and the depth of indentation are defined by the following equalities*

$$\delta(c, t, P) = c^{(2-\kappa)/\alpha}(P/P_1)^{\lambda/\alpha}\delta(1, t, P_1),$$
$$l(c, t, P) = c^{-\kappa/\alpha}(P/P_1)^{1/\alpha}l(1, t, P_1). \tag{55}$$

Remarks. As will be shown below, Corollary 1 (Borodich, 1989) gives theoretical background for the Meyer formula.

The formulas (55) follow from Theorem 3. They cannot be derived in a direct way using the similarity transformations of Theorem 4.

For linear elastic materials, $\kappa = 1$. Hence, one has $\delta \sim P^{\lambda/(\lambda+1)}$ and $l \sim P^{1/(\lambda+1)}$. This is in accordance with Galin's (1946) formulae (18) and (19) for isotropic materials and in the case of $\lambda = 2$ with the Willis (1966) solution for anisotropic elastic solids.

5.3 Punches Described by Parametric-Homogeneous Functions

In all the papers mentioned above, the similarity method was applied to Hertz type contact problems for convex punches described by homogeneous functions H_λ. However, natural body surfaces are rough and real contact regions are discrete. Usually, analysis of contact problem for rough surfaces is based on a multi-asperity approach. This means that first the problem is solved for one asperity, and then the solution is used to represent multi-asperity contact. This involves introducing various additional assumptions into the problem. It is difficult to trace the influence of these assumptions. So, it is important to study contact problems for non-convex punches in a direct way, i.e. to study them as Hertz type contact problems, without employing the multi-asperity approach. It was shown by Borodich (1993b) that Hertz type contact problems can be self-similar if functions describing contact surfaces are not only homogeneous but also parametric-homogeneous (PH). Let us recall the definitions of PH- and PQH-functions (Borodich, 1994, 1998a).

The function $B_\lambda : \mathbb{R}^n \to \mathbb{R}$ is called a PQH-function of degree λ and parameter p with weights $\boldsymbol{\alpha} = (\alpha_1, \ldots, \alpha_n)$ if there exists a positive parameter $p,\ p \neq 1$ such that it satisfies the following identity $B_\lambda(p^{k\alpha_1}x_1, \ldots, p^{k\alpha_n}x_n; p) = p^{k\lambda} B_\lambda(\mathbf{x}; p)$, and the parameter is unique in some neighbourhood.

PH-functions $b_\lambda : \mathbb{R}^n \to \mathbb{R}$ are a particular case of PQH-functions when $\alpha_1 = \ldots = \alpha_n$.

Evidently, if p is a parameter of a PH-function then p^m is also a parameter of the same function for any integer m. To avoid an ambiguous definition, we will take as the parameter the least $p,\ p > 1$. Actually, any $b_\lambda : \mathbb{R}_+ \to \mathbb{R}$ can be represented as $b_\lambda(x, p) = Ax^\lambda b_0(x, p)$. Hence, to describe the whole PH-function we should give the values of the appropriate $b_0(x, p)$ on the fundamental domain $(c, pc]$ where c is an arbitrary positive number.

The PH-functions are numerous and the following Weierstrass type fractal functions (f) are particular cases of such functions of degree λ

$$f(x) = \sum_{n=-\infty}^{\infty} p^{-\lambda n} H_o(p^n x) \tag{56}$$

where H_o is a bounded Hölder function of order b, and $0 < \lambda < b \le 1$ (H_o satisfies the Hölder condition, i.e. there is a nonnegative real constant C, such that $|H_o(x) - H_o(y)| \le C|x - y|^b$ for any real x and y).

A sine log-periodic function

$$b_0(x; p) = \sin\left(\frac{2\pi \ln x}{\ln p} + \Phi\right) \tag{57}$$

is another particular case of a PH-function. Here, Φ is a constant.

PH-functions can have both fractal and smooth graphs. In particular, the body shape can be described as a Hertzian body (elliptical paraboloid) with superimposed small roughness represented by a PH-function of zero degree.

$$b_2(x; p) = Ax^2\left[1 + \epsilon b_0(x; p)\right] \tag{58}$$

where A and ϵ are constants, and p is the scaling parameter of the punch shape. Figure 1 shows a graph of a PH-parabola (58) with roughness b_0 represented by a sine log-periodic function (57).

As in Hertz contact, we can prove (Borodich, 1994, 1998b)

Theorem 5. *Let the shape of a blunt punch be determined by a positive PH-function b_λ of degree $\lambda > 0$ and parameter p.*

In addition, let the operator of the constitutive relations $\mathcal{F}$ satisfy (52).

Assume further that for every value of compressing force P_0 on the half-interval $(P_1, p^\alpha P_1]$ the solution of the contact problem (1)-(4) and (5) is given by the functions $\sigma_{ij}(\mathbf{x}, t, P_0; p)$, $\epsilon_{ij}(\mathbf{x}, t, P_0; p)$, $u_i(\mathbf{x}, t, P_0; p)$, quantity $\delta(t, P_0; p)$, and region $G(t, P_0; p)$.

Then, the contact boundary-value problem for a force P is satisfied by

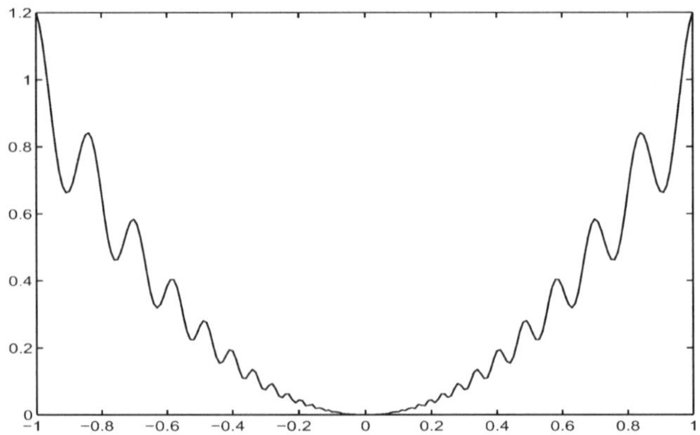

Fig. 1 A log-periodic parabola as a punch profile (after Borodich, 1998b; Borodich and Galanov, 2002).

$$u_i(\mathbf{x}, t, P; p) = p^{-k_0\lambda} u_i(p^{k_0}\mathbf{x}, t, P_0; p),$$
$$\epsilon_{ij}(\mathbf{x}, t, P; p) = p^{-k_0(\lambda-1)} \epsilon_{ij}(p^{k_0}\mathbf{x}, t, P_0; p),$$
$$\sigma_{ij}(\mathbf{x}, t, P; p) = p^{-\kappa k_0(\lambda-1)} \sigma_{ij}(p^{k_0}\mathbf{x}, t, P_0; p),$$
$$\delta(t, P; p) = p^{-k_0\lambda} \delta(t, P_0; p)$$

and the contact region $G(t, P_0; p)$ changes according to the homothetic transformation, i.e.,

$$[(x_1, x_2) \in G(t, P; p)] \Longleftrightarrow [(p^{k_0}x_1, p^{k_0}x_2) \in G(t, P_0; p)],$$

where $P_0 = p^{k_0\alpha} P, \quad k_0 \in \mathbb{Z}.$

Remark. It is evident that the solution to the contact problem has discrete self-similar properties, i.e., it is repeated in scaling form for any load $p^{k(\lambda+1)} P_0,\ k \in \mathbb{Z}$. Hence, it is possible to get the whole solution using the results of numerical simulation of the problem on a finite half-interval of external parameter (the so-called fundamental domain of the discrete group) only. In the above case, this fundamental domain is $(P_1, p^\alpha P_1]$.

Particular cases of Theorem 5 were considered earlier (see, e.g. Borodich, 1993b).

Theorem 6. *Let the shape of the punch be determined by a positive PH-function b_λ of degree $\lambda > 0$ and parameter p.*

In addition let the operator of the constitutive relations $\mathcal{F}$ satisfy (52).

Assume further that for every value of the depth of indentation δ_I on the half-interval $(\delta_1, p^\lambda \delta_1]$ the solution of the contact problem (1)- (7) is given by the functions $\sigma_{ij}(\mathbf{x}, \delta_I; p)$, $\epsilon_{ij}(\mathbf{x}, \delta_I; p)$, $u_i(\mathbf{x}, \delta_I; p)$, the force $P(\delta_I; p)$, and region $G(\delta_I; p)$.

Then, the contact boundary value problem for each depth δ is satisfied by

$$\begin{aligned} u_i(\mathbf{x}, \delta; p) &= p^{-k_0\lambda} u_i(p^{k_0}\mathbf{x}, \delta_I; p), \\ \epsilon_{ij}(\mathbf{x}, \delta; p) &= p^{-k_0(\lambda-1)} \epsilon_{ij}(p^{k_0}\mathbf{x}, \delta_I; p), \\ \sigma_{ij}(\mathbf{x}, \delta; p) &= p^{-\kappa k_0(\lambda-1)} \sigma_{ij}(p^{k_0}\mathbf{x}, \delta_I; p), \\ P(\delta; p) &= p^{-k_0\alpha} P(\delta_I; p) \end{aligned} \tag{59}$$

and the contact region $G(\delta_I; p)$ changes by a homothetic transformation, i.e.,

$$[(x_1, x_2) \in G(\delta; p)] \Longleftrightarrow [(p^{k_0}x_1, p^{k_0}x_2) \in G(\delta_I; p)],$$

where k_0 is taken such that $\delta_I = p^{k_0\lambda}\delta$.

These Transformations A and B of the contact problems for anisotropic, non-linear smooth bodiess were introduced by Borodich (1989, 1993a) and for fractal punches by Borodich (1993b). Later Transformations A and B and the similarity approach was discussed by Roux et al. (1993) in application to the contact of isotropic linearly elastic fractal bodies.

A numerical solution to a particular case of the contact problem for an isotropic linear elastic half-space when the surface roughness is described by a smooth log-periodic function, was studied numerically (Borodich and Galanov, 2002), i.e. the contact problem for rough punches is studied as a Hertz type contact problem without employing additional assumptions of the multi-asperity approach. The problem was solved only on the fundamental domain for the parameter of self-similarity because solutions for other values of the parameter can be obtained by renormalization of this solution. It was shown that the problem has some features of chaotic systems, namely the global character of the solution (the trend of the $P - \delta$ relation) is independent of fine distinctions between PH-functions $b_0(r; p)$ describing roughness in (58), while the local characteristics (the stress field) of the problem is sensitive to small perturbations of the punch shape.

6 Hardness Measurements and Depth-Sensing Techniques

Indentation testing is widely used for analysis and estimation of mechanical properties of materials. Historically, the first indentation tests were developed for hardness measurement. Later they were used for extracting the mechanical properties of materials. The estimations of the thin film mechanical properties can be affected by various factors (see, e.g. a discussion by Borodich et al., 2003).

6.1 Brief History of Hardness Measurements

The idea of hardness measurement can be traced back to Réaumur (1722) (see, e.g. review of Williams, 1942), who suggested comparing the relative hardness of two contacting materials. However, the analytical approach to the problem goes back to Hertz. In 1882 Hertz published another paper on contact problems where he suggested a way to evaluate the hardness of materials. He wrote:

> The hardness of a body is to be measured by the normal pressure per unit area which must act at the centre of a circular surface of pressure in order that in some point of the body the stress may just reach the limit consistent with perfect elasticity. (Hertz, 1882b)

Although Hertz contact theory (Hertz, 1882a) is of a great practical importance and is used in a number of contact models, this suggestion of measuring the hardness of a material by the initiation of plastic yield under an impressed hard ball was found to be impracticable (Johnson, 1985). Indeed, as early as 1909 Dinnik (1952) showed for a circular contact region, and later Belyaev (see §28 by Belyaev, 1924) for an elliptic contact region that according to Hertz contact theory, the point of maximum shearing stresses and consequently the point of first yield is beneath the contact surface and it is normally hidden from view. Hence, it is rather difficult to detect the first yield point experimentally.

Since that time, various experimental techniques have been developed for hardness measurement by indentation, and various definitions of hardness were also introduced. In 1900 Brinell delivered a lecture where he described existing experimental means for hardness measurement, and presented another simple test (Brinell test) based on indentation of hard balls. Brinell assumed the test could give a single numerical expression that may be used as a hardness number. However, soon after this Meyer (1908) showed that the hardness of a metal cannot truly be represented by a single number, and $P = ka^n$ where P is the load, k is an empirical coefficient, n is an exponent, and a is the radius of the projected impression after unloading.

The hardness H was defined originally as the ratio of the maximal load applied to the indenter to the area of the residual imprint after unloading

$$\text{Hardness} = \frac{\text{Load}}{\text{Area of imprint}}.$$

Brinell considered the area of the residual curved imprint, and the Brinell hardness is usually defined as

$$H_B = \frac{P}{A_s}, \quad A_s = \frac{\pi D}{2}\left(D - \sqrt{D^2 - 4a^2}\right).$$

where D is the diameter of the ball. Meyer suggested using the area of the impression projected on the initial contact plane. Hence, the Meyer hardness is defined as (see, e.g. Tabor, 1951)

$$H_M = \frac{P}{A_s}, \quad A_s = \pi a^2.$$

Thus, the Hertz linearized formulation of a boundary value problem may be applied to the Meyer approach, while it is not applicable to the Brinell test. A semi-analytical treatment of the Meyer test was given by Tabor (1951). Another treatment of the Meyer test (see (55)) based on the similarity approach, was given earlier (see Borodich, 1989, 1993a).

However, hardness is now often defined as the ratio of current contact force to the current contact area

$$\text{Hardness} = \frac{\text{Load}}{\text{Area of contact}}.$$

Here this definitionIn will be adopted.

Compared with spherical indenters, conical and pyramidal indenters have the advantage that geometrically similar impressions are obtained at different loads even in the non-linearized formulation (Smith and Sandlund, 1925; Mott 1956). Apparently, Ludwik (1908) was the first to use a diamond cone in a hardness test. In 1922 two other very popular indenters were introduced. Rockwell (1922) introduced a sphero-conical indenter (the Rockwell indenter), while Smith and Sandlund (1922, 1925) suggested using a square-base diamond pyramid (the Vickers indenter). These and other classic methods of measuring hardness are described in detail by Williams (1942), Mott (1956), and also in various standard textbooks. However, there is a difficulty in machining a four-sided indenter in such a way that the sides meet in a point and not as a chisel edge. This is why Berkovich and his research colleagues suggested three-sided indenters for micro-hardness tests (Khrushchev and Berkovich, 1950; Mott 1956).

6.2 *Depth-Sensing Techniques*

Further progress in micro- and nano-hardness came from the introduction of depth-sensing indentation, i.e. the continuously monitoring the displacement of the indenter into the sample surface for both loading and unloading branches. The idea of the continuous monitoring the displacement of the indenter was first introduced by Grodzinskii (1953). However, the modern depth-sensing indentation technique, based on the use of electronics, was introduced by Kalei (1968), who recorded load-depth diagrams for various metals and minerals. For example, the diagram was recorded for a chromium film of 1 μm thickness when the maximum depth of indentation was 150 nm. This revolutionary technique was developed very rapidly, first in the former Soviet Union and then world-wide. Modern sensors can accurately monitor the load in the micro-Newton range and the depth of indentation in the nanometer range.

Introduction of a method for determining of Young's modulus according to the indentation diagram was a very important step in the interpretation of indentation tests. The method was introduced by Bulychev et al. (1975, 1976). Evidently, the load-displacement diagram at loading reflects both elastic and plastic deformation of the material, while the unloading is taking place elastically. The boundary demarcating the elastic and plastic regions may be estimated only by numerical techniques, for example by the finite element method. Therefore, Bulychev et al. (1975) applied the elastic contact solution to the unloading path of the load-displacement diagram, assuming that the non-homogeneity of the residual stress field in a sample after plastic deformation may be neglected. The BASh (Bulychev-Alekhin-Shorshorov) equation for the stiffness S of the upper portion of the load-displacement curve at unloading is the following:

$$S = \frac{dP}{d\delta} = \frac{4\sqrt{A_s}}{\sqrt{\pi}}\vartheta^{-1} \tag{60}$$

where the contact area $A_s = \pi a^2$. We remind that $2\vartheta^{-1} = E^*$ is called the reduced elastic modulus.

The BASh relation was originally derived for spherical and conical indenters using exact solutions for these indenters collected by Lur'e (1955). However, later it was shown that the relation is valid for an arbitrary axisymmetric indenter (Pharr et al., 1992; Borodich and Keer, 2004b). Thus, the BASh relation is an example of a fundamental relation for depth-sensing indentation.

Note that the BASh relation is valid only for frictionless elastic contact. Borodich and Keer (2004a) developed the Mossakovskii approach to the adhesive problem, and derived a relation that is analogous to the BASh relation.

Let us show that the BASh relation (60) can be derived directly from the Galin solution, namely (8)-(10). To do this we employ the Leibnitz rule of differentiation of an integral by a parameter α

$$\frac{d}{d\alpha}\int_{L_1(\alpha)}^{L_2(\alpha)} F(x,\alpha)dx = \int_{L_1(\alpha)}^{L_2(\alpha)} \frac{dF(x,\alpha)}{d\alpha}dx + F(L_2,\alpha)\frac{dL_2}{d\alpha} - F(L_1,\alpha)\frac{dL_1}{d\alpha}.$$

For both equations (8) and (9), the parameter $\alpha = a$, the limits of integrations $L_1 = 0$ and $L_2 = a$, while $F(L_2,\alpha) = 0$. Hence, we have

$$\begin{aligned}\frac{dP}{da} &= 4\vartheta^{-1}\int_0^a \rho_1 \Delta f(\rho_1)\frac{d\sqrt{a^2-\rho_1^2}}{da}d\rho_1 \\ &= 4\vartheta^{-1}a\int_0^a \rho_1 \Delta f(\rho_1)\frac{1}{\sqrt{a^2-\rho_1^2}}d\rho_1,\end{aligned} \tag{61}$$

$$\frac{d\delta}{da} = \int_0^a \rho_1 \Delta f(\rho_1)\frac{d[\tanh^{-1}(\sqrt{1-\rho_1^2/a^2})]}{da}d\rho_1. \tag{62}$$

Substituting $v = \sqrt{1-\rho_1^2/a^2}$ into (61), one obtains

$$\frac{d[\tanh^{-1} v]}{da} = \frac{1}{1-v^2}\frac{\rho_1^2 a^{-3}}{\sqrt{1-\rho_1^2/a^2}} = \frac{1}{\sqrt{a^2-\rho_1^2}}. \tag{63}$$

By substituting this formula into (62) and comparing the result with (61), we obtain the BASh relation for frictionless contact

$$\frac{dP}{d\delta} = \frac{dP/da}{d\delta/da} = 4a\vartheta^{-1} = \frac{4\sqrt{A_s}}{\sqrt{\pi}}\vartheta^{-1}.$$

6.3 Adhesive (No-Slip) Indentation

By differentiating (43) with respect to a, one obtains that the slope of the $P - \delta$ curve is

$$\frac{dP}{d\delta} = \frac{P'(a)}{\delta'(a)} = \frac{4\mu \ln(3 - 4\nu)}{(1 - 2\nu)} a. \tag{64}$$

We can conclude that the BASh relation (60) should be corrected by the factor C in frictional contact:

$$S = \frac{dP}{d\delta} = 4C\vartheta^{-1}\frac{\sqrt{A_s}}{\sqrt{\pi}} \tag{65}$$

where for adhesive (no-slip) contact (Borodich and Keer, 2004a)

$$C = \frac{(1 - \nu)\ln(3 - 4\nu)}{1 - 2\nu}. \tag{66}$$

It follows from (66) that C decreases from $\ln 3 = 1.0986$ at $\nu = 0$ and takes its minimum $C = 1$ at $\nu = 0.5$. Taking into account that full adhesion preventing any slip within the contact region does not occur for real physical contact and there is some frictional slip at the edge of the contact region (see Galin, 1945; Spence, 1975), we can conclude that the values of (66) can be taken as the upper bound for the correction factor C in (65).

6.4 Similarity Considerations of 3D Indentation

As has been shown above, the self-similarity approach is valid for non-linear anisotropic materials in both the frictionless and frictional cases. Roughly speaking, if the stress-strain relation of the coating is $\sigma \sim \epsilon^{\kappa}$ where κ is the work-hardening exponent of the constitutive relationship, and the shape function is homogeneous, then the problem is self-similar. Following the recent approach by Borodich et al. (2003), we will apply the similarity approach to blunted indenters, which, like real indenters, have some deviation from their nominal shapes. Hence, it is important to derive theoretical formulae which are valid for general 3D schemes of nanoindentation by indenters of non-ideal shapes.

Let P_1 be some initial value of the external load, $l(P_1)$ and $\delta(P_1)$ be respectively the characteristic size of the contact region and the depth of indentation (displacement) at this load. Then l and δ at any other value of the load for monomial indenters and materials with power-law stress-strain relations can be re-scaled using (55).

Let us denote by P_1, $(A_s)_1$, l_1 and δ_1 respectively some initial load, the corresponding contact area, the characteristic size of the contact region and the displacement. Then (55) can be re-written as

$$\frac{l}{l_1} = c^{\frac{-\kappa}{\alpha}} \left(\frac{P}{P_1}\right)^{\frac{1}{\alpha}}, \quad \frac{\delta}{\delta_1} = c^{\frac{2-\kappa}{\alpha}} \left(\frac{P}{P_1}\right)^{\frac{\lambda}{\alpha}} \tag{67}$$

and as shown by Borodich et al. (2003), the rescaling formula for the contact area A is

$$\frac{A_s}{(A_s)_1} = c^{\frac{-2}{\lambda}} \left(\frac{\delta}{\delta_1}\right)^{\frac{2}{\lambda}}. \tag{68}$$

If one considers the same indenter then $c = 1$. It follows from (68) that if the indenter tip is described as a monomial function of degree λ, then $\delta \sim A_s^{\lambda/2}$ independently of the work hardening exponent κ.

For a fixed indenter, i.e. $c = 1$, the hardness is the following function of the depth of indentation:

$$\frac{H}{H_1} = \left(\frac{\delta}{\delta_1}\right)^{\frac{\kappa(\lambda-1)}{\lambda}}.$$

However, for an ideal conical or pyramid-shaped indenter $\lambda = 1$, and the hardness is constant.

The formulae (67) and (68) were obtained by assuming the homogeneity of material properties, and that the stress-strain relation remains the same for any depth of indentation. This is not always true. However, as we have seen, non-ideal indenter geometries can also affect the interpretation of the experimental results.

Acknowledgments

In 1974–1979 the author was a student at the Faculty of Mechanics and Mathematics of Moscow State University. He took courses of his narrow specialisation at the Theory of Plasticity Division, where he had the good fortune to meet Professor L.A. Galin. It is a great pleasure and honour for the author to contribute to this edition of Professor Galin's book. The author is very grateful to Professor G.M.L. Gladwell for his invitation to write this chapter.

References

Abramov, V.M. (1937) The problem of contact of an elastic half-plane with an absolutely rigid rough foundation. *Doklady AN SSSR*, **17**, 173–178 [in Russian].

Abramovitz, M. & Stegun, I.A. (Eds.) (1964) *Handbook of Mathematical Functions with Formulas, Graphs, and Mathematical Tables.* Washington, U.S. Govt. Print. Off.

Belyaev, N.M. (1924) Local stresses by compressing elastic solids. In: *Collection of papers "Engineering Structures and Structural Mechanics"*, Put', Leningrad, pp. 27–108 [in Russian].

Borodich, F.M. (1983) Similarity in the problem of contact between elastic bodies. *J. Appl. Math. Mech.*, **47**, 519–521.

Borodich, F.M. (1988) Use of the theory of similarity in the nonlinear problem of contact between an indenter and anisotropic metallic foundations. In: *Abstracts of Reports of All-Union Conference "Metal"-Programme's Fulfillers*, A.A. Bogatov et al. (Eds.), Krasnoyarskii Polytechnical Institute Press, Abakan, pp. 195–196 [in Russian].

Borodich, F.M. (1989) Hertz contact problems for an anisotropic physically nonlinear elastic medium. *Strength of Materials*, **21**, 1668–1676.

Borodich, F.M. (1990a) Hertz contact problems for an elastic anisotropic half-space with initial stresses. *Soviet Appl. Mech.*, **26**, 126–132.

Borodich, F.M. (1990b) On three-dimensional contact problems of the anisotropic plasticity and creep theories. In: *Proceedings of All-Union Institute on Light Alloys*, Moscow, pp. 1–8 [in Russian].

Borodich, F.M. (1990c) Three-dimensional problems of contact of blunt bodies with continuous media. Thesis (D.Sc.) Moscow State University, Moscow [in Russian].

Borodich, F.M. (1990d) On the three-dimensional problem of contact between a rigid punch and an anisotropic metallic foundation in the case of nonlinear creep. In: *Investigation in the Fields of Theory, Technology and Instrumentation of the Punching Industry*, Tula, pp. 16–22 [in Russian].

Borodich, F.M. (1993a) The Hertz frictional contact between nonlinear elastic anisotropic bodies (the similarity approach). *Int. J. Solids Structures*, **30**, 1513–1526.

Borodich, F.M. (1993b) Similarity properties of discrete contact between a fractal punch and an elastic medium. *Comp. R. Acad. Sci., Paris, Ser. 2*, **316**, 281–286.

Borodich, F.M. (1994) Some applications of the fractal parametric-homogeneous functions. *Fractals*, **2**, 311–314.

Borodich, F.M. (1998a) Parametric homogeneity and non-classical self-similarity. I. Mathematical background. *Acta Mechanica*, **131**, 27–45.

Borodich, F.M. (1998b) Parametric homogeneity and non-classical self-similarity. II. Some applications. *Acta Mechanica*, **131**, 47–67.

Borodich, F.M. and Galanov, B.A. (2002) Self-similar problems of elastic contact for non-convex punches. *JMPS*, **50**, 2441–2461.

Borodich, F.M., Keer, L.M. and Korach, C.S. (2003) Analytical study of fundamental nanoindentation test relations for indenters of non-ideal shapes. *Nanotechnology*, **14**, 803–808.

Borodich, F.M., and Keer, L.M. (2004a) Evaluation of elastic modulus of materials by adhesive (no-slip) nanoindentation *Proc. R. Soc. Ser. A*, **460**, 507–514.

Borodich F.M. and Keer L.M. (2004b) Contact problems and depth-sensing nanoindentation for frictionless and frictional boundary conditions. *Int. J. Solids Struct.*, **41**, 2479–2499.

Bradley, R.S. (1932) The cohesive force between solid surfaces and the surface energy of solids. *Phil. Mag*, **13**, 853–862.

Brinell, J.A. (1900) Mémoire sur les épreuves a bille en acier. In: *Communications présentées devant le Congrés international des méthodes d'essai des matériaux de construction, tenu à Paris du 9 au 16 juillet 1900*, **2**, pp. 83–94.

Bryant, M.D. and Keer, L.M. (1982) Rough contact between elastically and geometrically identical curved bodies. *Trans. ASME, J. Appl. Mech.*, **49**, 345–352.

Bulychev, S.I., Alekhin, V.P., Shorshorov, M.Kh., Ternovskii, A.P. and Shnyrev, G.D. (1975) Determination of Young's modulus according to indentation diagram. *Industrial Lab.*, **41**, 1409–1412.

Bulychev, S.I., Alekhin, V.P., Shorshorov, M.Kh. and Ternovskii, A.P. (1976) Mechanical properties of materials studied from kinetic diagrams of load versus depth of impression during microimpression. *Strength of Materials*, **8**, 1084–1089.

Carpick, R.W., Agraït, N., Ogletree, D.F., Salmeron, M. (1996) Measurement of interfacial shear (friction) with an ultrahigh vacuum atomic force microscope. *J. Vac. Sci. Technol. B*, **14**, 1289–1295.

Derjaguin, B. (1934) Untersuchungen über die Reibung und Adhäsion, IV. Theorie des Anhaftens kleiner Teilchen. *Kolloid Zeitschrift*, **69**, 155–164.

Derjaguin, B.V., Muller, V.M. and Toporov, Y.P. (1975) Effect of contact deformations on adhesion of particles. *J. Colloid Interface Sci.*, **53**, 314–326.

Dinnik, A.N. (1952) *Impact and Compression of Elastic Bodies*. In: *Collected Works*, Vol. 1, Izd. Acad. Nauk Ukr. SSR, Kiev [in Russian].

Erdelyi, A. (Ed.) (1954) *Tables of Integral Transforms*, Vol. 1, McGraw-Hill, New York.

Galanov, B.A. (1981a) Approximate solution to some problems of elastic contact of two bodies. *Mechanics of Solids*, **16**, 61–67.

Galanov, B.A. (1981b) Approximate solution of some contact problems with an unknown contact area under conditions of power law of material hardening. *Dokl. AN Urain. SSR, A*, **6**, 36–41 [in Russian and Ukrainian]

Galanov, B.A. (1982) Approximate method for solution of some contact problems with an unknown contact area for two creeping bodies. *Soviet Appl. Mechanics*, **18**, 49–55.

Galanov, B.A. and Grigor'ev, O.N. (1994) Adhesion and wear of diamond. Part I. Modelling. Technical Report. Institute Prob. Mat. Sci., Nat. Ac. Sci. Ukraine, Kiev, pp. 1–14 [in Russian].

Galanov, B.A. and Kravchenko, S.A. (1986) A numerical solution of the problem of contact between a rigid punch and a half-space in the case of nonlinear creep and an unknown contact area. *Soprotivlenie material. i teoriya sooruzhen.*, **49**, 68–71 [in Russian].

Galin, L.A. (1945) Indentation of a punch in presence of friction and adhesion. *J. Appl. Math. Mech. (PMM)*, **9**, 413–424 [in Russian].

Galin, L.A. (1946) Spatial contact problems of the theory of elasticity for punches of circular shape in planar projection. *J. Appl. Math. Mech. (PMM)*, **10**, 425–448 [in Russian].

Galin, L.A. (1953) *Contact Problems in the Theory of Elasticity*, Gostekhizdat, Moscow/Leningrad.

Gladwell, G.M.L. (1980) *Contact Problems in the Classical Theory of Elasticity*. Sijthoff and Noordhoff, Alphen aan den Rijn.

Goodman, L.E. (1962) Contact stress analysis of normally loaded rough spheres. *J. Appl. Mech.*, **29**, 515–522.

Grodzinski, P. (1953) Hardness testing of plastics. *Plastics*, **18**, 312–314.

Hertz, H. (1882a) Ueber die Berührung fester elastischer Körper. *J. reine angewandte Mathematik.*, **92**, 156–171. [English transl. Hertz, H. (1896) On the contact of elastic solids. In: *Miscellaneous Papers by H. Hertz*, D.E. Jones and G.A. Schott (Eds.), Macmillan, London, pp. 146–162.]

Hertz, H. (1882b) Ueber die Berührung fester elastischer Körper und über die Härte. *Verhandlungen des Vereins zur Beförderung des Gewerbefleißes. Berlin, Nov, 1882*. [English transl. Hertz, H. (1896) On the contact of elastic solids and on hardness. In: *Miscellaneous Papers by H. Hertz*, D.E. Jones and G.A. Schott (Eds.), Macmillan, London, pp. 163–183.]

Johnson, K.L. (1958) A note on the adhesion of elastic solids. *Brit. J. Appl. Phys.*, **9**, 199–200.

Johnson, K.L. (1985) *Contact Mechanics.* Cambridge University Press, Cambridge.

Johnson, K.L., Kendall, K. and Roberts, A.D. (1971) Surface energy and the contact of elastic solids. *Proc. R.. Soc. Lond. A*, **324**, 301–313.

Kalei, G.N. (1968) Some results of microhardness test using the depth of impression. *Mashinovedenie*, **4**(3), 105–107 [in Russian].

Keer, L.M. (1967) Mixed boundary-value problems for an elastic half-space. *Proc. Camb. Phil. Soc.*, **63**, 1379–1386.

Kendall, K. (2001) *Molecular Adhesion and Its Applications.* Kluwer Academic/Plenum Publishers, New York.

Khadem, R. and Keer, L.M. (1974) Coupled pairs of dual integral equations with trigonometric kernels. *Quart. Appl. Math.*, **31**, 467–480.

Khrushchov, M.M. and Berkovich, E.S. (1950) Experience in the application of electronic microscope for measurement of very small imprints obtained at microhardness test. *Izvestiya AN SSSR. Otd. Tekh. Nauk*, **11**, 1645–1649 [in Russian]

Kochin, N.E. (1940) Theory of a wing of finite span with circular form in plane. *J. Appl. Math. Mech. (PMM)*, **4**, 3–32 [in Russian].

Kolosov, G.V. (1914) Über einige Eigenschaften des ebenen Problems der Elastizitätstheorie. *Z. Math. Phys.*, **62**, 383–409.

Ludwik, P. (1908) *Die Kegeldruckprobe, ein neues Verfahren zur Härtebestimmung von Materialien.* Springer, Berlin.

Lure, A.I. (1955) *Three-Dimensional Problems of the Theory of Elasticity*, Gostekhizdat, Moscow. [English transl. Lur'e, A.I. (1964) *Three-Dimensional Problems of the Theory of Elasticity*, J.R.M. Radok (Ed.), Interscience Publishers, New York.]

Maugis, D. (2000) *Contact, Adhesion and Rupture of Elastic Solids*, Springer-Verlag, Berlin.

Meyer, E. (1908) Untersuchungen über Härteprüfung und Härte. *Physikalische Z.*, **9**, 66–74.

Mossakovskii, V.I. (1954) The fundamental mixed problem of the theory of elasticity for for a half-space with a circular line separating the boundary conditions. *J. Appl. Math. Mech. (PMM)*, **18**, 187–196 [in Russian].

Mossakovskii, V.I. (1963) Compression of elastic bodies under conditions of adhesion (axisymmetric case). *J. Appl. Math. Mech.*, **27**, 630–643.

Mott, B.A. (1956) *Micro-Indentation Hardness Testing.* Butterworths Sc. Pub., London.

Muskhelishvili, N.I. (1949) *Some Basic Problems of the Mathematical Theory of Elasticity*, Moscow. [English transl. by J.R.M. Radok, Noordhoff International Publishing, Leyden, 1977.]

Pobedrya, B.E. (1984) The deformation theory of plasticity of anisotropic media. *J. Appl. Math. Mech.*, **48**, 10–17.

Pharr, G.M. Oliver, W.C. and Brotzen, F.R. (1992) On the generality of the relationship among contact stiffness, contact area, and elastic modulus during indentation. *J. Mater. Res.*, **7**, 613–617.

Réaumur, R.A. (1722) L'art de convertir le fer forge en acier, et l'art d'adoucir le fer fondu, ou de faire des ouvrages de faire fondu, aussi finis que de fer forge. Michel Brunet, Paris, pp. 566–587. [Reprinted in part, with résumé by F. Cournot, in *Revue de Metallurgie, Memoires*, **19**, 447–468 (1922).]

Rockwell, S.P. (1922) Testing metals for hardness. *Trans. Amer. Soc. Steel Treat.*, **2**, 1013–1033.

Roux, S., Schmittbuhl, J., Vilotte, J.-P. and Hansen, A. (1993) Some physical properties of self-affine rough surfaces. *Europhys. Lett.*, **23**, 277–282.

Shtaerman, I. Ya. (1939) On the Hertz theory of local deformations resulting from the pressure of elastic solids. *Dokl. Akad. Nauk SSSR*, **25**, 360–362 [in Russian].

Smith, R.L. and Sandland, G.E. (1922) An accurate method of determining the hardness of metals, with particular reference to those of a high degree of hardness. *Proc. Inst. Mech. Eng.*, **1**, 623–641.

Smith, R.L. and Sandland, G.E. (1925) Some notes on the use of a diamond pyramid for hardness testing. *J. Iron and Steel Inst.*, **111**, 285–294.

Sneddon, I.N. (1965) The relation between load and penetration in the axisymmetric Boussinesq problem for a punch of arbitrary profile. *Int. J. Eng. Sci.*, **3**, 47–57.

Spence, D.A. (1968) Self similar solutions to adhesive contact problems with incremental loading. *Proc. R. Soc. London A*, **305**, 55–80.

Spence, D.A. (1975) The Hertz contact problem with finite friction. *J. Elast.*, **5**, 297–319.

Tabor, D. (1951) *The Hardness of Metals*. Oxford, Clarendon Press.

Williams, S.R. (1942) *Hardness and Hardness Measurements.* American Society for Metals, Cleveland.

Willis, J.R. (1966) Hertzian contact of anisotropic bodies. *J. Mech. Phys. Solids*, **14**, 163–176.

Further Developments of Galin's Stick-Slip Problem

Olesya I. Zhupanska

1 Introduction

Galin's classical work (1945, 1953) contact with partial slip was the first successful attempt to take into account friction in the problem of normal contact. As Galin was unable to find an exact solution of the formulated problem (he determined the stick-slip boundary approximately and pointed out the values of the friction coefficient for which total stick or total slip solutions should be used), the problem of contact with partial slip of a rigid punch with an elastic half plane has been challenged by many researchers. At the same time Galin's problem stimulated the development of solutions for other contact problems with friction that feature different punch geometries and different material responses.

In this chapter we discuss some of the studies on contact problems with friction that were motivated by Galin's seminal work. First we turn our attention to Galin's original formulation of the stick-slip problem for a rigid flat punch indenting an elastic half plane, and trace its development in the works of other authors who either pursued Galin's ideas in order to improve his approximate solution, or developed entirely independent solution procedures. Then we briefly dwell on formulations and solutions of stick-slip problems that concern other types of loads and indenter geometries. In the spirit of Galin's work, we focus on contributions with substantial analytical merit.

We give a detailed discussion of the works published in Russian; the ones published in English will receive a briefer overview (for the details the reader is referred to the original papers). We consider papers that appeared in the top Russian (Soviet) technical journals that can be reached by the readers around the world.

Note that to keep uniformity throughout this chapter we follow Galin's notation wherever possible. Therefore sometimes the results of other authors look somewhat different from the way they appeared in the original papers.

2 The Stick-slip Problem in Galin's Formulation

2.1 Refinements of Galin's Solution

First we briefly revisit the key points of Galin's solution procedure in order to illuminate the distinctive features of solutions by other authors. Galin considered the problem of indentation of a flat rigid punch into an elastic half plane in the presence of partial slip on the interface. He divided the contact region into stick and slip zones with an unknown boundary between them, and assumed that in the stick zone the tangential force was insufficient to displace the points of the elastic half plane relative to the punch, and in the slip zone the punch slipped relative to the elastic half plane. The essence of Galin's solution procedure lies in the introduction of a new function $s(z)$ (4.7.12) that maps some curvilinear quadrangle region S (Figure 4.7.1) onto the upper half of the complex plane, and in the formulation of a Riemann–Hilbert problem for the function $w_1(z)$ (defined by (2.2.28)). As mentioned in Section 4.7, the solution of the problem depends on the conformal mapping $s(z)$. Having difficulty mapping the original region S onto the upper half plane of the complex plane, Galin approximated S by another region using the following conformal mapping (4.7.34). He showed that his approximation of S (the hatched region in Figure 4.7.1) is quite good. Moreover, using this approximation he determined that the coordinate a of the stick-slip boundary is defined by the first equation of (4.7.39).

Finally, examining the location of the stick-slip boundary depending on the coefficient of friction, Galin resolved when the total stick and total slip solutions have to be used. Therefore, Galin's solution of the formulated boundary-value problem (4.7.1)–(4.7.6) is not exact. Furthermore, this solution includes neither explicit expressions for contact stresses nor explicit expressions for the potentials $w_1(z)$ and $w_2(z)$.

Galin mentions that to find an exact mapping of the curvilinear quadrangle S onto the upper half plane of the complex plan one needs to find the solutions of the following Fuchsian differential equation (linear differential equations that have regular singular points are known as Fuchsian differential equations) with respect to the function $s(z)$ of the desired conformal mapping:

$$s'' + \left(\frac{1-\alpha/\pi}{z} + \frac{1-\alpha/\pi}{z-1} + \frac{1-\alpha/\pi}{z-a}\right)s' + \frac{(1-\alpha/\pi)(1-2\alpha/\pi)(z-\lambda)}{z(z-1)(z-a)}s = 0, \tag{1}$$

where $\alpha = \angle B_1 A_1 E_1$ (see Figure 4.7.1), and λ is an accessory parameter that is unknown *a priori*. The Fuchsian equation (1) is also known as Heun's equation, see, e.g., Ronveaux (1995). It is known that conformal mappings of curvilinear polygons are governed by Fuchsian equations, see, e.g., Nehari (1952). It is also known that there is no general solution procedure for the Fuchsian equation (1) since the accessory parameter λ and the free point a are not known *a priori* and must be determined as part of the solution.

Primarily because of the difficulties associated with finding the parameters a and λ, Galin did not pursue the solution of equation (1), and used the conformal mapping (4.7.34) for the region that approximates the curvilinear quadrangle S.

In 1972, Mossakovskii and Biskup pursued Galin's idea of reducing the original stick-slip contact problem to a Fuchsian equation although they used a different solution procedure. The problem of determining the unknown parameters in the Fuchsian equation was circumvented by imposing some additional relationships between these parameters and the coefficients of the corresponding Riemann–Hilbert problem. Here we reproduce the key steps of Mossakovskii and Biskup's solution procedure along with some intermediate steps that were omitted in the exposition of the original paper.

Following Muskhelishvili (1963), Mossakovskii and Biskup represented the stresses and derivatives of displacements in the half plane in terms of a single complex valued function $\Phi(z)$:

$$2\mu\left(\frac{\partial u}{\partial x}+i\frac{\partial v}{\partial x}\right)=\kappa\Phi(z)+\Phi(\bar{z})-(z-\bar{z})\bar{\Phi}'(\bar{z}),$$

$$\sigma_{yy}-i\tau_{xy}=\Phi(z)-\Phi(\bar{z})+(z-\bar{z})\bar{\Phi}'(\bar{z}), \tag{2}$$

where as before $\kappa=3-4\nu$. Then the original boundary-value problem (4.7.1)–(4.7.6) for the case $a=b$ ($Q=0$) was converted to the Riemann–Hilbert problem for two piecewise analytic functions $\Phi(z)$ and $\bar{\Phi}(z)$ with discontinuous coefficients at the boundary:

$$\begin{aligned}
&\Phi^-(x)=\Phi^+(x),\ \bar{\Phi}^-(x)=\bar{\Phi}^+(x)\quad(-\infty<x<-1),\\
&(\rho+i)\Phi^-(x)-(\rho-i)\bar{\Phi}^-(x)-(\rho+i)\Phi^+(x)\\
&\quad+(\rho-i)\bar{\Phi}^+(x)=0\quad(-1<x<-a),\\
&\kappa\left(\Phi^-(x)-\bar{\Phi}^+(x)\right)+\Phi^+(x)-\bar{\Phi}^-(x)=0\quad(-1<x<-a),\\
&\kappa\left(\Phi^-(x)+\bar{\Phi}^+(x)\right)+\Phi^+(x)+\bar{\Phi}^-(x)=0\quad(-a<x<a),\\
&\kappa\left(\Phi^-(x)-\bar{\Phi}^+(x)\right)+\Phi^+(x)-\bar{\Phi}^-(x)=0\quad(-a<x<a),\\
&(\rho-i)\Phi^-(x)-(\rho+i)\bar{\Phi}^-(x)-(\rho-i)\Phi^+(x)\\
&\quad+(\rho+i)\bar{\Phi}^+(x)=0\quad(a<x<1),\\
&\kappa\left(\Phi^-(x)-\bar{\Phi}^+(x)\right)+\Phi^+(x)-\bar{\Phi}^-(x)=0\quad(a<x<1),\\
&\Phi^-(x)=\Phi^+(x),\ \bar{\Phi}^-(x)=\bar{\Phi}^+(x)\quad(1<x<\infty).
\end{aligned}\tag{3}$$

Equations (3) were obtained by substitution of the relations (2) and their conjugates into the boundary conditions (4.7.1)–(4.7.6).

To solve the Riemann–Hilbert problem (3), new functions $Y(z)$ and $\bar{Y}(z)$ were introduced by relationships

$$\Phi(z) = \frac{1}{\sqrt{z^2-1}} Y(z), \quad \bar{\Phi}(z) = \frac{1}{\sqrt{z^2-1}} \bar{Y}(z), \tag{4}$$

whereupon boundary conditions for $Y(z)$ and $\bar{Y}(z)$ take the form

$$\begin{aligned}
Y^-(x) &= -\frac{2\rho}{D} Y^+(x) + \frac{(\rho - i)(\kappa+1)}{D} \bar{Y}^+(x), \\
\bar{Y}^-(x) &= -\frac{(\rho + i)(\kappa+1)}{D} Y^+(x) + \frac{2\rho\kappa}{D} \bar{Y}^+(x) \quad (-l < x < a), \\
Y^-(x) &= \frac{1}{\kappa} Y^+(x), \\
\bar{Y}^-(x) &= \kappa \bar{Y}^+(x) \quad (-a < x < a), \\
Y^-(x) &= -\frac{2\rho}{D} Y^+(x) + \frac{(\rho + i)(\kappa+1)}{D} \bar{Y}^+(x), \\
\bar{Y}^-(x) &= -\frac{(\rho - i)(\kappa+1)}{D} Y^+(x) + \frac{2\rho\kappa}{D} \bar{Y}^+(x) \quad (a < x < l).
\end{aligned} \tag{5}$$

Here

$$D = -\rho(\kappa - 1) + i(k+1). \tag{6}$$

Additionally, the functions $Y(z)$ and $\bar{Y}(z)$ are holomorphic functions outside the contact area. Therefore problem (5) is a Riemann–Hilbert problem for two unknown functions with piecewise constant coefficients. This problem is reduced to the following Fuchsian differential equation for $Y(z)$:

$$\begin{aligned}
&\frac{d^2Y(z)}{dz^2} + \left[\frac{1/2 - \varphi/\pi}{z+1} + \frac{1/2 + \varphi/\pi}{z+a} + \frac{1/2 + \varphi/\pi}{z-a} + \frac{1/2 - \varphi/\pi}{z-1} \right] \frac{dY(z)}{dz} \\
&+ \left[-\frac{(1-a^2)\,\varphi/\pi}{z+1} + \frac{(1-a^2)\,\varphi/\pi}{z-1} + \lambda_a \right] \frac{Y(z)}{(z^2-1)(z^2-a^2)} = 0.
\end{aligned} \tag{7}$$

Here λ_a is an accessory parameter that is unknown *a priori* and should be determined as a part of the solution, and φ stands for

$$\varphi = \arctan \frac{\rho(\kappa - 1)}{\kappa + 1}. \tag{8}$$

The Fuchsian equation (7) was solved by Mossakovskii and Biskup numerically using the Runge–Kutta method. Unknowns a and λ_a were also determined numerically from additional conditions requiring that coefficients of increase for the integrals of the Fuchsian equation (7) for a circuit around a pair of arbitrary fixed points should coincide with coefficients of the corresponding Riemann–Hilbert problem (5). Distributions of contact stresses were calculated numerically. The location a of the stick-slip boundary was found to be in excellent agreement with the result obtained by Galin using the approximate conformal mapping technique (1945, 1953).

Another extension of Galin's work may be found in Mossakovskii et al. (1989) who considered the boundary-value problem (4.7.1)–(4.7.6) for the case $a = b$ ($Q = 0$) as well. Following Galin's work they introduced potentials $w_1(z)$, $w_2(z)$, formulated Riemann–Hilbert problems for these functions, and introduced the function $s(z)$ by means of (4.7.12). Then Galin's approximate conformal mapping was used

$$\varsigma = -\frac{s+i}{s-i} = \frac{\Phi_2}{\Phi_1} = -\frac{w_2 + iw_1}{w_2 - iw_1} \tag{9}$$

and new functions $Y_1(z)$ and $Y_2(z)$ were designated

$$Y_{1,2} = \sqrt{z^2-1}\Phi_{1,2}. \tag{10}$$

The original boundary-value problem was subsequently reduced to the Rieman-Hilbert problem for the product of functions, $Y_1 \cdot Y_2$, and the logarithm of the quotient of these functions, $F = \ln\left(Y_1/Y_2\right)$:

$$\begin{aligned}
&(Y_1(x)\cdot Y_2(x))^- = (Y_1(x)\cdot Y_2(x))^+ \quad (-\infty < x < -1 \text{ and } 1 < x < \infty),\\
&(Y_1(x)\cdot Y_2(x))^- = e^{2i\varphi}\,(Y_1(x)\cdot Y_2(x))^+ \quad (-1 < x < -a),\\
&(Y_1(x)\cdot Y_2(x))^- = (Y_1(x)\cdot Y_2(x))^+ \quad (-a < x < a),\\
&(Y_1(x)\cdot Y_2(x))^- = e^{2i\varphi}\,(Y_1(x)\cdot Y_2(x))^+ \quad (a < x < 1),
\end{aligned} \tag{11}$$

where φ is determined by (8), and

$$\begin{aligned}
&F^-(x) = F^+(x) \quad (-\infty < x < -1 \text{ and } 1 < x < \infty),\\
&F^-(x) + F^+(x) = -i\cdot 4\arctan\rho \quad (-1 < x < -a),\\
&F^-(x) - F^+(x) = 2\ln\kappa \quad (-a < x < a),\\
&F^-(x) + F^+(x) = i\cdot 4\arctan\rho \quad (a < x < 1).
\end{aligned} \tag{12}$$

In accordance with Muskhelishvili (1963), the solution of problem (11) is

$$\Phi_1(z)\cdot\Phi_2(z) = -\frac{P^2}{z^2-1}\left(\frac{z^2-1}{z^2-a^2}\right)^{\varphi/\pi}, \tag{13}$$

where P is the compressive force

$$\int_0^a p(x)dx = \frac{P}{2}. \tag{14}$$

The solution of problem (12) has the form

$$\frac{\Phi_1(z)}{\Phi_2(z)} = \exp\left\{c\int_z^\infty \frac{dz}{\sqrt{(z^2-1)(z^2-a^2)}}\right\}, \tag{15}$$

where c is an unknown parameter determined by either of the following relationships:

$$c = -i \ln \kappa \Bigg/ \int_a^1 \frac{dx}{\sqrt{(1-x^2)(x^2-a^2)}},$$

$$c = 2i \arctan \rho \Bigg/ \int_1^\infty \frac{dx}{\sqrt{(x^2-1)(x^2-a^2)}}. \tag{16}$$

The analytical expressions for the contact stresses are as follows:

$$\sigma_{yy} = \mathrm{Re}\big(\Phi_1^{M-}(z) - \Phi_1^{M+}(z)\big),$$
$$\tau_{xy} = \mathrm{Im}\big(\Phi_1^{M-}(z) - \Phi_1^{M+}(z)\big). \tag{17}$$

Here the functions $\Phi_1^M(z)$ and $\Phi_2^M(z)$ are obtained immediately from (13) and (15):

$$\Phi_1^M(z) = \frac{iP}{2\pi\sqrt{z^2-1}} \left(\frac{z^2-1}{z^2-a^2}\right)^{\varphi/2\pi} \times \exp\left\{\frac{c}{2}\int_z^\infty \frac{dz}{\sqrt{(z^2-1)(z^2-a^2)}}\right\},$$

$$\Phi_2^M(z) = \frac{iP}{2\pi\sqrt{z^2-1}} \left(\frac{z^2-1}{z^2-a^2}\right)^{\varphi/2\pi} \times \exp\left\{-\frac{c}{2}\int_z^\infty \frac{dz}{\sqrt{(z^2-1)(z^2-a^2)}}\right\}. \tag{18}$$

The unknown coordinate a that defines the location of the stick-slip boundary is taken from Galin's solution. Therefore, the contribution of the paper to the solution of the Galin problem may be seen in the derivation of the contact stresses. Indeed, as acknowledged by Mossakovskii, Biskup, and Mossakovskaia, their work is just a further development of Galin's method.

2.2 Solution of the Galin Stick-Slip Problem Due to Antipov and Arutyunyan

A completely independent approach to Galin's stick-slip problem was proposed by Antipov and Arutyunyan (1991), who gave an exact solution to the problem. Their method is based on the reduction of the original boundary-value problem to a particular Riemann problem for two functions, which itself is reduced to an infinite system of algebraic equations. Solution to this system may be obtained with any prescribed accuracy.

Below we describe the solution procedure developed by Antipov and Arutyunyan (1991). Following the original paper, the stick-slip problem for the contact of a rigid punch and an elastic half plane is reformulated in polar coordinates, namely

$$\begin{aligned}
&u_r|_{\theta=0} = \delta_t \quad (0 < r < a),\\
&u_\theta|_{\theta=0} = \delta_n \quad (0 < r < 1),\\
&(\tau_{r\theta} + \rho\sigma_{\theta\theta})|_{\theta=0} = 0 \quad (a < r < 1),\\
&\sigma_{\theta\theta}|_{\theta=0} = \tau_{r\theta}|_{\theta=0} = 0 \quad (1 < r < \infty),\\
&u_\theta|_{\theta=-\pi/2} = \tau_{r\theta}|_{\theta=-\pi/2} = 0 \quad (0 < r < \infty).
\end{aligned} \tag{19}$$

Here, due to the symmetric setup, the problem is written only for a quarter of the plane. Additionally, the equilibrium condition implies that

$$\int_0^1 \sigma_{\theta\theta}|_{\theta=0}\, dr = \frac{P}{2}. \tag{20}$$

The authors then introduced the following new unknown functions

$$\begin{aligned}
\chi_1(r) &= \sigma_{\theta\theta}(r, 0),\\
\chi_2(r) &= \tau_{r\theta}(r, 0) + \rho\sigma_{\theta\theta}(r, 0),\\
\psi_1(r) &= \frac{E}{(1+\nu)} \frac{\partial u_r(r, 0)}{\partial r},\\
\psi_2(r) &= \frac{E}{(1+\nu)} \frac{\partial u_\theta(r, 0)}{\partial r},
\end{aligned} \tag{21}$$

where E is the Young modulus, and ν is Poisson's ratio. The equilibrium condition (20) immediately becomes

$$\int_0^a \chi_1(r) dr = \frac{P}{2}. \tag{22}$$

Then, the Mellin integral transform was applied to the elastostatic equations, equation of continuity, and Hooke's law. For plain strain this leads to the following equations:

$$\begin{aligned}
&\Sigma_{\theta z}^{\mathrm{IV}} + 2(z^2+1)\Sigma_{\theta z}'' + (z^2-1)^2\Sigma_{\theta z} = 0 \quad (-\pi < \theta < 0),\\
&(z-1)\mathrm{T}_{r\theta z} = \Sigma_{\theta z}',\\
&z(z-1)\eta_{rz} = \frac{1+\nu}{E}\left((1-\nu)\Sigma_{\theta z}'' + (\nu z + 1 - \nu)(z-1)\Sigma_{\theta z}\right),\\
&z(z^2-1)\eta_{\theta z} = \frac{1+\nu}{E}\left((1-\nu)\Sigma_{\theta z}''' + (2z^2 - \nu z^2 + 2\nu z - z + 1 - \nu)\Sigma_{\theta z}'\right).
\end{aligned} \tag{23}$$

Here

$$\begin{aligned}
\Sigma_{\theta z} &= \int_0^\infty \sigma_{\theta\theta} r^z dr, \quad \mathrm{T}_{r\theta z} = \int_0^\infty \tau_{r\theta} r^z dr,\\
\eta_{rz} &= \int_0^\infty \frac{\partial u_r}{\partial r} r^z dr, \quad \eta_{\theta z} = \int_0^\infty \frac{\partial u_\theta}{\partial r} r^z dr.
\end{aligned} \tag{24}$$

By complementing boundary conditions (19) with equations (23) they obtained the following boundary value problem:

$$\Sigma_{\theta z}|_{\theta=0} = \Phi_1^-(z),$$
$$\left(\rho\Sigma_{\theta z} + (z-1)^{-1}\Sigma'_{\theta z}\right)\Big|_{\theta=0} = a^{z+1}\Phi_2^-(z),$$
$$\left((1-\nu)\Sigma''_{\theta z} - (\nu z + 1 - \nu)(z-1)\Sigma_{\theta z}\right)|_{\theta=0} = z(z-1)a^{z+1}\Phi_1^+(z),$$
$$\left((1-\nu)\Sigma'''_{\theta z} + (2z^2 - \nu z^2 + 2\nu z - z + 1 - \nu)\Sigma'_{\theta z}\right)\Big|_{\theta=0} = z(z^2-1)\Phi_2^+(z),$$
$$\Sigma'_{\theta z}|_{\theta=-\pi/2} = \Sigma'''_{\theta z}|_{\theta=-\pi/2} = 0. \tag{25}$$

Here

$$\phi_1^-(z) = \int_0^1 \chi_1(r)r^z dr, \quad \phi_2^-(z) = \int_0^1 \chi_2(ar)r^z dr,$$
$$\phi_1^+(z) = \int_0^1 \psi_1(ar)r^z dr, \quad \phi_2^+(z) = \int_0^1 \psi_2(r)r^z dr. \tag{26}$$

The problem (25) was reduced to the homogeneous Riemann problem

$$a^{z+1}\phi_1^+(z) = K_1(z)\phi_1^-(z) - \frac{1}{2}(\kappa+1)a^{z+1}\tan(\pi z/2)\,\phi_2^-(z),$$
$$\phi_2^+(z) = K_0(z)\phi_1^-(z) - \frac{1}{2}(\kappa-1)a^{z+1}\phi_2^-(z), \tag{27}$$

where

$$K_0(z) = \frac{1}{2}(\kappa-1)\,\mathrm{cotan}\,(\pi z/2) + \frac{1}{2}(\kappa-1)\rho,$$
$$K_1(z) = \frac{1}{2}(\kappa-1) + \frac{1}{2}(\kappa+1)\rho\tan(\pi z/2) \tag{28}$$

and $\kappa = 3 - 4\nu$.

The first key point in Antipov and Arutyunyan's solution procedure is the reduction of the homogeneous Riemann problem (27) to a special form, convenient for the follow-up solution technique proposed by the authors. To this end, the following factorization of the function $K_0(z)$ was performed

$$K_0(z) = K_0^+(z)K_0^-(z). \tag{29}$$

Here

$$K_0^+(z) = -\frac{(\kappa+1)}{2\sin\varphi}\frac{\Gamma(-z/2)}{\Gamma(1-\varphi/\pi - z/2)}, \quad K_0^-(z) = -\frac{\Gamma(1+z/2)}{\Gamma(\varphi/\pi + z/2)}, \tag{30}$$

where φ is determined by (8). Substitution of (29) into (27) yields

$$-\kappa\frac{\phi_2^-(z)}{K_0^-(z)} = K_0^+(z)\phi_1^+(z) - a^{-z-1}\frac{K_1(z)\phi_2^+(z)}{K_0^-(z)},$$

$$\frac{\phi_2^+(z)}{K_0^+(z)} = K_0^-(z)\phi_1^-(z) - (\kappa-1)a^{z+1}\frac{\phi_2^-(z)}{2K_0^+(z)}. \tag{31}$$

Here function $K_1(z)/K_0^-(z)$ is a meramorphic function in the region D^+: $\mathrm{Re}(z) < \gamma_0 \in (-\varepsilon, 0)$ $(0 < \varepsilon < 1)$ with poles $z = -2\varphi/\pi - 2j$ and $z = -1 - 2j$ $(j = 0, 1, \ldots)$. The function $[K_0^+(z)]^{-1}$ is a meramorphic function in the region D^-: $\mathrm{Re}(z) > \gamma_0 \in (-\varepsilon, 0)$ $(0 < \varepsilon < 1)$ with poles $z = -2\varphi/\pi + 2 + 2j$ $(j = 0, 1, \ldots)$.

The second key step in Antipov and Arutyunyan's solution procedure is the introduction of new unknown functions

$$\Psi_0^\pm(z) = \sum_{j=0}^{\infty}\frac{A_j^\pm}{z + 2\varphi/\pi \mp 2j \mp 1 - 1}, \qquad \Psi_1^-(z) = \sum_{j=0}^{\infty}\frac{B_j}{z+1+2j}, \tag{32}$$

where

$$A_j^+ = \operatorname*{Res}_{z=-2\varphi/\pi+2+2j}\left\{-(\kappa-1)a^{z+1}[2K_0^+(z)]^{-1}\phi_2^-(z)\right\},$$

$$A_j^- = \operatorname*{Res}_{z=-2\varphi\pi-2j}\left\{-a^{-z-1}[K_0^-(z)]^{-1}K_1(z)\phi_2^+(z)\right\},$$

$$B_j = \operatorname*{Res}_{z=-1-2j}\left\{-a^{-z-1}[K_0^-(z)]^{-1}K_1(z)\phi_2^+(z)\right\}. \tag{33}$$

Application of the Liouville theorem allowed the authors to write the solution of the Riemann problem (31) in terms of unknown functions $\Psi_0^+(z)$, $\Psi_1^+(z)$ and $\Psi_1^-(z)$, namely

$$\phi_1^-(z) = \frac{C + \Psi_0^+(z)}{K_0^-(z)} - (\kappa-1)a^{z+1}\frac{\Psi_0^-(z) + \Psi_1^-(z)}{2\kappa K_0^+(z)},$$

$$\phi_1^+(z) = \frac{\Psi_0^-(z) + \Psi_1^-(z)}{K_0^+(z)} + a^{-z-1}\frac{K_1(z)\left(C + \Psi_0^+(z)\right)}{K_0^-(z)},$$

$$\phi_2^-(z) = -\frac{K_0^-(z)\left(\Psi_0^-(z) + \Psi_1^-(z)\right)}{\kappa},$$

$$\phi_2^-(z) = K_0^+(z)\left(C + \Psi_0^+(z)\right), \tag{34}$$

where C is an arbitrary constant. Eventually the problem is reduced to the determination of the unknown functions $\Psi_0^+(z)$, $\Psi_1^+(z)$ and $\Psi_1^-(z)$. Substitution of (32), (33), and (30) into (34) yields the following infinite system of algebraic equations

$$A_{n*}^- = a^{2\varphi/\pi-1+2n}\delta_{on}^+\left(1 - \sum_{j=0}^{\infty}\frac{A_{j*}^+}{2(n+j+1)}\right),$$

$$B_{n*} = a^{2n}\delta^{+}_{1n}\left(1 - \sum_{j=0}^{\infty}\frac{A^{+}_{j*}}{2(n+j-\varphi/\pi)+3}\right),$$

$$A^{+}_{n*} = a^{2n+3-2\varphi/\pi}\delta^{-}_{on}\left(\sum_{j=0}^{\infty}\frac{A^{-}_{j*}}{2(n+j+1)} + \sum_{j=0}^{\infty}\frac{B_{j*}}{2(n+j-\varphi/\pi)+3}\right) \quad (35)$$

with respect to new unknowns $A^{\pm}_{n*}$, B_{n*}, that are related to old variables as $A^{\pm}_{n} = CA^{\pm}_{n*}$, $B_n = CB_{n*}$. The following designations were applied in (35)

$$\delta^{+}_{0n} = -\frac{2\kappa(\kappa+1)}{\pi(\kappa-1)}\frac{\Gamma^2(n+\varphi/\pi)}{n!^2},$$

$$\delta^{+}_{1n} = \frac{(\kappa+1)^3}{4\pi(\kappa-1)\sin^2\varphi}\frac{\Gamma^2(n+1/2)}{\Gamma^2(n+3/2-\varphi/\pi)},$$

$$\delta^{-}_{0n} = -\frac{2(\kappa-1)\sin^2\varphi}{\pi(\kappa+1)\kappa}\frac{\Gamma^2(n+2-\varphi/\pi)}{n!^2}. \quad (36)$$

The system (36) was solved by the asymptotic method. The desired coefficients $A^{\pm}_{n*}$, B_{n*} are sought in the form of expansions with respect to the parameter a

$$A^{-}_{n*} = a^{2n-1+2\varphi/\pi}\sum_{j=0}^{\infty}a^{-}_{nj}a^{2j}, \quad A^{+}_{n*} = a^{2n+2}\sum_{j=0}^{\infty}a^{+}_{nj}a^{2j},$$

$$B_{n*} = a^{2n}\sum_{j=0}^{\infty}b_{nj}a^{2j}, \quad (37)$$

which lead to rapidly (exponentially) converging asymptotic expansions in a for the coefficients $A^{\pm}_{n}$, B_n. The system (35) still contains an unknown coordinate a of the stick-slip boundary and constant C that have to be found as a part of the solution. The constant C was obtained as follows: substitution of (22) into (25) gives $\phi^{-}_{1}(0) = P/2$ and subsequent substitution of $\phi^{-}_{1}(0)$ along with (21) and (32) into the first equation (34) produces

$$C = \frac{P}{2l\Gamma(\varphi/\pi)}\left(1 + \sum_{j=0}^{\infty}\frac{A^{+}_{j*}}{2(\varphi/\pi-1-j)}\right)^{-1}. \quad (38)$$

To determine the unknown coordinate a of the stick-slip boundary, the authors bring into consideration the stress intensity factor, defined as

$$K(a) = \lim_{r\to a-0}(a-r)^{1-\varphi/\pi}\chi_2(r) \quad (39)$$

and require that $K(a) = 0$, e.g., $\tau_{r\theta}(a,0) + \rho\sigma_{\theta\theta}(a,0) = 0$, in order to guarantee finiteness of the contact stresses in the vicinity of the stick-slip boundary. With use of the result (34) it was shown that

$$\phi_2^-(z) \sim -\frac{z^{-\varphi/\pi}}{2^{1-\varphi/\pi}\kappa} C \sum_{j=0}^{\infty} \left(A_{j*}^- + B_{j*} \right), \quad z \to \infty, \quad z \in D^-,$$

which immediately leads to

$$K(a) = -C(\kappa\Gamma(\varphi/\pi))^{-1}(a/2)^{1-\varphi/\pi} \sum_{j=0}^{\infty}(A_{j*}^- + B_{j*}).$$

Therefore the coordinate a of the stick-slip boundary is determined by condition

$$\sum_{j=0}^{\infty}(A_{j*}^- + B_{j*}) = 0. \tag{40}$$

Thus, to complete the solution of the original boundary-value problem (19) one had to solve the infinite system of algebraic equations (35) and satisfy two additional conditions (38) and (40). Besides, contact stresses are evaluated by means of the inverse Mellin integral transform.

In this way, reduction of the original boundary-value problem to a Riemann problem of special form allowed the authors to deduce an exact solution in the form of an infinite series which itself is a solution of an infinite system of linear algebraic equations.

2.3 *Solution to the Galin Stick-Slip Problem Due to Spence*

A completely different treatment of Galin's problem was given by Spence (1973). Although he considered not only a flat punch, but also polynomial rigid indenters, here we focus our attention on a flat rigid punch indenting an elastic half space by a normal force P with the boundary conditions (4.7.1)–(4.7.6) ($a = b, Q = 0$). Spence denoted the surface values of the normal and shear stresses by

$$\sigma_{yy}\big|_{y=0} = -\frac{P}{a} p_0(x), \qquad \tau_{xy}\big|_{y=0} = \frac{P}{a} q_0(x) \tag{41}$$

and used the following result due to Muskhelishvili (1963):

$$\frac{1-\nu}{\pi}\int_{-1}^{1}\frac{p_0(t)dt}{t-x} + \frac{1}{2}(1-2\nu)\,q_0(x) = -\frac{\mu}{P}\frac{dv}{dx}$$

$$\frac{1-\nu}{\pi}\int_{-1}^{1}\frac{q_0(t)dt}{t-x} - \frac{1}{2}(1-2\nu)\,p_0(x) = \frac{\mu}{P}\frac{du}{dx}. \tag{42}$$

The boundary conditions can be rewritten as

$$
\begin{aligned}
&\frac{\mu}{P}\frac{dv}{dx} = 0 \quad -1 < x < 1,\\
&\frac{\mu}{P}\frac{du}{dx} = 0 \quad -a < x < a,\\
&p_0(x) = q_0(x) = 0 \quad -\infty < x < 1, \quad 1 < x < \infty,\\
&p_0(x) = \rho q_0(x) \quad a < x < 1,\\
&p_0(x) = -\rho q_0(x) \quad -1 < x < a.
\end{aligned}
\tag{43}
$$

Substituting representations (42) into the boundary conditions (43) and taking into account the symmetry of $p_0(x)$ and the antisymmetry of $q_0(x)$, Spence reduced the stick-slip boundary-value problem to a dual system of integral equations with respect to the unknown contact stresses

$$
\begin{aligned}
&\frac{2x}{\pi}\int_0^1 \frac{p_0(t)dt}{t^2 - x^2} + \frac{1-2\nu}{2-2\nu} q_0(x) = 0 \quad (0 < x < 1),\\
&\frac{2}{\pi}\int_0^1 \frac{tq_0(t)dt}{t^2 - x^2} - \frac{1-2\nu}{2-2\nu} p_0(x) = 0 \quad (0 < x < a).
\end{aligned}
\tag{44}
$$

As the first approximation to the solution of the system (44) Spence considered the case when the influence of friction on the normal stress is neglected (such approximation to the contact problems with friction was first suggested by Goodman (1962) and often is called Goodman's approximation). In this case, the equations in (44) become uncoupled and the system admits an analytic solution. In particular, the distribution of the normal stress is the same as for smooth frictionless contact:

$$
p_0(x) = (1/\pi)\left(1 - x^2\right)^{-\frac{1}{2}}. \tag{45}
$$

The coordinate a of the stick-slip boundary was determined from the requirement that the contact shear stress is continuous and bounded across the stick-slip boundary:

$$
\begin{aligned}
&\int_0^1 \frac{dt}{\sqrt{\left(1-t^2\right)\left(1-a^2t^2\right)}} \Big/ \int_0^1 \frac{dt}{\sqrt{\left(1-t^2\right)\left(1-(1-a^2)t^2\right)}}\\
&= \rho(2-2\nu)/(1-2\nu),
\end{aligned}
\tag{46}
$$

By taking into account (45) and (46), Spence deduced that the distribution of normalized shear stress in the stick zone:

$$
\begin{aligned}
q_0(x) = -\frac{\rho}{\pi}&\left(1 \Big/ \int_0^1 \frac{dt}{\sqrt{\left(1-t^2\right)\left(1-a^2t^2\right)}}\right)\\
&\left(\int_0^x \frac{dt}{\sqrt{\left(1-t^2\right)\left(1-a^2t^2\right)}} \Big/ \sqrt{1-x^2}\right) \quad (0 < x < a).
\end{aligned}
\tag{47}
$$

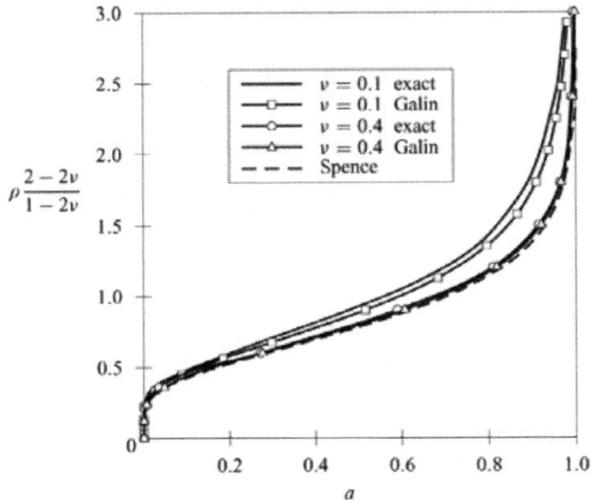

Fig. 1 Stick zone size as a function of the friction coefficient and Poisson's ratio.

The sign "–" appears in the right-hand side of (47) because the friction coefficient has opposite signs in Galin's and Spence's formulations of the boundary conditions.

Spence also tackled the original coupled stick-slip problem where the effect of friction on the distribution of normal contact stress is preserved. Initially he reduced the singular integral equations (44) to Volterra equations. The system of Volterra equations was eventually solved numerically.

Moreover, Spence noticed that the equation for the determination of the stick-slip boundary (46) for uncoupled solution is similar to Galin's equation (4.7.39). Indeed, the left-hand sides of both equations are identical. Spence compared the accuracy of both solutions, (46) and (4.7.39), relative to the numerical solution obtained for the system of coupled equations (44). He indicated that, in general, Galin's approximation (equation (4.7.39)) gives better accuracy than the approximation due to (46), especially when Poisson's ratio is small. Nevertheless, due to significant simplifications, the assumption of independence of the normal stress on the shear stress has become common and is now widely used in the contact mechanics literature (see, for example, Hills and Sackfield, 1987; Johnson, 1985), whereas Galin's sophisticated and more rigorous, but lacking physical interpretation argument has less favorable attention nowadays. Figure 1 shows variation of the stick zone size with parameter $\rho(2-2\nu)/(1-2\nu)$ at different values of Poisson's ratio found from exact solution of the problem as well as determined by Galin's and Spence's approximations.

Now we turn our attention to stick-slip problems that differ from the Galin's original formulation in either applied load, geometry, or material response.

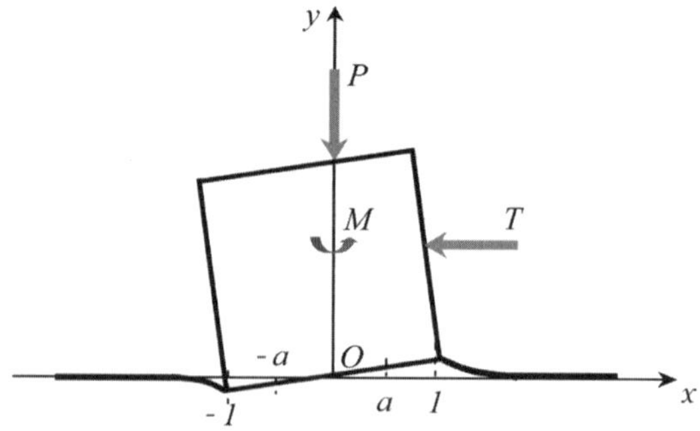

Fig. 2 An elastic half plane indented when a moment is applied.

3 Extensions of the Stick-slip Contact Problem

3.1 Arbitrary Load

Mossakovskii et al. (1983) generalized Galin's problem and considered indentation with stick and slip of a rigid flat punch into an elastic half plane when, in addition to normal and tangential forces, external forces with a couple of moment M are applied, and the base of the punch forms an angle ε with the x-axis (Figure 2).

The boundary conditions for the displacement components on the surface of the elastic half plane take the form

$$\begin{aligned} &v(x) = \varepsilon x \quad (-1 < x < -a), \\ &u(x) = 0, \quad v(x) = \varepsilon x \quad (-a < x < a), \\ &v(x) = \varepsilon x \quad (a < x < 1). \end{aligned} \tag{48}$$

The boundary conditions for stresses remain the same as in Galin's stick-slip problem.

The authors constructed the solution of the problem (48) as a superposition of two known solutions: the first was Muskhelishvili's solution (1953) of the contact problem of full stick for an inclined rigid punch and an elastic half plane; the second was a solution of the Galin stick-slip problem given by Mossakovskii and Biskup (1972) with some modifications in the solution procedure for the Fuchsian equation (7). Eventually the complex valued function $\Phi(z)$ that represents stresses and derivatives of displacements in the half plane (2) was found in the form

$$\Phi(z) = -\frac{iM}{\pi(\lambda_a - 1 + 2a_{-1}^2)}\left[1 - (z - a_{-1})\left(\frac{1}{z} - \frac{\lambda_a - 1}{2z^3} + \frac{a_{-1}}{z^2} + \dots\right)\right], \tag{49}$$

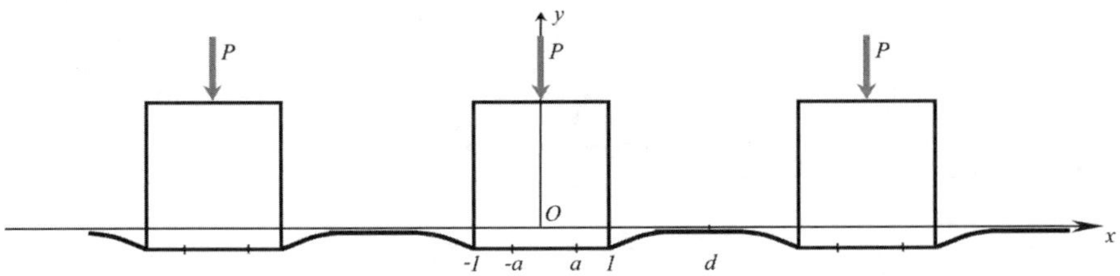

Fig. 3 A periodic system of rigid punches in the contact with an elastic half plane.

where λ_a is the accessory parameter of the Fuchsian equation (7) and the coefficient a_{-1} was determined numerically from the increase in the integrals of the Fuchsian equation (7) for a circuit around the point at infinity.

3.2 A Periodic System of Punches

Antipov (2000) considered a periodic system of rigid punches pressed into an elastic half plane by normal forces applied to each punch. He assumed that in each punch's contact region there is a stick and a slip zone (Figure 3).

Due to such a periodic structure, the boundary-value problem was formulated only for the half-strip $0 < x < d, -\infty < y < 0$ and reduced to a system of singular integral equations with respect to unknown normal and shear stresses in the contact area $-1 < x < 1$. The solution procedure for this system is based on reduction to a matrix Wiener–Hopf problem. After factorization, which is precisely the same as in (30), the Wiener–Hopf functional equations take a form similar to (31). Follow up solution procedure is due to Antipov and Arutyunyan (1991) as described above. The computations were carried out for Poisson's ratio $\nu = 0.3$ and different values of the friction coefficient. The results showed that the stick zone size and, correspondingly, the stress intensity factor at the edges of the contact area increase as the distance between punches decrease.

3.3 Self-Similarity Approach in Stick-Slip Contact Problems

Spence (1968) originated a self-similarity approach to contact problems. The key feature of his approach is the self-similarity argument that yields geometrically similar stress and displacement fields at each step of the application of progressive load for any indenter having profile $y = -A\,|x|^n$ (2D case) or $z = -Br^n$ (axisymmetric case). For such indenters the contact and stick-slip boundaries do not remain stationary under progressively increasing load, and have to be determined from the solution of the problem. Spence's innovative idea changed the treatment of con-

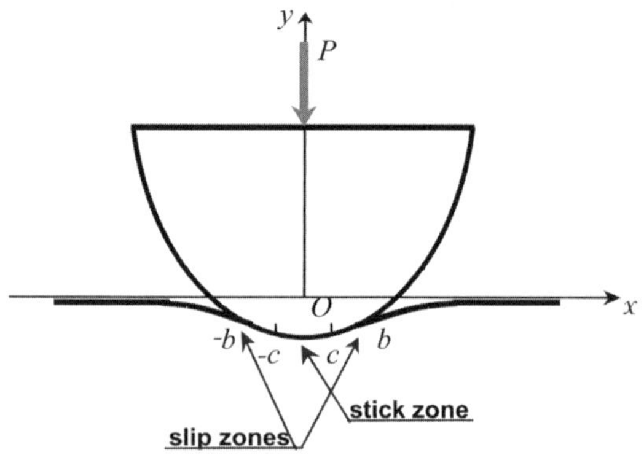

Fig. 4 An elastic half plane indented by a rigid power law indenter.

tact problems that involve contact boundaries moving under progressive load, and allowed one to avoid an incremental step-by-step computation of contact stresses performed simultaneously with the increase of the contact area (see, for example, Mossakovskii, 1954, 1963; Goodman, 1962).

In stick-slip problems for power law indenters (Figure 4) the self-similarity argument leads to a constant proportion between stick and slip zones at every step of progressive loading, and mathematically it implies the following boundary condition on the lateral displacement u in the stick zone

$$u(x)|_{y=0} = C^* \, |x|^n \quad (-c/a < x < c/a) \tag{50}$$

for 2D case (3D case is similar). Here C^* is a non-positive constant that is unknown *a priori* and has to be determined as a part of the solution.

Dependence (50) ensures a constant displacement at each point of the stick zone for progressively increasing load. In other words, the lateral displacement at any given point of the stick zone does not change when the boundary of the contact area changes.

In this setting, Spence considered problems with full stick (1968), and stick-slip (1973, 1975). He found that, under normal indentation, the size of the slip zone(s) is the same for all power-law indenters. This follows from the possibility of transforming the equations and boundary conditions for power law indenters into those for a flat-ended punch. Particularly in the 2D case the solution for the power law indenter is recovered by quadratures

$$p(x) = nx^{n-1} \int_x^1 \frac{p_0(t)dt}{t^n}, \quad q(x) = nx^{n-1} \int_x^1 \frac{q_0(t)dt}{t^n} \quad (0 < x < 1). \tag{51}$$

Here the notation is as in (41). Moreover, since the ratio between stick and slip zone(s) is the same for all power law indenters, the unknown stick-slip boundary is found from the solution of the corresponding problem for the flat punch.

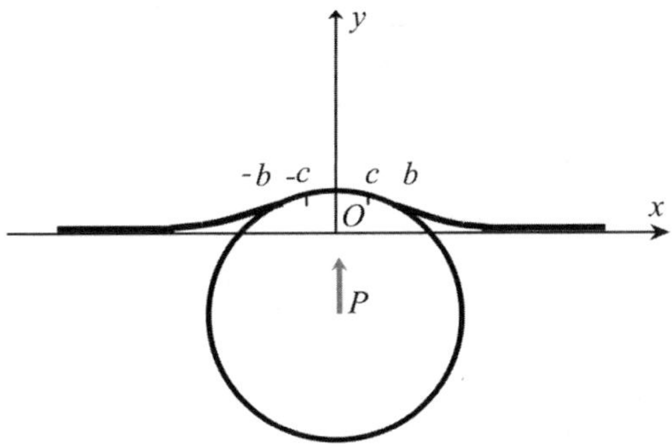

Fig. 5 An elastic half space indented by a rigid cylinder.

Spence (1975) also showed that similarity considerations are still applicable in axisymmetric contact problems. In other words, it is sufficient to solve the problem for a flat punch, and after that to use quadrature, as in (51), to obtain the solution of the original stick-slip contact problem for a power law indenter. Spence constructed an iterative numerical solution using a dual system of Volterra equations with respect to unknown contact stresses, and calculated contact stress distributions for indentation by a flat punch and by a sphere. As in the 2D problem, he obtained a simple equation for the determination of the stick-slip radius a for the uncoupled problem when the effect of friction on the normal contact stress is ignored:

$$\frac{1}{2c}\ln\left(\frac{1+c}{1-c}\right) \Big/ \int_0^1 \frac{dt}{\sqrt{(1-t^2)\left(1-(1-c^2)t^2\right)}} = \rho(2-2\nu)/(1-2\nu). \quad (52)$$

As expected, the approximation (52) provides the most inaccurate results when Poisson's ratio approaches zero.

A completely independent analytical solution procedure for 2D stick-slip contact problems was suggested by Zhupanska and Ulitko (2005), who gave an exact solution to the normal contact with friction of a rigid cylinder with an elastic half space (Figure 5).

The problem was formulated as a stick-slip problem with Spence's self-similarity condition that implied the following behavior of the horizontal strain in the stick zone $-c < x < c$:

$$\left.\frac{\partial u}{\partial x}\right|_{y=0} = 2C^* |x| \qquad (-c < x < c), \quad (53)$$

where C^* is an unknown constant that has to be determined as a part of the solution. The boundary condition for the normal displacement v has the form

$$v|_{y=0} = \frac{b^2 - x^2}{2R} \qquad (-b < x < b). \quad (54)$$

The boundary conditions for stresses are the same as in Galin's stick-slip problem. This is a nonlinear mixed boundary value problem of planar elasticity for conditions of plane strain. The key points of the authors' solution procedure were: (i) the use of bipolar coordinates (α, β) that are related to the Cartesian coordinates (x, y) by

$$x = a\frac{\sinh\alpha}{\cosh\alpha + \cos\beta},$$

$$y = a\frac{\sin\beta}{\cosh\alpha + \cos\beta} \quad (-\infty < \alpha < \infty, -\pi \leq \beta \leq \pi); \tag{55}$$

and (ii) the use of the Papkovich–Neuber general solution of the elastostatic equations. It is apparent that the original boundary conditions in the bipolar coordinates are reformulated in three infinite regions, namely $\beta = \pi, \infty < \alpha < 0, \beta = 0, -\infty < \alpha < \infty$, and $\beta = \pi$, $0 < \alpha < \infty$, which substantially facilitates the solution procedure.

The elastic stresses and displacements on the surface of the half space have the following representation by two harmonic Papkovich–Neuber functions Φ_2 and Φ_3

$$\left.\frac{\partial u}{\partial x}\right|_{y=0} = \varepsilon_x = \left.\frac{\partial \Phi_3}{\partial y}\right|_{y=0}, \quad \left.\frac{\sigma_{yy}}{2\mu}\right|_{y=0} = \left.\frac{\partial}{\partial y}[2(1-\nu)\Phi_2 - \Phi_3]\right|_{y=0},$$

$$\left.\frac{\partial v}{\partial x}\right|_{y=0} = \left.\frac{\partial}{\partial x}[(3-4\nu)\Phi_2 - \Phi_3]\right|_{y=0},$$

$$\left.\frac{\tau_{xy}}{2\mu}\right|_{y=0} = \left.\frac{\partial}{\partial x}[(1-2\nu)\Phi_2 - \Phi_3]\right|_{y=0}, \tag{56}$$

where μ is the shear modulus, and the Fourier integral representations for the Papkovich–Neuber functions Φ_2 and $\Phi_3$in bipolar coordinates have the form

$$\Phi_2(\alpha, \beta) = C_2 \ln\left(\frac{\cosh\alpha/2 + \sin\beta/2}{\cosh\alpha/2 - \sin\beta/2}\right) + \frac{1}{\sqrt{2\pi}}\int_{-\infty}^{\infty}[A_2(\lambda)\cosh\lambda\beta + B_2(\lambda)\sinh\lambda\beta]\, e^{-i\lambda\alpha}d\lambda,$$

$$\Phi_3(\alpha, \beta) = (1-2\nu)C_2 \ln\left(\frac{\cosh\alpha/2 + \sin\beta/2}{\cosh\alpha/2 - \sin\beta/2}\right) + \frac{1}{\sqrt{2\pi}}\int_{-\infty}^{\infty}[A_3(\lambda)\cosh\lambda\beta + B_3(\lambda)\ \sinh\lambda\beta]\, e^{-i\lambda\alpha}d\lambda, \tag{57}$$

where C_2 is an unknown constant that should be a part of the solution, and $A_{2,3}(\lambda)$, $B_{2,3}(\lambda)$ are unknown densities. This allowed the authors to reduce the boundary-value problem to a singular integral equation with respect to the unknown normal stress σ in the slip zones (versus a dual system of singular integral equations found in other treatments of problems with partial slip):

$$\frac{1}{2\pi}\int_{\alpha_0}^{\infty}\frac{\sigma(\xi)}{2\mu}\left[\int_{-\infty}^{\infty}\frac{\sinh\pi\lambda\cosh\pi(\lambda-i\gamma)}{\cosh\pi(\lambda+\theta)\cosh\pi(\lambda-\theta)}e^{-i\lambda(\alpha-\xi)}d\lambda\right.$$

$$\left.+\int_{-\infty}^{\infty}\frac{\sinh\pi\lambda\cosh\pi(\lambda+i\gamma)}{\cosh\pi(\lambda+\theta)\cosh\pi(\lambda-\theta)}\times e^{-i\lambda(\alpha+\xi)}d\lambda\right]\frac{d\xi}{\cosh\xi-1}$$

$$=\frac{i}{D}\left\{\frac{a}{R}\frac{\partial^2}{\partial\alpha^2}\left(\frac{\cos\theta\alpha}{\sinh\frac{\alpha}{2}}\right)+2(1-\nu)\frac{C_2}{a}\frac{\cos\theta\alpha}{\sinh\frac{\alpha}{2}}+\frac{\cosh\pi\theta}{2\pi}C^*a\right.$$

$$\left.\times\left[\frac{\sin\theta\alpha}{\cosh\frac{\alpha}{2}}+\frac{\partial}{\partial\alpha}\int_0^{\infty}\left(\frac{\sin\theta(\alpha+\eta)}{\cosh\frac{\alpha+\eta}{2}}-\frac{\sin\theta(\alpha-\eta)}{\cosh\frac{\alpha-\eta}{2}}\right)\frac{d\eta}{\cosh\eta+1}\right]\right\},$$

$$\alpha_0\le\alpha<\infty, \tag{58}$$

where

$$\cosh\pi\theta=\kappa=\frac{2(1-\nu)}{\sqrt{3-4\nu}},$$

$$D=\sqrt{4(1-\nu)^2+\rho^2(1-2\nu)^2},$$

$$\tan\pi\gamma=\rho\frac{1-2\nu}{2(1-\nu)} \tag{59}$$

and ρ is the coefficient of dry friction between the half space and cylinder. The value of the constant α_0

$$\alpha_o=\operatorname{arctan}h\left(c/b\right), \tag{60}$$

depends on the size ratio of the stick ($-\infty<\alpha<\infty, \beta=0$) and the slip ($\alpha_0<|\alpha|<\infty, \beta=\pi$) zones and, therefore, has to be determined from the solution of the problem.

In equation (58), the influence of the shear stress on the normal displacement in the contact area is preserved. An analytical solution of the equation was constructed using the Wiener–Hopf technique (Noble, 1958) via the following procedure. First, equation (58) was reduced to the form

$$\int_0^{\infty}k(t-\tau)\varphi(\tau)d\tau=F(\alpha_0+t), 0<t,\tau<\infty. \tag{61}$$

Here

$$\alpha=\alpha_0+t,\quad \frac{\sigma(\alpha_0+\tau)}{2\mu}\frac{1}{\cosh(\alpha_0+\tau)-1}=\varphi(\tau),$$

$$k(t-\tau)=\frac{1}{2\pi}\int_{-\infty}^{\infty}\frac{\sinh\pi\lambda\cosh\pi(\lambda-i\gamma)}{\cosh\pi(\lambda+\theta)\cosh\pi(\lambda-\theta)}e^{-i\lambda(t-\tau)}d\lambda$$

$$=-\frac{i}{\pi D}\left[(1-2\nu)\rho\pi\delta(t-\tau)+(1-\nu)\frac{\cos\theta(t-\tau)-\rho\sin\theta(t-\tau)}{\sinh\frac{t-\tau}{2}}\right],$$

$$
\begin{aligned}
F(\alpha_0+t) = &-\frac{i}{\pi}\Bigg[\pi \sin\pi\gamma\delta(2\alpha_0+t+\tau)\\
&-\frac{1-\nu}{D}\frac{\cos\theta(2\alpha_0+t+\tau)+\rho\sin\theta(2\alpha_0+t+\tau)}{\sinh(\alpha_o+\frac{t+\tau}{2})}\Bigg]\\
&+\frac{i}{D}\Bigg\{\frac{a}{R}\frac{\partial^2}{\partial t^2}\left(\frac{\cos\theta(\alpha_0+t)}{\sinh\frac{\alpha_0+t}{2}}\right)+2(1-\nu)\frac{C_2}{a}\frac{\cos\theta(\alpha_0+t)}{\sinh\frac{\alpha_0+t}{2}}+\frac{\cosh\pi\theta}{2\pi}C^*a\\
&\times\Bigg[\frac{\sin\theta(\alpha_0+t)}{\cosh\frac{\alpha_0+t}{2}}+\frac{\partial}{\partial t}\left(\int_0^\infty\left(\frac{\sin\theta(\alpha_0+t+\eta)}{\cosh\frac{\alpha_0+t+\eta}{2}}-\frac{\sin\theta(\alpha_0+t-\eta)}{\cosh\frac{\alpha_0+t-\eta}{2}}\right)\right.\\
&\left.\times\frac{d\eta}{\cosh\eta+1}\right)\Bigg]\Bigg\}. \qquad (62)
\end{aligned}
$$

In the integral equation (61), the integral with the singular kernel $k(t-\tau)$ is isolated on the left-hand side of the equation, while the right-hand side retains an integral with the regular kernel $m(2\alpha_0+t+\tau)$. The integral on the right-hand side of equation (61) accounts for the mutual influence of the identical slip zones on the distribution of the normal contact stresses. The integral equation (61) is not exactly of the Wiener–Hopf type, since, along with the singular integral in the left-hand side, it also contains the regular integral with respect to the unknown function in the right-hand side. Yet, the Wiener–Hopf technique can be employed for solving this equation. Namely, by letting the regular integral with unknown σ in the right-hand side of (61) be temporarily regarded as a known function. This integral vanishes when the size of slip zones is relatively small ($\alpha_0 \to +\infty$), in which case equation (61) reduces to a singular integral equation with a difference kernel.

Applying to (61) the Fourier integral transform over the interval $-\infty < x < \infty$, we obtain a Wiener–Hopf equation in the form

$$\Phi_+(z)K(z) = \Psi_-(z) + F_+(z). \qquad (63)$$

Here

$$\Phi_+(z) = \frac{1}{\sqrt{2\pi}}\int_0^\infty \varphi(t)e^{izt}dt, \qquad (64)$$

$$\Psi_-(z) = \frac{1}{\sqrt{2\pi}}\int_{-\infty}^0 e^{izt}dt\int_0^\infty k(t-\tau)\varphi(\tau)d\tau \qquad (65)$$

are the new unknown functions, analytic in the upper and lower half planes of the complex plane respectively. The kernel $K(z)$ admits an explicit decomposition in the form (Ulitko, 2000).

$$\frac{K(z)}{z} = \frac{\sinh\pi z\cosh\pi(z-i\gamma)}{\cosh\pi(z+\theta)\cosh\pi(z-\theta)}\frac{1}{z} = \frac{K_+(z)}{K_-(z)}. \qquad (66)$$

Here $K_+(z)$ and $K_-(z)$ are analytic functions in the upper and lower half planes of the complex plane, respectively:

$$K_+(z) = \frac{\Gamma(1/2 - iz - i\theta)\Gamma(1/2 - iz + i\theta)}{\Gamma(1 - iz)\Gamma(1/2 - iz - \gamma)},$$

$$K_-(z) = \frac{\Gamma(1 + iz)\Gamma(1/2 + iz + \gamma)}{\Gamma(1/2 + iz + i\theta)\Gamma(1/2 + iz - i\theta)} \tag{67}$$

and the function $F_+(z)$ is analytic in the upper half of the complex plane. Further factorization of the product $F_+(Z)K_-(z)$ as

$$F_+(z)K_-(z) = f_+(z) + f_-(z) \tag{68}$$

leads to the solution for the function $\Phi_+(z)$ in the form

$$\Phi_+(z) = \frac{f_+(z)}{zK_+(z)}, \tag{69}$$

with the extra conditions

$$f_+(0) = 0, \qquad f_+\left(-\frac{i}{2} + i\gamma\right) = 0, \tag{70}$$

that ensure analyticity of the function $\Phi_+(z)$ for $\text{Im}(z) > -1$, and consequently, finiteness and continuity of the contact stresses across the stick-slip boundary. The function $f_+(z)$ is defined as

$$f_+(z) = \frac{i}{\sqrt{2\pi}}\left[e^{-(\frac{1}{2}-i\theta)\alpha_0}\sum_{k=0}^{\infty}\frac{L_k}{k+\frac{1}{2}-iz-i\theta} + e^{-(\frac{1}{2}+i\theta)\alpha_0}\sum_{k=0}^{\infty}\frac{\bar{L}_k}{k+\frac{1}{2}-iz+i\theta}\right],$$

$$L_k = \frac{\Gamma(k+\frac{3}{2}-i\theta)\Gamma(k+1+\gamma-i\theta)}{\Gamma(k+1-2i\theta)}\frac{e^{-k\alpha_0}}{k!}\left\{\left(k+\frac{1}{2}-i\theta\right)^2\frac{a}{R} + 2(1-\nu)\frac{C_2}{a}\right.$$
$$-i(-1)^k C^* a\frac{\cosh\pi\theta}{2\pi}\left[1 + 2\left(k+\frac{1}{2}-i\theta\right) + 2\left(k+\frac{1}{2}-i\theta\right)^2\left((-1)^k\frac{\pi}{\cosh\pi\theta} - q_k\right)\right]$$
$$\left. + (1-i\rho)e^{-(k+\frac{1}{2}-i\theta)\alpha_0}\frac{\cos\pi\gamma}{\sqrt{2\pi}}Z_k\right\}. \tag{71}$$

The solution (68) is defined apart from the unknowns Z_k that have to be determined from an infinite algebraic system obtained by substitution of the solution (69) for $\Phi_+(z)$ into

$$Z_n = \Phi_+\left(\theta + i(n + 1/2)\right). \tag{72}$$

Thus, to complete the solution of equation (61) one has to solve the infinite system (72) for Z_k, and to satisfy two additional conditions (70). It is worth repeating that the infinite series appears in the solution due to the presence of a regular part in the kernel of the singular integral equation along with the singular difference part.

This paper also contains an analysis of the solution from a mechanical viewpoint, where distributions of contact stress, strain and displacement fields are derived and corresponding graphs for various material parameters are plotted.

The advantage of Zhupanska and Ulitko's solution procedure becomes even more evident in the limiting case of full stick between the cylinder and half space. It leads to a simple analytical solution to the problem by means of the Fourier integral transform. Finally, we note that this solution procedure is quite general, and allows for construction of analytical solutions to stick-slip contact problems for any power law indenter.

The self-similarity argument was exploited in the solutions of contact problems with finite friction not only in the case of elastic but also visco-elastic and elasto-plastic material response. Particularly, we mention contribution to the self-similarity approach by Borodich (1993), Borodich and Galanov (2002), and Borodich and Keer (2004).

Acknowledgement

I would like to extend my gratitude to Professor G.M.L. Gladwell for inviting me to contribute to the new edition of Galin's classical book, and to comment on the developments that stemmed from Galin's stick-slip contact problem. Like for many other researchers, my own endeavors in this field have been greatly inspired by Galin's seminal work.

References

Antipov, Y.A. (2000) Galin's problem for a periodic system of stamps with friction and adhesion. *International Journal of Solids and Structures*, **37**, 2093–2125.

Antipov, Y.A., Arutyunyan, N.K. (1991) Contact problems of the theory of elasticity with friction and adhesion. *PMM Journal of Applied Mathematics and Mechanics*, **55**(6), 887–901 [in Russian].

Borodich, F.M. (1993) The Hertz frictional contact between nonlinear elastic anisotropic bodies (the similarity approach). *International Journal of Solids and Structures*, **30**(11), 1513–1526.

Borodich, F.M., Galanov, B.A. (2002) Self-similar problems of elastic contact for non-convex punches. *Journal of the Mechanics and Physics of Solids*, **50**, 2441–2461.

Borodich, F.M., Keer L.M. (2004) Contact problems and depth-sensing nanoindentation for frictionless and frictional boundary conditions. *International Journal of Solids and Structures*, **41**, 2479–2499.

Galin, L.A. (1945) Pressure of a punch with friction and cohesion domains. *PMM Journal of Applied Mathematics and Mechanics*, **9**, 413–424 [in Russian].

Galin, L.A. (1953) *Contact Problems of the Theory of Elasticity*. Gostehizdat, Moscow (First Russian edition).

Goodman, L.E. (1962) Contact stress analysis of normally loaded rough spheres. *Transactions of the ASME, Journal of Applied Mechanics*, **29** (9), 515–522.

Johnson, K.L. (1985) *Contact Mechanics*, Cambridge University Press, Cambridge.

Hills, D.A., Sackfield, A. (1987) The stress field induced by normal contact between dissimilar spheres. *Transactions of the ASME, Journal of Applied Mechanics*, **54**, 8–14.

Mossakovskii, V.I. (1954) The fundamental general problem of the theory of elasticity for a half space with a circular curve determining boundary conditions, *PMM Journal of Applied Mathematics and Mechanics*, **18**(2), 187–196 [in Russian].

Mossakovskii V.I. (1963) Compression of elastic bodies under conditions of adhesion (axisymmetric case). *PMM Journal of Applied Mathematics and Mechanics*, **27**(3), 418–427 [in Russian].

Mossakovskii, V.I., Biskup, A.G. (1972) Pressing-in of die when friction and cohesion are present. *Doklady Akademii Nauk SSSR*, **206**(5) 1068–1070 [in Russian].

Mossakovskii, V.I., Biskup, A.G., Mossakovskaia, L.V. (1983) Further development of Galin's problem with dry friction and adhesion. *Doklady Akademii Nauk SSSR*, **271**(1), 60–64 [in Russian].

Mossakovskii, V.I., Biskup, A.G., Mossakovskaia, L.V. (1989) On one method of solution of two-dimensional contact problems with dry friction and adhesion. *Doklady Akademii Nauk SSSR*, **308**(3), 561–564 [in Russian].

Muskhelishvili, N.I. (1963) *Some Basic Problems of the Mathematical Theory of Elasticity*, Noordhoff, Groningen, the Netherlands.

Nehari, Z. (1952) *Conformal Mapping*, McGraw-Hill, New York.

Noble, B. (1958) *Methods Based on the Wiener–Hopf Technique*, Pergamon, London.

Ronveaux, A. (1995) *Heun's Differential Equations*, Oxford University Press, Oxford.

Spence, D.A. (1968) Self-similar solutions to adhesive contact problems with incremental loading. *Proceedings of the Royal Society of London, Series A*, **305**, 81–92.

Spence, D.A. (1973) An eigenvalue problem for elastic contact with finite friction. *Proceedings of the Cambridge Philosophical Society*, **73**, 249–268.

Spence, D.A. (1975) The Hertz contact problem with finite friction. *Journal of Elasticity*, **5**, 297–319.

Ulitko, A.F. (2000) Exakte Lösung des Kontaktproblems für zwei Zylinder unter Berücksichtigung der Reibung. *Zeitschrift für Angewandte Mathematik und Mechanik* **80**(7), 435–455.

Zhupanska, O.I., Ulitko, A.F. (2005) Contact with friction of a rigid cylinder with an elastic half space. *Journal of the Mechanics and Physics of Solids*, **53**, 975–999.